INTERNATIONAL CENTRE FOR MECHANICAL SCIENCES

COURSES AND LECTURES - No. 224

RAREFIED GAS FLOWS
THEORY AND EXPERIMENT

EDITED BY
W. FISZDON
INSTITUTE OF FUNDAMENTAL
TECHNOLOGICAL RESEARCH
WARSAW

SPRINGER-VERLAG WIEN GMBH

Originally published by Springer-Verlag Wien New York in 1981

ISBN 978-3-211-81595-3 ISBN 978-3-7091-2898-5 (eBook)
DOI 10.1007/978-3-7091-2898-5

PREFACE

The theoretical description and analysis of the behaviour of gases is based on models which are not uniformly adequate and useful for this material, that is composed of a very large number of discrete particles.

At one end, in the case of dense gases, where very small Knudsen numbers are involved, the continuum model has proved its validity. At the other hand, when dealing with rarefied gases, characterized by large Knudsen numbers, the statistical mechanics model is the most appropriate.

As far as the latter case is concerned, until the middle of this century the aim was to develop our knowledge and to reach the possibility of deducing the transport properties of rarefied gases and of correlating them with their molecular interaction laws, as well as of deducing their molecular description, based on Boltzmann's equation, and of correlating it with the continuous one, exemplified by the Navier-Stokes equations.

It is only in the late fifties that it became evident that in many problems of practical interest, related with the motion of rarefied gases, it was necessary to take into account some of the properties due to their discrete, molecular nature, and thus the kinetic theory of gases, or rather the dynamics of rarefied gases, was established.

This volume contains the set of five lecture series which covered selected aspects of the rarefied gas dynamics and gave a coherent picture of this field of mechanics.

– C. Cercignani's lectures introduce Boltzmann's equation and describe the methods used for solving it;

– J.P. Guiraud considers the fundamental mathematical problems concerning the existence and uniqueness of the solution of Boltzmann's equation in various spaces;

– W. Fiszdon provides an example of the application of the kinetic theory to the solution of the simple, one-dimensional well-defined problem of the shock-wave structure in an infinite space;

– S. Nocilla's lectures deal with the very involved problem of gas-surface interaction, whose solution is required in all problems concerning a gas confined in spaces bounded by solid surfaces;

– J.J. Smolderen offers an insight into the experimental methods used in rarefied gas dynamics and presents the experimental results obtained.

In publishing this volume in the by now well-established series of CISM Lecture Notes, I hope that the young – and perhaps also the not so young – scientists who wish to get acquainted with the problems of rarefied gasdynamics will find this work useful and the time spent in reading it rewarding.

W. Fiszdon

LIST OF CONTRIBUTORS

CERCIGNANI, C., Politecnico di Milano, Milano, Italy.

FISZDON, W., Institute of Fundamental Technological Research, Warsaw, Poland.

GUIRAUD, J.P., Université P. & M. Curie, Paris, France.

NOCILLA, S., Politecnico di Torino, Torino, Italy.

SMOLDEREN, J.J., von Karman Institute for Fluid Dynamics, Rhode St. Genese, Belgium.

CONTENTS

Experimental Methods and Results in Rarefied Gas Dynamics
by J.J. Smolderen

The Structure of Plane Shock Waves
by W. Fiszdon

METHODS OF SOLUTION OF THE BOLTZMANN EQUATION FOR RAREFIED GASES

Carlo Cercignani
Politecnico di Milano

CHAPTER I

THE BOLTZMANN EQUATION

1. Classical Dynamics and Boltzmann Equation

According to the molecular theory of matter, a macroscopic volume of gas (say, 1 cm^3) is a system of a very large number (say, 10^{20}) of molecules moving in a rather irregular way. In principle, we may assume, ignoring quantum effects, that the molecules are particles (mass points or other systems with a small number of degrees of freedom) obeying the laws of classical mechanics. We may also assume that the laws of interaction between the molecules are perfectly known so that, in principle, the evolution of the system is computable, provided suitable initial data are given. If the molecules are, e.g., mass points the equations of motion are:

$$\dot{\xi}_i = X_i \qquad \dot{x}_i = \xi_i \tag{1.1a}$$

or

$$\ddot{x}_i = X_i \tag{1.1b}$$

where x_i is the position vector of the i-th particle (i= 1, . . . N) and ξ_i its velocity vector; both x_i and ξ_i are functions of the time variable t and the dots denote, as usual, differentiation with respect to t. Here X_i is the force acting upon the i-th particle divided by the mass of the particle. Such a force will in general be the sum of the resultant of external forces (e.g. gravity or, if the observer is not inertial, apparent forces such as centrifugal or Coriolis forces) and the forces describing the action of the other particles of the system on the i-th particle. As we said before, the expression of such forces must be given as a part of the description of the mechanical system.

In order to compute the time evolution of the system, one would have to solve the 6N first-order differential equations, Eq. (1.1a) in the 6N unknowns constituting the components of the 2N vectors $(\mathbf{x}_i, \xi_i)$ $(i = 1, \ldots, N)$. A prerequisite for this is the knowledge of the 6N initial conditions:

$$\mathbf{x}_i(0) = \mathbf{x}_i^0 \quad ; \quad \dot{\mathbf{x}}_i(0) = \xi_i(0) = \xi_i^0 \tag{1.2}$$

where the components of $\mathbf{x}_i^0$ and ξ_i^0 are 6N given constants which describe the initial state of the system.

However, solving the above initial value problem for a number of particles of a realistic order of magnitude (say, $N \simeq 10^{20}$) is an impossible and useless task, for the following reasons:

1) We have to know the initial data $\mathbf{x}_i^0$ and ξ_i^0 i.e. the positions and velocities of all the molecules at $t = 0$ and obtaining these data appears difficult, even in principle. In fact, it would involve the simultaneous measurement of the positions and velocities of all the molecules at $t = 0$.

2) The information on the initial data, if available in spite of the above remark, is enormous and the duration of a human life would be insufficient to utter a faint fraction of these data (assume that one may give the six data for each particle in one second and observe that there are less than 10^8 seconds in a year, i.e. less than 10^{10} seconds in the lifetime of a human being).

3) Even if we could obtain the data and feed them in a computer, it seems impossible to imagine a computer capable of solving so many equations (Think of the number of computer cards required to supply the initial data!).

4) No matter how accurately we measure or give the initial data, the latter cannot be infinitely accurate; e.g., we shall never consider more than 100 decimal figures in our computations, thus introducing truncation errors of order 10^{-100}.

As a consequence, one should consider the evolution, not of a system, but of an ensemble of identical systems whose initial data differ from each other by quantities of the order of the accepted errors. It is possible to see that, for a system of about 10^{20} molecules, truncation errors of the order of 10^{-100} in the computations would make it impossible to compute the motion of these molecules for more than one millionth of a second.

5) Even if we could work with infinitely many figures (!), we should include all the particles of the Universe in our computations. In fact according to Borel the displacement of 1 gram of matter by 1 cm on a not too distant star (say, Sirius)

would produce a change of force larger than 10^{-100} times a typical force acting on the molecule and we would fall back to the difficulty mentioned under 4), unless we want to include all the particles of the Universe (!) in our computations.

6) Even if we overcome the above difficulties and compute the subsequent evolution of the considered system, this detailed information would be useless because knowledge of where the single molecules are and what their velocities are is information which, in this form, does not tell us what we really want to know, e.g., the pressure exerted on a wall by a gas at a given density and temperature.

The conclusion is that the only significant and useful results are those about the behavior of many systems in the form of statistics, i.e. information about probable distributions. This information can be obtained by averaging over our ignorance (meaning the incapability of macroscopic bodies to detect certain microscopic details of another macroscopic body) or the errors introduced by neglecting the influence of other bodies in the Universe.

As a result, only averages can be computed and are what matter, provided they are related to such macroscopic quantities as pressure, temperature, stresses, heat flow, etc. This is the basic idea of statistical mechanics.

The first kind of averaging which appears in any treatment of mechanics based upon statistical ideas is, as suggested by the above consideration, over our ignorance of initial data. However, other averaging processes or limiting procedures are usually required, to take into account the interactions of particles in a statistical fashion. These interactions also include the interaction of the molecules of a fluid with the solid boundaries, which bound the region where the fluid flows and are also formed of molecules.

When we deal with statistical mechanics, therefore, we talk about probabilities instead of certainties: i.e., in our description, a given particle will not have a definite position and velocity, but only different probabilities of having different positions and velocities. In particular, this is true for the kinetic theory of gases, i.e. the statistical mechanics of gas molecules, and the theories of transport of particles (neutrons, electrons, photons, etc.). Under suitable assumptions, the information required to compute averages for these systems can be reduced to the solution of an equation, the so-called Boltzmann equation. In the case of neutrons the equation is frequently called transport equation, while the name of transfer equation is in use for the case of photons (radiative transfer).

In order to discuss the behavior of a system of N mass points satisfying Eqs. (1.1a), it is highly convenient to introduce the so called phase space, i.e. a $6N$ –

dimensional space where the Cartesian coordinates are the 3N components of the N position vectors $\mathbf{x}_i$ and the 3N components of the N velocities ξ_i.

In this space, the state of a system at a given time t, if known with absolute accuracy, is represented by a point whose coordinates are the 6N values of the components of the position vectors and velocities of the N particles (frequently, the momenta of the particles are used in place of their velocities, but the difference will not matter for our purposes). Let us introduce the 6N – dimensional vector $\mathbf{z}$ which gives the position of the representative point in phase space; clearly, the components of $\mathbf{z}$ are orderly given by the 3N components of the N three-dimensional vectors $\mathbf{x}_i$ and the 3N components of the N three--dimensional vectors ξ_i. The evolution equation for $\mathbf{z}$ is from Eqs. (1a).

$$\dot{\mathbf{z}} = \frac{d\mathbf{z}}{dt} = \mathbf{Z} \tag{1.3}$$

where $\mathbf{Z}$ is a 6N-dimensional vector, whose components are orderly given by the 3N components of the N three-dimensional vectors ξ_i and the 3N components of the N three-dimensional vectors $\mathbf{X}_i$. Given the initial state, i.e. a point $\mathbf{z}_o$ in phase space, Eq. (1.3) determines $\mathbf{z}$ at subsequent times (provided the conditions for existence and uniqueness of the solution are satisfied).

If the initial data are not known with absolute accuracy, we must introduce a probability density $P_o(\mathbf{z})$ which gives us the distribution of probability for the initial data and we can try to set up the problem of computing the probability density at subsequent times, $P(\mathbf{z}, t)$. In order to achieve this, we must find an evolution equation for $P(\mathbf{z}, t)$; this can be easily done, as we shall see, provided the forces are known, i.e. if the only uncertainty is on the initial data.

An intuitive way of deriving the equation satisfied by $P(\mathbf{z}, t)$ is the following. We replace each mass point by a continuous distribution with density proportional to the probability density; in such a way, the system of mass points is replaced by a sort of fluid with density proportional to P and velocity $\dot{\mathbf{z}} = \mathbf{Z}$. Hence conservation of mass will give:

$$\frac{\partial P}{\partial t} + \operatorname{div}(P\mathbf{Z}) = 0 \tag{1.4}$$

where, as usual, for any vector **u** of the phase space, we write

$$\text{div } \mathbf{u} = \sum_{i=1}^{6N} \frac{\partial u_i}{\partial z_i} \equiv \frac{\partial}{\partial \mathbf{z}} \cdot \mathbf{u} \tag{1.5}$$

Eq.(1.4) is the Liouville equation; note that the components of **z** are independent variables.

But:

$$\text{div}(P\mathbf{Z}) = \mathbf{Z} \cdot \text{grad } P + P \text{ div } \mathbf{Z} \tag{1.6}$$

where, as usual, grad $P \equiv \partial P/\partial \mathbf{z}$ is the vector of components $\partial P/\partial z_i$. Hence P satisfies the equation:

$$\frac{\partial P}{\partial t} + \mathbf{Z} \cdot \text{grad } P + P \text{ div } \mathbf{Z} = 0 \tag{1.7}$$

Usually, div **Z** = 0. In fact, since $\mathbf{x}_i$ and $\boldsymbol{\xi}_i$ are independent variables:

$$\text{div } \mathbf{Z} = \sum_{i=1}^{N} \left(\frac{\partial}{\partial \mathbf{x}_i} \cdot \boldsymbol{\xi}_i + \frac{\partial}{\partial \boldsymbol{\xi}_i} \cdot \mathbf{X}_i \right) = \sum_{i=1}^{N} \frac{\partial}{\partial \boldsymbol{\xi}_i} \cdot \mathbf{X}_i \tag{1.8}$$

If the force per unit mass is velocity-independent, then also $(\partial/\partial\boldsymbol{\xi}_i) \cdot \mathbf{X}_i = 0$, and div = 0 as announced. Note, however, that for some velocity dependent forces $(\partial/\partial\boldsymbol{\xi}_i) \cdot \mathbf{X}_i = 0$, the most notable case being that of the Lorentz force acting on a charged particle in a magnetic field. We shall always consider forces such that div **Z** = 0 (typically, velocity-independent forces). Hence we write the Liouville equation in the following form:

$$\frac{\partial P}{\partial t} + \mathbf{Z} \cdot \frac{\partial P}{\partial \mathbf{z}} = 0 \tag{1.9}$$

Eq. (1.9) can be of course rewritten in terms of the variables x_i, ξ_i :

$$(1.10) \qquad \frac{\partial P}{\partial t} + \sum_{i=1}^{N} \xi_i \cdot \frac{\partial P}{\partial x_i} + \sum_{i=1}^{N} X_i \cdot \frac{\partial P}{\partial \xi_i} = 0$$

where $\partial P/\partial x_i$ is the gradient in the three-dimensional space of the positions of the i-th particle, $\partial P/\partial \xi_i$ the gradient in the three-dimensional space of velocities of the i-th particle.

Once P is known, it is possible to deduce, by integration, the reduced (s-body) distribution functions:

$$(1.11) \qquad P^{(s)} = \int P \prod_{i=s+1} dx_i \, d\xi_i$$

where integration extends to the entire 3(N-s)-dimensional space described by the vectors x_i, ξ_i $(s+1 \leq i \leq N)$. The function $P^{(s)}(x_i, \xi_i, t)$ gives the probability density that, at time t, the i-th mass point $(1 \leq i \leq s)$ has position between x_i and $x_i + dx_i$ and velocity between ξ_i and $\xi_i + d\xi_i$, as one uses to say in a shortened but suggestive way.

If general nonequilibrium phenomena are to be considered, a direct use of the Liouville equation is out of question when N has an order of magnitude comparable with the number of atoms contained in a macroscopic volume of gas. In order to be able to say something, it is essential to work with the reduced distribution functions $P^{(s)}$, corresponding to low values of s.

An equation for $P^{(1)}$ was derived by Boltzmann in 1872; this equation can be written as follows:

$$(1.12) \qquad \frac{\partial P^{(1)}}{\partial t} + \xi \cdot \frac{\partial P^{(1)}}{\partial x} = Q(P^{(1)}, P^{(1)})$$

where $Q(P^{(1)}, P^{(1)})$ is a complicated expression which we shall consider in more detail below and, to simplify, we assumed that no external forces are exerted on our system (except at the boundary).

The first problem which arises in connection with equation (1.12) is whether it is compatible with the Liouville equation, Eq. (1.10); as a matter of fact, as we said above, $P^{(1)}$ may be computed (in principle), once the solution P of Eq.(1.10) is known. It is a well known fact that a naive interpretation of this compatibility leads to paradoxes, as was elucidated by the controversies which opposed Boltzmann on one side, Zermelo and Loschmidt on the other (1,2). Such paradoxes are centered upon the fact that the Liouville equation is time-reversible (it goes into itself when we change t into $-t$, ξ_i into $-\xi_i$), whilst the Boltzmann equation describes an irreversible behavior, as is shown by the famous H theorem to be discussed in section 3.

Accordingly, if the Boltzmann equation is to be justified, this can be achieved only in a suitable sense, to be defined. To be precise, let us consider, for the sake of simplicity, a system of N rigid spheres (which, of course, is a limiting case of a system of points with strongly repulsive forces). Let σ denote the diameter of the spheres. One may suggest the following conjecture: when $N \to \infty$ and $\sigma \to 0$ in such a way that $N\sigma^2$ remains finite, the behavior of the system is described by the Boltzmann equation for a certain set of initial data and $t > 0$ (and not for $t < 0$) (3).

One must, of course, define the meaning of the limit of a system described by certain variables when the number of these variables tends to infinity. This is not exceedingly difficult if one considers only solutions of the Liouville equation which depend symmetrically upon the N sextuples of variables $(\mathbf{x}_i, \xi_i)$ and adopting a definition which imposes obvious conditions upon $P^{(s)}$ with a fixed s when $N \to \infty$. It is then possible (4), as we shall see, to "prove" the compatibility of the Boltzmann equation with the Liouville equation.

The first step is to deduce the following hierarchy of equations:

$$\frac{\partial P_N^{(s)}}{\partial t} + \Sigma\, \xi_i \cdot \frac{\partial P_N^{(s)}}{\partial \mathbf{x}_i} = (N-s)\sigma^2 \sum_{i=1}^{s} \int [P^{(s+1)'} - P_N^{(s+1)}] |V_i \cdot n| \, dn \, d\xi_* \tag{1.13}$$

$$(s = 1,2,\ldots,N)$$

where a subscript N has been added to $P^{(s)}$ in order to emphasize the dependence upon N, the integration with respect to the unit vector n is extended to the hemisphere $V_i \cdot n > 0$ the arguments of $P_N^{(s+1)}$ are the same as those of $P_N^{(s)}$ (i.e.

$t, x_j, \xi_i, 1 \leqslant j \leqslant s$) plus $x_i + n\sigma$ and ξ_i and the arguments of $P_N^{(s+1)'}$ are the same as those of $P_N^{(s-1)}$ except for $x_i - n\sigma$ which is replaced by $x_i + n\sigma$ and ξ_i, ξ_* which are replaced by ξ_i', ξ_*' given by:

$$\text{(1.14)} \qquad \begin{aligned} \xi_i' &= \xi_i - n(n \cdot V_i) \\ \xi_*' &= \xi_* + n(n \cdot V_i) \end{aligned} \qquad (V_i = \xi_i - \xi_*)$$

Eq. (1.13) can be derived under the only assumptions of symmetrical dependence of $P_N^{(s)}$ upon the particles and sufficient smoothness of $P_N^{(s)}$ (delta-like singularities are not allowed (4). In addition a surface integral extended to the boundary of the region where the system is enclosed is assumed to be zero, which is rigorously correct if the boundary specularly reflects the particles.

In particular, for $s = 1$ Eq. (1.13) gives:

$$\text{(1.15)} \qquad \frac{\partial P_N^{(1)}}{\partial t} + \xi_1 \cdot \frac{P_N^{(1)}}{\partial x_1} = (N-1)\,\sigma^2 \int [P_N^{(2)'} - P_N^{(2)}]\,|\,V_1 \cdot n\,|\, dn d\xi_*$$

This equation shows that the time evolution of the one-particle distribution function, $P_N^{(1)}$, depends upon the two-particle distribution function, $P_N^{(2)}$. In order to have a closed form equation for $P_N^{(1)}$, it is necessary to express $P_N^{(2)}$ in terms of $P_N^{(1)}$; a simple intuitive way of doing this is to assume the absence of correlation, i.e. to write:

$$\text{(1.16)} \qquad P_N^{(2)}(x_1, \xi_1, x_*, \xi_*, t) = P_N^{(1)}(x_1, \xi_1, t)\; P_N^{(1)}(x_*, \xi_*, t)$$

This relation can be obtained (2) in the case of thermal equilibrium for $N \to \infty$. If we accept it even in the case of non-equilibrium and insert Eq. (3.2) into Eq. (3.1) an equation involving $P_N^{(1)}$ alone is found. This is essentially the "*stosszahlansatz*" used by Boltzmann (6) to derive the equation for $P_N^{(1)}$, which is the Boltzmann equation, written above, Eq. (1.12).

We have no right, however, to postulate Eq. (1.16) because $P_N^{(2)}$, is determined by another equation (Eq. (1.13) with $s = 2$) involving $P_N^{(3)}$, and the latter in turn, by another equation, Eq. (1.13) with $s = 3$, involving $P_N^{(4)}$, etc. The least requirement is, therefore, to show that Eq. (1.16) is not in contrast with the equation regulating the time evolution of $P_N^{(s)}$ $(s \geq 2)$. Now, we cannot prove this statement, at least if we take it literally. In fact, Eq. (1.16) means that the states of the two molecules considered are statistically uncorrelated. Now, this makes sense for any two randomly chosen molecules of the gas, since they do not interact when they are far apart, and therefore behave independently. In particular, this seems true for two molecules which are going to collide, because they are just two random molecules whose paths happen to cross; but the same statistical independence is far from being true for two molecules which have just collided. We note, however, that Eq. (1.15) involves $P_N^{(2)}$ for molecules that are entering a collision, because when proving Eq. (1.13), one eliminates the values of $P_N^{(s+1)}$ corresponding to after--collision states (4). This remark is important, but problems still arise because Eq. (1.13) for $P_N^{(s)}$ is valid provided $\mathbf{x}_i$ $(i = 1, \ldots, s)$ is outside the sets $|\mathbf{x}_i - \mathbf{x}_j| \leq \sigma$ $(j = 1, \ldots s;\ j \neq i)$; the volume of these sets grows linearly with s, being proportional to $s\sigma^3$. These sets are, however, negligible in the limit $\sigma \to 0$, $N \to \infty$ for a fixed s (or even if we let s grow with N $(s \leq N)$, provided $N\sigma^3 \to 0$ as is the case for a perfect gas). We conclude that Boltzmann's *ansatz,* Eq. (1.16), is not true in a literal sense, but could become true for $N \to \infty$, $\sigma \to 0$ provided we specify that Eq.(1.16) is valid almost everywhere, i.e. ceases to be valid in exceptional sets of zero measure (among which the set of after-collision states).

Accordingly, we must prove that Eq. (1.16) (for $N \to \infty$, $\sigma \to 0$) is not in contrast with the equations governing the time evolution of $P_N^{(s)}$ $(s \geq 2)$. We shall prove more, i.e. that the factorization property:

$$P^{(s)} = \prod_{i=1}^{s} P^{(1)}(\mathbf{x}_i, \xi_i, t) \tag{1.17}$$

(where

$$P^{(s)} = \lim_{N \to \infty} P^{(s)}\ ) \tag{1.18}$$

is not in contrast with Eqs. (1.13) provided $\sigma \to 0$ in such a way that $N\sigma^2$ is bounded (hence $N\sigma^3 \to 0$). In order to prove this, we shall assume that the limit shown in Eq. (1.18) exists for any finite s and the resulting function $P^{(s)}$ is sufficiently smooth.

Then, if we fix s and let $N \to \infty$, $\sigma \to 0$ in Eq. (1.13) in such a way that $N\sigma^2$ is bounded, we obtain:

$$(1.19)\qquad \frac{\partial P^{(s)}}{\partial t} + \sum_{i=1}^{s} \xi_i \cdot \frac{\partial P^{(s)}}{\partial x_i} = (N\sigma^2)\, \Sigma \int [P^{(s+1)'} - P^{(s+1)}]\, |\, V_i \cdot n\, |\, dn d\xi_*$$

$$(s = 1, 2, 3, \ldots)$$

where the arguments in $P^{(s+1)'}$ and $P^{(s+1)}$ are the same as above, except for the fact that, of course, the arguments $x_i - n\sigma$, and $x_i + n\sigma$, reduce to x_i. Eqs. (1.19) give a complete description of the time evolution of a Boltzmann gas, provided the initial value problems is well set for this system of equations.

A particular solution of Eqs. (1.19) can be found in the form given by Eq. (1.17) provided the one-particle distribution function satisfies

$$(1.20)\qquad \frac{\partial P}{\partial t} + \xi \cdot \frac{\partial P}{\partial x} = (N\sigma^2) \int (P'P'_* - PP_*)\, |\, V \cdot n\, |\, dn d\xi_*$$

where we wrote ξ and x in place of ξ_1 and x_1, P in place of $P^{(1)}$ while P_*, P', P'_* denote that the argument ξ appearing in P is to be replaced by ξ_*, ξ', ξ'_* respectively. The above statement is straightforwardly verified by substituting Eq. (1.17) into Eq. (1.19) provided Leibnitz's rule for differentiating a product is used when evaluating the time derivative of $P^{(s)}$.

Hence, if the system of Eqs. (1.19) admits a unique solution for a given initial datum, we conclude that the solution corresponding to a datum satisfying the "chaos assumption":

$$(1.21)\qquad P^{(s)} = \prod_{i=1}^{s} P^{(1)}(x_i, \xi_i, 0) \qquad (t = 0)$$

will remain factored for all subsequent times and the one-particle distribution function $P = P^{(1)}$ will satisfy the Boltzmann equation. Therefore the factorization assumption, Eq. (1.17) is not inconsistent with the dynamics of rigid spheres in the limit $N \to \infty$, $\sigma \to 0$ ($N\sigma^2$ bounded) and leads to the Boltzmann equation.

One may ask where irreversibility, which, as we shall see, follows from equation (1.20), crept into the latter equation. It was in a seemingly innocent step made when deducing Eq. (1.13); in the proof (4), one uses the laws of elastic impact between spheres to eliminate the state after a collision in favor of the state before the collision. This is correct, because we want to use the equations to predict the future from the past, but, had we acted in the opposite way (which is exactly on the same foot from a purely mathematical viewpoint) and then repeated the same steps as above, we would have obtained an "anti-Boltzmann" equation, identical to Eq. (1.20) except for a minus sign in front of the right hand side! In other words, when deriving the Boltzmann equation, we chose a precise meaning for the inequality $t > 0$, which is a purely conventional statement in the dynamics of a conservative system. The answer to the irreversibility paradox is, therefore that the Boltzmann equation cannot describe the time-reversed process, because we renounced to describe the after-collision states which become the before-collision states in the reversed process.

Before concluding this section, let us look more carefully at the equation which we have obtained, Eq. (1.20). We note that it is a nonlinear, integral, partial differential, functional equation, where the specification "functional" refers to the fact that the unknown function P appears in the integral term not only with the arguments ξ (the current velocity variable) and ξ_* (the integration variable) but also with the arguments ξ' and ξ'_*. The latter variables are related to ξ and ξ_* by the condition of being transformed into ξ and ξ_* by the effect of a collision, according to Eqs. (2.12), i.e.:

$$\begin{aligned} \xi' &= \xi - n(n \cdot V) \\ \xi'_* &= \xi_* + n(n \cdot V) \end{aligned} \qquad (V = \xi - \xi_*) \tag{1.22}$$

The integral in the right hand side of Eq. (1.20); which is called the collision term, is extended to all the values of ξ_* and the hemisphere $|n| = 1$, $V \cdot n > 0$. We observe that it could be equivalently extended to the whole unit sphere and divided by 2, because changing **n** into $-\mathbf{n}$ does not alter the integrand.

Frequently, when dealing with the Boltzmann equation, one introduces a different unknown f which is related to P by:

$$f = NmP = MP \tag{1.23}$$

where N is the number of molecules, m the mass of a molecule and M the total mass. The meaning of f is an (expected) mass density in the phase space of a single particle, i.e. the (expected) "mass per unit volume" in the six-dimensional space described by $(\mathbf{x}, \boldsymbol{\xi})$. We note that because of the normalization condition

$$\int P \, d\mathbf{x} \, d\boldsymbol{\xi} = 1 \tag{1.24}$$

we have

$$\int f \, d\mathbf{x} \, d\boldsymbol{\xi} = M \tag{1.25}$$

It is clear that in terms of f we have:

$$\frac{\partial f}{\partial t} + \boldsymbol{\xi} \cdot \frac{\partial f}{\partial \mathbf{x}} = \frac{\sigma^2}{m} \int (f' f'_* - f f_*) \, |\mathbf{V} \cdot \mathbf{n}| \, d\mathbf{n} \, d\boldsymbol{\xi}_* \tag{1.26}$$

wheres, $f_* = f(\boldsymbol{\xi}_*)$, $f'_* = f(\boldsymbol{\xi}'_*)$, $f' = f(\boldsymbol{\xi}')$. This is the form of the Boltzmann equation for a gas of rigid spheres which will be used in the following.

The above considerations could be repeated if an external force per unit mass, $\mathbf{X}$, acts on the molecules, the only influence of this force being that one should add a term $\mathbf{X} \cdot \partial f/\partial \boldsymbol{\xi}$ the left-hand side of Eq. (1.26). Since we shall usually consider cases when the external action on the gas is exerted through solid boundaries (surface forces), we shall not usually write the abovementioned term describing body forces; it should be kept in mind, however, that such simplification implies neglecting, *inter alia*, gravity.

A Boltzmann equation can be also obtained for molecules interacting with at-distance force, provided they are short-range. Then the range σ takes the place of the molecular diameter in the limiting process described above. Then although things are less straightforward, an equation similar to Eq. (1.26) follows. The only change is that the factor

$$\sigma^2 \mathbf{V} \cdot \mathbf{n} \; d\mathbf{n} = \sigma^2 V \cos\theta \sin\theta \, d\theta \, d\epsilon \tag{1.27}$$

where θ is the angle between $\mathbf{n}$ and $\mathbf{V}$, ϵ the azimuth of $\mathbf{n}$ in a system of polar coordinates with $\mathbf{V}$ as polar axis, is replaced by

$$V r dr \, d\epsilon = V r \frac{\partial r}{\partial \theta} d\theta d\epsilon = B(\theta, V) \, d\theta \, d\epsilon \tag{1.28}$$

where r is the so called impact parameter related to θ by two-body dynamics (2, 3, 5, 7). In particular, for molecules interacting with a power law potential, $B(\theta, V)$ factorizes into the product of a fractional power of V and a function of θ. For the so called Maxwellian molecules (inverse fifth power force) $B(\theta, V)$ reduces to a function of θ alone.

We remark that the proof given above is not rigorous from the viewpoint of mathematical analysis. A rigorous proof has been announced by O.H. Lanford III (8) for the case of rigid spheres under suitable restrictive assumptions on P_N.

2. Boltzmann Equation and Continuum Fluid Dynamics

In this lecture we shall consider some of the basic properties of the Boltzmann equation and the relationship of the latter to continuum mechanics, in particular, with macroscopic fluid-dynamics. Let us start with a study of the right hand side of Eq.(1.26), which, according to the remarks made at the end of the first lecture, can be generalized as follows:

$$\frac{\partial f}{\partial t} + \xi \cdot \frac{\partial f}{\partial x} = Q(f, f) \tag{2.1}$$

where (Q(f,f) is a quadratic expression defined by (see Eq. (1.28)):

$$Q(f, f) = \frac{1}{m} \int (f' f'_* - f f_*) \, B(\theta, V) \, d\xi_1 \, d\epsilon d\theta \tag{2.2}$$

The operator Q acts on the velocity-dependence of f; it describes the effect of interactions, and is accordingly called the collision operator. Q(f, f), i.e. the integral in Eq. (2.2) is called the collision integral or, simply, the collision term. We shall study some properties which make the manipulation of Q possible in many problems of basic character in spite of its complicated form. Actually, we shall study here a slightly more general expression, the bilinear quantity

$$Q(f, g) = \frac{1}{2m} \int (f' g'_* + f'_* g' - f g_* - f_* g) \, B(\theta, V) \, d\xi_* \, d\epsilon d\theta \tag{2.3}$$

It is clear that when $g = f$, Eq. (2.3) reduces to Eq. (2.2); in addition,

$$Q(f, g) = Q(g, f) \tag{2.4}$$

Our first aim is to study some manipulations of the eightfold integral

$$\int Q(f, g) \, \varphi(\xi) d\xi = \frac{1}{2m} \int (f' g'_* + f'_* g' - f g_* - f_* g) \, \varphi(\xi) \, B(\theta, V) \, d\xi d\xi_* d\theta d\epsilon \tag{2.5}$$

where the integrals with respect to ξ are extended to the whole velocity space and $\varphi(\xi)$ is any function of ξ such that the indicated integrals exist.

We now perform the interchange of variables $\xi' \to \xi'_*$, $\xi_* \to \xi$ (which implies also $\xi' \to \xi'_*$, $\xi'_* \to \xi'$ because of Eqs. (1.22)). Then, since both $B(\theta, V)$ and the quantity within parentheses transform into themselves, and the Jacobian of the transformation is obviously unity, we have

$$\int Q(f,g)\,\varphi(\xi)d\xi = \frac{1}{2m}\int (f'g'_* + f'_*g' - fg_* - f_*g)\,\varphi(\xi_*)\,B(\theta,V)\,d\xi d\xi_*\,d\theta d\epsilon \tag{2.6}$$

This equation is identical to Eq. (2.5) except for having $\varphi(\xi_*)$ in place of $\varphi(\xi)$. Now we consider another transformation of variables in Eq. (2.5): $\xi \to \xi'$ and $\xi_* \to \xi'_*$ (here, as above, the unit vector $\mathbf{n}$ in Eq. (1.22) is considered as fixed). It is an easy matter to prove that the absolute value of the Jacobian of this transformation is unity, i.e. (2, 3, 7, 8) $d\xi d\xi_* = d\xi' d\xi'_*$ and Eq. (6.5) becomes

$$\int Q(f,g)\,\varphi(\xi)\,d\xi = \frac{1}{2m}\int (f'g'_* + f'_*g' - fg_* - f_*g)\,\varphi(\xi)\,B(\theta,V)\,d\xi'\,d\xi'_*\,d\theta d\epsilon \tag{2.7}$$

where now, since ξ' and ξ'_* are integration variables, we must express ξ and ξ_* by means of the relations inverting Eqs. (1.22) which are:

$$\xi = \xi' - \mathbf{n}(\mathbf{n}\cdot\mathbf{V}')$$
$$\xi_* = \xi'_* + \mathbf{n}(\mathbf{n}\cdot\mathbf{V}') \tag{2.8}$$

where $\mathbf{V}' = \xi' - \xi'_*$ is related to $\mathbf{V} = \xi - \xi_*$ by:

$$\mathbf{V}' = \mathbf{V} - 2\mathbf{n}(\mathbf{n}\cdot\mathbf{V}) \tag{2.9}$$

and, consequently:

$$\mathbf{V}'\cdot\mathbf{n} = -\mathbf{V}\cdot\mathbf{n} \tag{2.10}$$

and the hemisphere $V \cdot n > 0$ corresponds to $V' \cdot n < 0$; we may change, however, n into $-n$, without altering the expressions of ξ, ξ_* and integrate over the hemisphere $V' \cdot n > 0$. We can also change the names of integration variables and call ξ, ξ_* what we called ξ', ξ'_* before. Then, because of Eqs. (1.22) and (2.8), we can consistently call ξ' and ξ'_* what we called ξ and ξ_* before, and write Eq. (2.7) as follows:

$$(2.11) \quad \int Q(f,g)\varphi(\xi)d\xi = \frac{1}{2m}\int (fg_* + f_*g - f'g'_* - f'_*g')\varphi(\xi')B(\theta,V)d\xi d\xi_* d\theta d\epsilon$$

where $B(\theta, V)$ is not affected by the change, since Eq. (2.9) implies $V' = V$. We can rewrite Eq. (2.11) as follows:

$$(2.12) \quad \int Q(f,g)\varphi(\xi)d\xi = -\frac{1}{2m}\int (f'g'_* + f'_*g' - f g_* - f_*g)\varphi(\xi')B(\theta,V)d\xi d\xi_* d\theta d\epsilon$$

This equation is identical to Eq. (2.5) except for a minus sign and having $\varphi(\xi'_*)$ in place of $\varphi(\xi)$.

Finally, let us interchange ξ and ξ_* in Eq. (2.12) as we did in Eq. (2.5) to obtain Eq. (2.6). The result is:

$$(2.13) \quad \int Q(f,g)\varphi(\xi)d\xi = -\frac{1}{2m}\int (f'g'_* + f'_*g' - fg_* - f_*g)\varphi(\xi'_*)B(\theta,V)d\xi d\xi_* d\theta d\epsilon$$

which is identical to Eq. (2.5) except for a minus sign and for having $\varphi(\xi'_*)$ in place of $\varphi(\xi)$.

We have thus obtained four different expressions for the same quantity: Eqs. (2.5), (2.6), (2.12), (2.13). We can now obtain more expressions by taking appropriate linear combinations of the four basic ones; we are particularly interested in the combination which is obtained by adding the above four expressions and dividing by four. The result is:

$$\int Q(f,g)\varphi(\xi)d\xi = \frac{1}{8m}\int (f'g'_* + f'_*g' - fg_* - f_*g)(\varphi + \varphi_* - \varphi' - \varphi'_*)B(\theta,V)d\xi d\xi_* d\theta d\epsilon \tag{2.14}$$

This equation expresses a basic property of the collision term, which will be frequently used in the following. In the particular case of $g = f$ Eq. (2.14) reads:

$$\int Q(f,f)\varphi(\xi)d\xi = \frac{1}{4m}\int (f'f'_* - ff_*)(\varphi + \varphi_* - \varphi' - \varphi'_*)B(\theta,V)d\xi d\xi_* d\theta d\epsilon \tag{2.15}$$

We now observe that the integral appearing in Eq. (2.14) is zero, independent of the particular f and g, if

$$\varphi + \varphi_* = \varphi' + \varphi'_* \tag{2.16}$$

is valid almost everywhere in velocity space. Since the integral appearing in the left hand side of Eq. (2.13) is the average change of the function $\varphi(\xi)$ in unit time by the effect of the collisions, the functions satisfying Eq. (2.16) are usually called "collision invariants". We now have the property that, if $\varphi(\xi)$ is assumed to be continuous, then Eq. (2.16) is satisfied if and only if

$$\varphi(\xi) = a + b\cdot\xi + c\,\xi^2 \tag{2.17}$$

where a and c are constant scalars and b a constant vector. The functions $\psi_0 = 1$, $(\psi_1, \psi_2, \psi_3) = \xi$, $\psi_4 = \xi^2$ are usually called the elementary collision invariants; thus a general collision invariant is a linear combination of the five ψ's.

In order to prove the above statement that Eq. (2.16) is satisfied if and only if $\varphi(\xi)$ has the form shown in Eq. (2.17), one may argue, on physical grounds, that the existence of a further collision invariant, linearly independent of the five ψ's would lead to a relation between ξ', ξ'_* on one side and ξ, ξ_* on the other involving just one parameter rather than the two parameters as dictated by the mechanics of a collision. For strictly mathematical proofs, we refer the reader to Grad (9), Carleman (10) or a forthcoming book of the author (5).

As a first, important application of the above results, we shall now investigate the existence of positive functions f which give a vanishing collision integral:

$$Q(f,f) = \int (f' f'_* - f f_*) B(\theta, V) d\xi_* d\theta d\epsilon = 0 \tag{2.18}$$

We want to show that such functions exist and are all given by

$$f(\xi) = \exp(a + b \cdot \xi + c\xi^2) \quad (c < 0) \tag{2.19}$$

where a, b, c have the same meaning as in Eq. (2.17). In order to show that this statement is true, we prove a preliminary result which will also be important later, i.e. that no matter what the distribution function is, the following inequality (Boltzmann's inequality) holds:

$$\int \log f \, Q(f,f) d\xi \leqslant 0 \tag{2.20}$$

and the equality sign applies if, and only if, f is given by Eq. (2.19). Now it is seen that the first statement is a simple corollary of the second one: in fact, if Eq. (2.18) is satisfied then multiplying it by log f and integrating gives Eq. (2.20) with the equality sign, which implies Eq. (2.19) if the second statement applies. *Vice versa,* if Eq. (2.19) holds, then, because of Eq. (2.16), applied to the function $\varphi = \log f$, $f' f'_* = f f_*$ and Eq. (2.18) is satisfied.

Let us prove, therefore, that Eq. (2.20) always holds for $f > 0$ and the equality sign implies, and is implied by, Eq. (2.19). If we use Eq. (2.15) with $\varphi = \log f$ we have:

$$\begin{aligned} \int \log f \, Q(f,f) d\xi &= \frac{1}{4m} \int (f' f'_* - f f_*) \log (f f_* / f' f'_*) B(\theta,V) d\xi d\xi_* d\theta d\epsilon \\ &= \frac{1}{4m} \int f' f'_* (1 - \lambda) \log \lambda \, B(\theta,V) d\xi d\xi_* d\theta d\epsilon \end{aligned} \tag{2.21}$$

where

$$\lambda = ff_*/(f'f'_*) \tag{2.22}$$

Now $f'f'_* > 0$, $B \geqslant 0$ (the equality sign applying only at $\theta = 0$); also, for any $\lambda \geqslant 0$ we have

$$(1 - \lambda) \log \lambda \leqslant 0 \tag{2.23}$$

and the equality sign applies if, and only if, $\lambda = 1$ (note that $(1 - \lambda)$ and $-\log \lambda$ are negative and positive together and both are zero if and only if $\lambda = 1$). If we use (Eq. (2.23), Eq. (2.21) implies Eq. (2.20) and the equality sign applies if, and only if, $\lambda = 1$, i.e.:

$$ff_* = f'f'_* \tag{2.24}$$

applies almost everywhere. But taking the logarithms of both sides of this equation, we obtain that $\varphi = \log f$ satisfies Eq. (2.16), i.e. $\varphi = \log f$ is given by Eq. (2.17); hence f is given by Eq. (2.19), as was to be shown.

We note that in Eq. (2.19) c must be negative, since f must be integrable over the whole velocity space. If we put $c = -\alpha$, $b = 2\alpha v$, where v is another constant vector, Eq. (2.19) can be written as follows:

$$f(\xi) = A \exp[-\alpha(\xi - v)^2] \tag{2.25}$$

where A is a constant related to a, α, v^2 (α, v, A constitute a new set of arbitrary constants). Eq. (2.25) is the familiar Maxwellian distribution; it is different from the usual form, because Eq. (2.25) describes a gas which is not at rest (for $v \neq 0$). However, Eq. (2.25) reduces to the usual form, if we change the reference frame to one moving with velocity v with respect to the frame for which Eq. (2.25) holds, and express the quantities A and α suitably in terms of internal energy and mass density. This interpretation will be shown to be correct at the end of this lecture.

Our next concern will be to show that the Boltzmann equation is not in contrast with the general conservation equations of continuum mechanics; on the contrary, the latter follow from the Boltzmann equation, thus showing that the Boltzmann gas is a continuum (remember that it is the limit of a particle model). In order to achieve this result, we must first show how to evaluate the macroscopic quantities once the distribution function is given.

The first quantity is the density in physical space ρ (x, t), which is nothing else than the integral of the density in the one-particle phase space f(x, ξ, t) with respect to all possible velocities

$$\rho(\mathbf{x}, t) = \int f d\xi \tag{2.26}$$

Because of the probabilistic meaning of f, the density ρ is the expected mass per unit volume at **x**, t or the product of the molecular mass m by the probability density of finding a molecule at **x**, t, i.e. the (expected) number density n(**x**, t)

$$n(\mathbf{x}, t) = \rho(\mathbf{x}, t)/m = \int P d\xi \tag{2.27}$$

where P is the probability density related to f by Eq. (1.23).

The mass velocity **v** is given by the average of the molecular velocity ξ

$$\mathbf{v} = \frac{\int \xi P d\xi}{\int P d\xi} = \frac{\int \xi f d\xi}{\int f d\xi} \tag{2.28}$$

where the integral in the denominator is due to the fact that P is not normalized to unity when we consider **x** as fixed and integrate only with respect to ξ . Because of Eq. (2.26), Eq. (2.28) can also be written as follows

$$\rho \mathbf{v} = \int \xi f d\xi \tag{2.29}$$

or, using components:

$$\rho v_i = \int \xi_i f d\xi \tag{2.30}$$

The mass velocity $\mathbf{v}$ is what we can directly perceive of the molecular motion by means of macroscopic observations; it is zero for the steady state of a gas enclosed in a specularly reflecting box at rest. Each molecule has its own velocity ξ which can be decomposed into the sum of $\mathbf{v}$ and another velocity

$$\mathbf{c} = \xi - \mathbf{v} \tag{2.31}$$

which describes the random deviation of the molecular velocity from the ordered motion with velocity $\mathbf{v}$. The velocity $\mathbf{c}$ is usually called the peculiar velocity or the random velocity; it coincides with ξ when the gas is macroscopically at rest. We note that, because of Eqs. (2.31), (2.30) and (2.26), we have:

$$\int c_i f d\xi = \int \xi_i f d\xi - v_i \int f d\xi = \rho v_i - \rho v_i = 0 \tag{2.32}$$

The quantity ρv_i which appears in Eq. (2.30) can be interpreted as the momentum density or, alternatively, as the mass flow (in the i direction). Other quantities which will be needed in the following are the momentum flow, the energy density and the energy flow. Since momentum is a vectory quantity, we have to consider the flow of the j component of momentum in the i direction; this is given by:

$$\int \xi_i (\xi_j f) d\xi = \int \xi_i \xi_j f d\xi \tag{2.33}$$

where we use the general fact that if a quantity has a density G in phase space (in this case $G = \xi_j f$), the associated flow through a surface S (i.e. the amount of that quantity that goes through S per unit surface and unit time) is given by $\int G \xi_n dt\, dS\, d\xi/(dS\, dt) = \int G \xi_n d\xi$, where integration is extended to all the possible velocities and dS denotes the area of a surface element, ξ_n the component of ξ along the normal to such element. Eq. (2.33) shows that the momentum flow is described by a symmetric tensor of second order. It is to be expected that in a macroscopic description only a part of the microscopically evaluated momentum flow will be identified as such, because the integral in Eq. (2.33) will be in general different from zero even if the gas is macroscopically at rest (absence of macroscopic momentum flow). In order to find out how the above momentum flow will appear in a macroscopic description, we have to use the splitting of ξ into mass velocity v and peculiar velocity c, according to Eq. (2.31). We have:

$$\begin{aligned}\int \xi_i \xi_j f d\xi &= \int (v_i + c_i)(v_i + c_j) f d\xi = v_i v_j \int f d\xi + v_i \int c_j f d\xi + \\ &+ v_j \int c_i f d\xi + \int c_i c_j f d\xi = \rho v_i v_j + \int c_i c_j f d\xi\end{aligned} \tag{2.34}$$

where Eq. (2.26) and (2.32) have been used. Thus the momentum flow decomposes into two parts, one of which is recognized as the macroscopic momentum flow (momentum density times velocity), while the second part is a hidden momentum flow due to the random motion of the molecules. How will this second part manifest itself in a macroscopic description? If we take a fixed region of the gas and observe the change of momentum inside it, we find that (in the absence of external body forces) the change can only in part be attributed to the matter which enters and leaves the region, leaving a second part which has no macroscopic explanation unless we attribute it to the action of a force exerted on the boundary of the region of interest by the contiguous regions of the gas. In other words the integral of $\int c_i c_j f d\xi$ appears as a contribution to the stress tensor (and, indeed, the only contribution to the stress tensor if the gas is a Boltzmann gas, for which the actual actions exerted by the molecules of a region on the molecules of another are neglected). We shall therefore write

$$p_{ij} = \int c_i c_j f d\xi \tag{2.35}$$

(a complete identification is correctly justified by the fact that, as we shall see later,

p_{ij} plays, in the macroscopic equations to be derived from the Boltzmann equation, the same role as the stress tensor in the conservation equations derived from macroscopic considerations).

An analogous decomposition is to be introduced for the energy density and energy flow. The energy density is given by $1/2 \int \xi^2 f d\xi$ and we have only to take $j = i$ and sum from $i = 1$ to $i = 3$ in Eq. (2.34) to deduce

$$\frac{1}{2} \int \xi^2 f d\xi = \frac{1}{2} \rho v^2 + \frac{1}{2} \int c^2 f d\xi \tag{2.36}$$

Again the first term in the right-hand side will be macroscopically identified with the kinetic energy density, while the second term will be ascribed to an "internal energy" of the gas. Therefore, if we introduce the internal energy per unit mass e we have for the density of internal energy per unit volume ρe :

$$\rho e = \frac{1}{2} \int c^2 f d\xi \tag{2.37}$$

We note that a relation exists between the internal energy density and the spur or trace (i.e. the sum of the three diagonal terms) of the stress tensor. In fact, Eqs. (2.37) and (2.35) give

$$p_{ii} = \int c^2 f d\xi = 2\rho e \tag{2.38}$$

(Here and in the following we shall use the convention of summing over repeated subscripts from 1 to 3, unless otherwise stated; in other words $p_{ii} \equiv \sum_{i=1}^{3} p_{ii}$). The spur divided by 3 gives the isotropic part of the stress tensor; it is therefore convenient to identify $p = p_{ii}/3$ with the gas pressure, at least in the case of equilibrium. The identification is also correct for non-equilibrium situations in the considered case of a monatomic perfect gas, but is generally incorrect. Therefore

$$p = 2\rho e \tag{2.39}$$

Eq. (2.39) is called the state equation of the gas and allows us to express any of the three quantities p, ρ, e in terms of the remaining two. Thermal equilibrium considerations show (5) that, for a monatomic perfect gas, e is a function of temperature, i.e. of an index which has the property of taking the same value for two systems in contact with each other in a state of equilibrium. Eq. (2.39) shows that p/ρ is constant at constant temperature for rarefied monatomic gases. It is this property which identifies such gases as the perfect gases obeying Boyle's law:

$$p = \rho R T \tag{2.40}$$

where T is the absolute temperature and R a constant (depending on the molecular mass, since $R = k/m$ where k is Boltzmann's universal constant). Eqs. (2.39) and (2.40) give the identification

$$e = \frac{3}{2} R T \tag{2.41}$$

We now have to investigate the energy flow; the total energy flow is obviously given by

$$\int \xi_i \left(\frac{1}{2} \xi^2 f\right) d\xi = \frac{1}{2} \int \xi_i \xi^2 \, d\xi \tag{2.42}$$

Using Eq. (2.31) gives:

$$\frac{1}{2} \int \xi_i \xi^2 f d\xi = \frac{1}{2} \int (v_i + c_i)(v^2 + 2c_j v_j + c^2) f d\xi =$$

$$= \frac{1}{2} v_i v^2 \int f d\xi + v_i v_j \int c_j f d\xi + \frac{1}{2} v_i \int c^2 f d\xi + \tag{2.43}$$

$$+ \frac{1}{2} v^2 \int c_i f d\xi + v_j \int c_i c_j f d\xi + \frac{1}{2} \int c_i c^2 f d\xi$$

i.e., using Eqs. (2.26), (2.32), (2.37), (2.35)

$$\frac{1}{2}\int \xi_i \xi^2 f d\xi = v_i(\frac{1}{2}\rho v^2 + e) + v_j p_{ij} + \frac{1}{2}\int c_i c^2 f d\xi \tag{2.44}$$

We now have three terms: the first one is obviously the energy flow due to macroscopic convection, and the second one is macroscopically interpreted as due to the work done by the stresses in unit time.

The third term is another kind of energy flow; the additional term is usually called the heat flux vector, and is denoted by **q**:

$$q_i = \frac{1}{2}\int c_i c^2 f d\xi \tag{2.45}$$

As for the case of the stress tensor, the identification is justified, as we shall see later, by the fact that **q** plays the same role as the flux vector in the macroscopic equations. However, the name "heat flux" is somewhat misleading, because there are situations when $q_i \neq 0$ and the temperature is practically constant everywhere; in this case one has to speak of a heat flux at constant temperature. The name "nonconvective energy flow" would be more appropriate for **q** but is not used.

The above discussion links the distribution function with the quantities used in the macroscopic description; in particular, e.g., p_{ij} can be used to evaluate the drag on a body moving inside the gas and **q** to evaluate the heat transfer from a hot body to a colder one when they are separated by a region filled by the gas.

In order to complete the connection, we now derive five differential equations, satisfied by the macroscopic quantities considered above, as a simple mathematical consequence of the Boltzmann equation; these equations are usually called the conservation equations, since they can be physically interpreted as expressing conservation of mass, momentum and energy.

In order to obtain these equations we consider the Boltzmann equation:

$$\frac{\partial f}{\partial t} + \xi_i \frac{\partial f}{\partial x_i} + X_i \frac{\partial f}{\partial \xi_i} = Q(f, f) \tag{2.46}$$

where for the sake of the generality we have introduced the body force term which is usually left out. We multiply both sides of Eq. (2.46) by the five collision invariants ψ_α ($\alpha = 0,1,2,3,4$) defined above and integrate with respect to ξ in accord with Eqs. (2.15) and (2.16) with $\varphi = \psi_\alpha$:

$$\int \psi_\alpha \, Q(f, f)\, d\xi = 0 \qquad (\alpha = 0, 1, 2, 3, 4) \tag{2.4}$$

for any f. Therefore for any f satisfying Eq. (2.46):

$$\frac{\partial}{\partial t}\int \psi_\alpha f\, d\xi + \frac{\partial}{\partial x_i}\int \xi_i \psi_\alpha f\, d\xi + X_i \int \psi_\alpha \frac{\partial f}{\partial \xi_i} = 0 \tag{2.48}$$

$$(\alpha = 0, 1, 2, 3, 4)$$

provided X_i does not depend on ξ.

If we take successively $\alpha = 0,1,2,3,4$ and use Eqs. (2.26), (2.29), (2.34)–(2.37), (2.44) and (2.45), we obtain

$$\frac{\partial \rho}{\partial t} + \frac{\partial}{\partial x_i}(\rho v_i) = 0$$

$$\frac{\partial}{\partial t}(\rho v_j) + \frac{\partial}{\partial x_i}(\rho v_i v_j + p_{ij}) = \rho X_j \tag{2.49}$$

$$\frac{\partial}{\partial t}\left[\rho\left(\frac{1}{2}v^2 + e\right)\right] + \frac{\partial}{\partial x_i}\left[\rho v_i\left(\frac{1}{2}v^2 + e\right) + p_{ij} v_j + q_i\right] = \rho X_i v_i$$

where we have also used the following relations:

$$\int \partial f/\partial \xi_i \, d\xi = 0 \,; \quad \int \xi_i \, \partial f/\partial \xi_i \, d\xi = -\int \delta_{ij} f\, d\xi = -\rho\delta_{ij} \,;$$

$$\frac{1}{2}\int \xi^2 \, \partial f/\partial \xi_i \, d\xi = -\int \xi_i f\, d\xi = -\rho v_i \tag{2.50}$$

which follow by partial integration and the conditions $\lim_{\xi\to\infty}(\psi_\alpha f) = 0$ which are required in order that all the integrals considered in the equations of this section exist. In the above $\delta_{ij} = 1$ for $i = j$, $\delta_{ij} = 0$ for $i \neq j$. Eqs. (2.49) are the basic equations of continuum mechanics, in particular of macroscopic gas dynamics; as they stand, however, they constitute an empty scheme, since there are five equations for 13 quantities (if Eq. (2.38) is taken into account). In order to have useful equations, one must have some expressions for p_{ij} and q_i in terms of ρ, v_i, e. Otherwise, one has to go back to the Boltzmann equation and solve it; and once this has been done, everything is done and Eqs. (2.49) are useless!

In any macroscopic approach to fluid dynamics, one has to postulate, either on the basic experiments or by plausible arguments, some phenomenological relations (the so-called "constitutive equations") between p_{ij}, q_i on one hand and ρ, v_i, e on the other. In the case of a gas, or, more generally, a fluid, there are two models which are well known: the Euler (or ideal) fluid:

$$p_{ij} = p\,\delta_{ij} \; ; \qquad q_i = 0 \tag{2.51}$$

and the Navier-Stokes-Fourier (or viscous and thermally conducting) fluid:

$$\begin{aligned} p_{ij} &= p\,\delta_{ij} - \mu\left(\frac{\partial v_i}{\partial x_j} + \frac{\partial v_j}{\partial x_i}\right) - \lambda \frac{\partial v_\kappa}{\partial x_\kappa}\,\delta_{ij} \\ q_i &= -\,\kappa\, \partial T/\partial x_i \end{aligned} \tag{2.52}$$

where μ and λ are the viscosity coefficients (usually one neglects the so called bulk viscosity; then $\lambda = -(2\mu)/3$) and k is the heat conduction coefficient (μ, λ and k can be functions of density ρ and temperature T).

No such relations are to be introduced in the microscopic description; the single unknown f contains all the information about density, velocity, temperature, stresses and heat flux! Of course, this is possible because f is a function of seven variables instead of four; the macroscopic approach (five functions of four variables) is simpler than the microscopic one (one function of seven variables) and is to be preferred whenever it can be applied. Therefore one of the tasks of a theory based on the Boltzmann equation is to deduce, for a gas in ordinary conditions, some approximate macroscopic model (in particular Eqs. (2.52) with μ, λ, k expressed in terms of molecular constants) and find out what the limits of application of this model are. For a long time this was considered to be the only task of the theory (7; 10). There are, however, regimes of such rarefaction that no general macroscopic theory in the usual sense is possible (constitutive equations such as Eq. (2.51) and (2.52) are not valid); in this case the Boltzmann equation must be solved and not used only to justify the macroscopic equations.

We note that if we apply Eqs. (2.26), (2.28), (2.37) to the Maxwellian given by Eq. (2.25) we find that the constant v appearing in the latter equation is actually the mass velocity, while

$$\alpha = 3(4e)^{-1} = (2RT)^{-1}, \quad A = \rho\left(\frac{4}{3}\pi e\right)^{-3/2} = \rho(2\pi RT)^{-3/2} \tag{2.53}$$

Furthermore,

$$p_{ij} = p\delta_{ij} = \frac{2}{3}\rho e\delta_{ij} \qquad q_i = 0 \tag{2.54}$$

i.e. a gas with a Maxwellian distribution satisfies the constitutive equations of the Euler fluid, Eq. (2.51).

3. Boundary Value Problems and H-Theorem

In this section we shall consider some aspects of the boundary value problems related to the Boltzmann equation. In particular, we shall prove an extension of the H-theorem of Boltzmann, which shows that the Boltzmann equation contains a description of time irreversible phenomena. As usual, we shall write the Boltzmann equation in the following form:

$$\frac{\partial f}{\partial t} + \xi \cdot \frac{\partial f}{\partial x} = Q(f, f) \qquad (3.1)$$

The boundary conditions to be matched with Eq. (31) are as important as the equation itself and were firstly investigated by Maxwell in 1879 in an Appendix to his last paper where he proposed the boundary conditions which bear his name. A systematic study of the form of the boundary conditions began only recently (2, 12-17). Under suitable assumptions, one can write the boundary conditions at a solid wall in the following form:

$$|\xi \cdot n| f(\xi) = \int_{\xi' \cdot n < 0} R(\xi' \to \xi) f(\xi') |\xi' \cdot n| d\xi' \qquad (\xi \cdot n < 0) \qquad (3.2)$$

where n is the normal to the solid wall pointing into the gas and $R(\xi' \to \xi)$ a scattering kernel which gives the probability density that a molecule impinging on the wall with velocity ξ' $(\xi' \cdot n < 0)$ will re-emerge with velocity ξ. The kernel $R(\xi' \to \xi)$ is not completely arbitrary because it must satisfy the following requirements:

1) The kernel is non-negative:

$$R(\xi' \to \xi) \geqslant 0 \qquad (3.3)$$

2) The kernel is normalized:

$$\int_{\xi \cdot n > 0} R(\xi' \to \xi) \, d\xi = 1 \qquad (\xi' \cdot n < 0) \qquad (3.4)$$

3) The kernel satisfies a reciprocity law:

$$f_w(\xi') \, |\xi' \cdot n| R(\xi' \to \xi) = f_w(\xi) |\xi \cdot n| R(-\xi \to -\xi') \qquad (3.5)$$
$$(\xi' \cdot n' < 0, \quad \xi \cdot n > 0)$$

where f_w is the Maxwellian corresponding to the temperature T_W and velocity u_w of the wall (assumed to be zero in Eqs. (3.2), (3.3), (3.4).

Eq. (3) is a simple consequence of the meaning of $R(\xi' \to \xi)$. Eq. (4) expresses the fact that no molecules remain trapped within the wall for a significant time interval. Eq. (5) has a more subtle meaning related to the assumptions that the wall is in local thermal equilibrium which is not disturbed by the gas molecules and the microscopic equations are time reversible (2, 13-17), Eqs. (4) and (5) imply

(3.6) $$\int_{\xi' \cdot n < 0} R(\xi' \to \xi) \mid \xi' \cdot n \mid f_w(\xi') d\xi' = \mid \xi \cdot n \mid f_w(\xi)$$

which ensures that, if the impinging molecules have distribution f_w, such distribution is not disturbed by the interaction with the wall. The set of Eqs. (4) and (6) is much less restrictive than the set of Eqs. (4) and (5). Eqs. (3), (4) and (6), however, are sufficient to prove (17) that

(3.7) $$\int f_w \, \xi \cdot n \; C(f/f_w) d\xi' \lessdot 0$$

where C is a convex function of its argument, f is evaluated at a point of the wall and integration is extended to the full ranges of values of the components of ξ, the values of f for $\xi \cdot n > 0$ being related to those for $\xi \cdot n < 0$ through Eq. (2). In particular, if we take $C(g) = g \log g - g$ and use Eq. (4), we obtain the inequality

(3.8) $$\int \xi \cdot n \; f \log f \; d\xi \leqslant - \frac{1}{RT_W} \mid q \cdot n \mid_{\text{solid}}$$

where q is the heat flux vector from the solid to the gas, evaluated inside the solid ($q \cdot n$ undergoes a jump at the interface when the gas slips on the wall (17). Eq. (8) is substantially due to Darrozès and Guiraud (12), a more detailed proof is given in Ref. (17)

We want to show that if we define

(3.9) $$\mathcal{H} = \int f \log f \; d\xi$$

$$\mathcal{H}_i = \int \xi_i \, f \log f \, d\xi \tag{3.10}$$

where f is any function which satisfies the Boltzmann equation, Eq. (2.46), then

$$\frac{\partial \mathcal{H}}{\partial t} + \frac{\partial \mathcal{H}_i}{\partial x_i} \leq 0 \tag{3.11}$$

Eq. (3.11) is an obvious consequence of the inequality in Eq. (2.20) and of the fact that f satisfies the Boltzmann equation; it is sufficient to multiply both sides of Eq. (2.46) by $1 + \log f$ and integrate over all possible velocities ξ, taking into account $d(f \log f) = (1 + \log f)df$, Eq. (2.20) and the fact that 1 is a collision invariant (vanishing of f for $\xi \to \infty$ is also understood).

We can now interpret Eq. (3.11) by rewriting it as follows:

$$\frac{\partial \mathcal{H}}{\partial t} + \frac{\partial \mathcal{H}_i}{\partial x_i} = \mathcal{S} \,, \qquad \mathcal{S} = \int \log f \; Q(f, f) d\xi \leq 0 \tag{3.12}$$

and introducing

$$H = \int_R \mathcal{H} \, dx \tag{3.13}$$

where R is a region filled by gas. If we had $\rho = 0$, then H would be a conserved quantity like mass, energy and momentum, since we can interpret the vector $\mathcal{H} = (\mathcal{H}_1, \mathcal{H}_2, \mathcal{H}_3)$ as the flow of H (note that here $\mathcal{H}$ is not the magnitude of the vector $\mathcal{H}$ but the density corresponding to the flow $\mathcal{H}$). Since $\rho \neq 0$ in general, but $\rho \neq 0$ we can say that molecular collisions act as a negative source for the quantity H. We also know that the source ρ is zero if and only if the distribution function is Maxwellian. Finally, we note that, as we did for momentum flow (Sect. 2), $\mathcal{H}$ can be split into a macroscopic (convective) flow of H, $\mathcal{H}u$ and a microscopic flow of H, $\mathcal{H} - \mathcal{H}u$.

If we integrate both sides of Eq. (3.12) with respect to x over the region R we have, if the boundary ∂R of R moves with velocity $\underset{\sim}{u}_o$.

$$\frac{dH}{dt} - \int (\mathcal{H} \cdot n - \mathcal{H} u_o \cdot n) dS = \int \mathcal{S} \, dx \leq 0 \tag{3.14}$$

where dS is a surface element of the boundary ∂R and n the inward normal. The second term in the integral comes from the fact that, if the boundary is moving, when forming the time derivative of H we have to take into account that the region of integration changes with time.

From Eq. (3.12) and (3.14) we deduce two classical forms of the celebrated H-theorem of Boltzmann:

a) If the gas is homogeneous ($\partial f/\partial x = 0$) and hence $\partial \mathcal{H}_i/\partial x_i = 0$ in Eq. (3.11) the quantity $\mathcal{H}$ (which is a function of time only) never increases with time and is steady if and only if the distribution function is Maxwellian (in fact $d\mathcal{H}/dt = 0$ implies $\rho = 0$ and because of the results of Sect. 2, f is Maxwellian)

b) If the gas is enclosed in a region such that the integral appearing in the left hand side of Eq. (9.6) is nonpositive:

$$\int_{\partial R} (\mathcal{H} - \mathcal{H}\, u_0) \cdot n \, dS \leqslant 0 \ , \tag{3.15}$$

the quantity H (which is a function of time only) never decreases with time and is steady if and only if the distribution function is Maxwellian.

The second form of the H-theorem covers more general situations but is somewhat unsatisfactory, because it can be applied only after having checked that Eq.(3.15) is satisfied. E.g., Eq. (3.15) is satisfied (with the equality sign), if the molecules are specularly reflected at the boundary ∂R (in the reference frame in which ∂R is at rest); this is the only case considered in the traditional treatments (9).

We can however obtain a sharp and significant form of the H-theorem if the boundary ∂R is a solid wall at which Eq. (3.2) holds. It is sufficient to note that

$$\begin{aligned} \int (\mathcal{H} - \mathcal{H}\, u_0) \cdot n \, dS &= \int (\xi - u_0) \cdot n \; f \log f \; d\xi \; dS \\ &= \int \xi_r \cdot n \; f \log f \; d\xi_r \; dS \end{aligned} \tag{3.16}$$

where $\xi_r = \xi - u_0$ is the molecular velocity in a reference frame moving with the wall. In this reference frame, Eq. (3.8) holds. Hence

$$\int (\mathcal{H} - \mathcal{H}\,u_o) \cdot dS \leqslant -\frac{1}{R}\int \frac{(q\cdot n)_{solid}}{T_o}\, dS \tag{3.17}$$

and Eq. (3.14) gives

$$\frac{dH}{dt} \leqslant -\frac{1}{R}\int \frac{(q\cdot n)_{solid}}{T_o}\, dS = -\frac{1}{R}\int \frac{d^*Q}{T_o} \tag{3.18}$$

where T_o may vary from point to point along ∂R and d^*Q is the heat transferred from the body to the gas at temperature T_o per unit time.

Hence we obtain the following

THEOREM (Generalized H-theorem). If the gas is bounded by nonporous solid walls at which a linear boundary condition, Eq. (3.8), holds, the quantity H given by Eq. (3.13) satisfies Eq. (3.18). In particular, if at no point of the boundary heat is flowing *from* the gas to the solid (i.e. at no point of ∂R $d^*Q < 0$) then H always decreases with time and can be constant only if the distribution function is Maxwellian.

Boltzmann's H-theorem is of basic importance because it shows that the Boltzmann equation has a basic feature of irreversibility: the quantity H always decreases even when it is not released to the surroundings [equality sign in Eq. (3.15)], or [Eq. (3.18)] when no energy exchange takes place between the gas and the surroundings.

This circumstance seems to be in conflict with the fact that the molecules constituting the gas follow the laws of classical mechanics which are time reversible. Accordingly, given at $t = t_o$ a motion with velocities $V_1, \ldots V_N$, we can always consider the motion with velocities $-V_1, \ldots, -V_N$ (and the same positions as before) at $t = t_o$; the backward evolution of the latter state will be equal to the forward evolution of the original one. Therefore if $dH/dt < 0$ in the first case, we shall have $dH/d(-t) < 0$ in the second case; i.e. $dH/dt > 0$, which contradicts Boltzmann's H theorem. This is Loschmidt's paradox (to simplify, the gas is assumed to be enclosed in a specularly reflecting box; otherwise the same objection applies to Eq. (3.2). The derivation of the Boltzmann equation given in Section 1 shows the solution of the paradox: for a more detailed discussion see Refs. 2, 3,5.

CHAPTER II

APPROXIMATING THE BOLTZMANN EQUATION

4. Flow Regimes and the Linearized Boltzmann Equation

If we want to solve the Boltzmann equation for realistic non-equilibrium situations, we must rely upon approximation methods, in particular perturbation procedures.

Two procedures of this kind have been long known, but they largely bypass both the significance and the difficulties of the Boltzmann equation. In order to describe them in a simple way, we observe that, if we denote by t a typical time, by L a typical length, by w a typical molecular velocity, then

$$\frac{\partial f}{\partial t} = O(t^{-1} f) \; ; \qquad \xi \cdot \frac{\partial f}{\partial x} = O(wL^{-1} f) \tag{4.1}$$

$$Q(f, f) = O(w\ell^{-1} f) \tag{4.2}$$

where ℓ is the mean free path, i.e. the average length of the free flight of a molecule between two successive interactions.

The combination $w\ell^{-1}$ can be considered as a natural measure of the collision frequency and its inverse $w^{-1}\ell$ as defining a mean free time θ. This discussion shows the existence of two basic nondimensional numbers in the Boltzmann equation, θ/t and ℓ/L. To simplify the discussion the length and time scales can be taken to be comparable in the sense that $\ell/L \cong \theta/t$. Accordingly, in a first approach, one can consider the left hand side of the Boltzmann equation as a single term of order of magnitude wL^{-1}, in such a way that the simple nondimensional number

$$Kn = \frac{\ell}{L} \tag{4.3}$$

expresses the relative magnitudes of the left and right hand side of the Boltzmann equation. It is clear that the so defined Knudsen number, Kn, ranges from 0 to ∞, $\text{Kn} \to 0$ corresponding to a fairly dense gas and $\text{Kn} \to \infty$ to a free molecular flow (i.e., a flow where molecules have negligible interactions with each other).

Two kinds of perturbation methods suggest themselves, one for $\text{Kn} \to 0$, the other for $\text{Kn} \to \infty$. The resulting expansions are known respectively as the Hilbert and Knudsen expansion (2).

Both these expansions, in their simplest form, present difficulties, due to the fact that the length and time scales are not always comparable, or there is more than one length scale. Accordingly, modified procedures have been proposed, the Chapman-Enskog expansion for $\text{Kn} \to 0$ and the successive collisions method for $\text{Kn} \to \infty$. The resulting expansions are useful in the so-called near-continuum (or slip) regime ($\text{Kn} \to 0$) and nearly free regime ($\text{Kn} \to \infty$). Accordingly an intermediate regime remains untouched by the above mentioned procedures, because it cannot be described in terms of either a higher order continuum theory or small corrections to a picture of essentially non-interacting particles. This regime is called the transition region and its treatment requires the full use of the Boltzmann equation (or at least sufficiently accurate models of the latter; see sect. 5).

We note that we can have situations where the flow is completely in the transition regime, and situations where one distinguishes various zones, one of which is neither continuum or free-molecular, E.g., in the flow past a solid body, even if the dimensions of the body are small with respect to the mean free path, several mean free paths from the body a continuum-like behavior takes over; the region close to the body belongs to the transition regime (note that the free molecular flow never appears, because the molecules arriving at the body surface have a distribution function which is influenced by the collisions they suffered one mean-free path away from the body).

If we want to investigate the transition regime, we have either to give up the idea of using perturbation methods or look for some other parameter, different from the Knudsen number, to be regarded as small in suitable conditions. According to the above discussion, the Boltzmann equation essentially contains one non-dimensional parameter (the Knudsen number); we have, therefore, to look for a new "small" parameter in the initial and boundary conditions, not in the Boltzmann equation itself. In this way we are led to the linearized Boltzmann equation (2,5):

$$\frac{\partial h}{\partial t} + \xi \cdot \frac{\partial h}{\partial x} = Lh \tag{4.4}$$

where h is the perturbation of the distribution function f about a maxwellian f_0:

$$f = f_0(1 + h) \tag{4.5}$$

and L is the so-called linearized collision operator simply related to $Q(f, f)$.

Equation (4.4) can be applied when the initial and boundary data for mass velocity and temperature can be considered close to constant values.

The linearized Boltzmann equation has the advantage of allowing a study of the transition regime without the difficulties related to the nonlinear character of the Boltzmann equation. The disadvantage is that, when linearizing, we lose the possibility of describing certain interesting features of a definitely nonlinear character, such as shock waves, and, in general, supersonic flows.

5. Model Equations

One of the major shortcomings in dealing with the Boltzmann equation is the complicated nature of the collision integral, in both the full nonlinear version, Q(f, f), and the linearized form, Lh. No wonder, therefore, that alternative, simpler expressions have been proposed for the collision term; they are known as collision models and any Boltzmann-like equation, where the Boltzmann collision integral is replaced by a collision model, is called a model equation or a kinetic model.

The idea behind this replacement is that the large amount of details of the two-body interactions (which are contained in the collision term and reflected, e.g., in the details of the spectrum of the linearized operator) are not likely to influence significantly the values of many experimentally measured quantities; i.e., unless very refined experiments are devised, the fine structure of the collision operator Q(f f), is blurred into a coarser description based upon a simpler operator J(f), which retains only the qualitative and average properties of the true collision operator.

The most widely known collision model (18) is usually called the Bhatnagar, Gross and Krook (BGK) model, although P. Welander (19) proposed it independently at about the same time. The idea behind the BGK model (retained by more sophisticated models) is that the main features of the collision operator are the following

a) the true collision term Q(f, f) satisfies the conservation equations; hence the collision model J(f) must satisfy:

$$\int \psi_\alpha J(f)\, d\xi = 0 \qquad (\alpha = 0, 1, 2, 3, 4) \tag{5.1}$$

b) the collision term expresses the tendency to a Maxwellian distribution (H-theorem).

The simplest way of taking this second feature into account is to imagine that each collision changes the distribution function $f(\xi)$ by an amount proportional to the departure of f from a Maxwellian $\phi(\xi)$; i.e., if ν is a constant with respect to ξ, we introduce the following collision model:

$$J(f) = \nu\,[\Phi(\xi) - f(\xi)] \tag{5.2}$$

The Maxwellian $\Phi(\xi)$ has five disposable scalar parameters $(\rho, \mathbf{v}, T)$; these are however, fixed by eq. (5.1) which implies

$$\int \psi_\alpha \Phi(\xi)\, d\xi = \int \psi_\alpha f(\xi)\, d\xi \tag{5.3}$$

i.e., at any space point and time instant, $\Phi(\xi)$ must have exactly the same density, velocity and temperature of the gas, given by the distribution function $f(\xi)$. Since the latter will, in general, vary with time and space coordinates, the same will be true for the parameters of $\Phi(\xi)$, which is accordingly called the local Maxwellian. The "collision frequency" ν is not restricted at this level and has to be fixed by means of additional considerations; we note, however, that ν can be a function of the local state of the gas and hence vary with both time and space coordinates.

We observe that the nonlinearity of the proposed $J(f)$ is much worse than the nonlinearity of the true collision term $Q(f, f)$; in fact, the latter is simply quadratic in f, while the former contains f in both the numerator and the denominator of an exponential (the $\mathbf{v}$ and T appearing in Φ are moments of f).

The main advantage in using the BGK operator is that, for any given problem, one can deduce integral equations for the macroscopic variables ρ, v, T; these equations are strongly nonlinear but simplify some iteration procedures and make the treatment of interesting problems feasible on a high speed computer. Another advantage of the BGK model is offered by its linearized form:

$$Lh = \nu\left(\sum_{\alpha=0}^{4} \psi_\alpha(\psi_\alpha, h) - h\right) \tag{5.4}$$

where h is the perturbation of the distribution function considered in section 4 and (g, h) denotes $\int f_0 g h \, d\xi_0$. The collision invariants ψ_α are here orthonormalized in such a way that

$$(\psi_\alpha, \psi_\beta) = \delta_{\alpha\beta} \qquad (\alpha, \beta = 0, 1, 2, 3, 4) \tag{5.5}$$

It is obvious that eq. (5.4) has a structure definitely simpler than the true linearized collision operator.

A problem which is easily solved with the nonlinear BGK model is the relaxation to equilibrium in the spatially homogeneous case. An arbitrary distribution function, $g(\xi)$, depending only on the velocity vector ξ, is given and we want to find its time evolution according to kinetic theory; the problem cannot be solved analytically with the full Boltzmann equation, but is trivial with the BGK model. As a matter of fact, we have

$$\frac{\partial f}{\partial t} = \nu(\Phi - f) \qquad f(\xi, 0) = g(\xi) \tag{5.6}$$

where the space derivatives can be omitted because the distribution function will remain homogeneous in physical space at any time. But Eqs. (5.1) together with Eq. (5.6), give:

$$\frac{\partial \rho}{\partial t} = 0, \qquad \frac{\partial \mathrm{v}}{\partial t} = 0, \qquad \frac{\partial T}{\partial t} = 0, \tag{5.7}$$

i.e., ρ, v, T are time independent (mass, momentum and energy are locally conserved in the homogeneous case). Then Eq. (5.6) is a very simple first-order differential equation with constant coefficients and constant source term: the solution is:

$$f(\xi, t) = g(\xi)e^{-\nu t} + (1 - e^{-\nu t})\ \Phi(\xi) \tag{5.8}$$

where the ρ, v, T in Φ can be calculated as moments of $g(\xi)$. The interpretation of Eq. (5.8) is quite simple: it describes an exponential approach to the equilibrium distribution $\Phi(\xi)$ with relaxation time $1/\nu$.

Another simple property of the BGK model is that it admits an easily proved H-theorem; i.e. for any distribution function f we have

$$\int \log f\ J(f)\ d\xi \leq 0 \tag{5.9}$$

and the equality sign applies only if f is a Maxwellian. In fact, we have:

$$\int \log f\ J(f)\ d\xi = \int \nu \log f\ (\Phi - f)\ d\xi$$

$$= \int \nu \log\left(\frac{f}{\Phi}\right)(\Phi - f)d\xi + \int \log \Phi \, J(f) \, d\xi \qquad (5.10)$$

Since $\log \Phi$ is a linear combination of the collision invariants, Eq. (5.1) shows that the last integral is zero. On the other hand:

$$(\Phi - f)\log\left(\frac{f}{\Phi}\right) \leq 0 \qquad (5.11)$$

and the equality sign applies only if $f = \Phi$; hence Eq. (5.9) follows.

The BGK model contains the most basic features of the Boltzmann collision integral but presents some shortcomings. Some of them can be avoided by suitable modifications, at the expense, however, of the simplicity of the model. A first modification (20, 21) can be introduced in order to allow the collision frequency ν to depend on the molecular velocity, instead of being locally constant; this modification is suggested by the circumstance that rigid sphere molecules, all the potentials with finite range, and power law potentials with angular cutoff (except Maxwell's molecules) present a collision frequency which varies with the molecular velocity and this variation is expected to be important at high molecular velocities. Formally, the modification is quite simple; we have only to allow ν to depend on ξ (more precisely on c) in Eq. (5.2), while requiring that Eq. (5.1) still holds. All the basic formal properties (including the H-theorem) are retained, but the density, velocity and temperature which now appear in the Maxwellian Φ are not the local density, velocity and temperature, but some ficticious local parameters, related to 5 moments of f weighted with $\nu(c)$; this follows from the fact that Eq. (5.1) now gives

$$\int \nu(c) \, \Phi \, \psi_\alpha \, d\xi = \int \nu(c) \, f \, \psi_\alpha \, d\xi \qquad (5.12)$$

instead of Eq. (5.3). A consequence of this fact is that the solution of the initial value problem is no longer as simple as before.

A different kind of correction to the BGK model (22, 23) is obtained, when we want to adjust the model to give the same Navier-Stokes equations as the full Boltzmann equation; in fact, as we shall see in the next chapter, the BGK model gives the value $Pr = 1$ for the Prandtl number, a value which is not in agreement with both the true Boltzmann equation and the experimental data for a monatomic gas (which agree in giving $Pr = 2/3$).

One of the unsatisfactory features of the BGK model as discussed above is that it is not derived from the Boltzmann equation by any kind of systematic procedure, but just guessed at on the basis of some qualitative information. The same statement applies to the two variants which have been briefly discussed.

Accordingly we shall investigate systematic procedures for deducing models of increasing accuracy. A satisfactory treatment is available only for the linearized operator, but something can be said also about nonlinear kinetic models.

The simplest model for the linearized operator is given by the linearized version of the BGK model, Eq. (5.4). This model has the following properties

(5.13) $$L\psi_\alpha = 0 \qquad (\alpha = 0, 1, 2, 3, 4)$$

(5.14) $$(h, Lh) \leq 0 \qquad (= 0, \text{ only if } h = \sum_{\alpha=0}^{4} c_\alpha \psi_\alpha)$$

(5.15) $$(g, Lh) = (Lg, h)$$

Putting $g = \psi_\alpha$ in Eq. (5.15) and using Eq. (5.13) we also get

(5.16) $$(\psi_\alpha, Lh) = 0 \qquad (\alpha = 0, 1, 2, 3, 4)$$

The three basic properties expressed by Eqs. (5.13)–(5.15) should be retained by any significant model. Eq. (5.13) and (5.15) for the BGK model are almost evident, while Eq. (5.14) follows from the fact that Eq. (5.4) can also be written as follows:

(5.17) $$Lh = \nu(Ph - h) = -\nu(I - P)h$$

where P is the projection operator onto the five-dimensional space $\mathfrak{J}$ spanned by the ψ's and I the identity operator (accordingly, $I-P$ is the projector onto $\mathfrak{W}$ the orthogonal complement of $\mathfrak{J}$); as a matter of fact, Eq. (5.17) implies:

$$(h, Lh) = -\nu(h, (I - P)h) = -\nu \,||(I - P)h||^2 \leq 0$$

and equality obviously holds only if $(I-P)h = 0$. Here $\| f \|$ denotes the norm in the Hilbert space $\mathcal{H}$ with scalar product (F, g).

A systematic procedure for improving the linearized BGK model, and characterizing the latter as the first step in a hierarchy of models approximating the collision operator for Maxwell molecules, with arbitrary accuracy, was proposed by Gross and Jackson (24). These authors started from the expansion of the collision operator for Maxwell molecules (see sect. 1) into a series of eigenfunctions which can be written as follows:

$$Lh = \sum_{R=0}^{\infty} \lambda_R \psi_R (\psi_R, h) \tag{5.18}$$

Here the ψ_R are products of Sonine polynomials times spherical harmonics and R is a label chosen in such a way that the collision invariants ψ_α correspond to $R = 0,1,2,3,4$. A systematic procedure for approximating L consists in partially destroying the fine structure in the spectrum of L by collapsing all the eigenvalues corresponding to $R > n$ into a single eigenvalue, which we shall denote by $-\nu_N$ (remember that $\lambda_R < 0$). This amounts to replacing L by an approximate operator L_N defined as follows:

$$L_N h = \sum_{R=0}^{N} \lambda_R \psi_R(\psi_R, h) - \nu_N \sum_{R=N} \psi_R(\psi_R, h) \tag{5.19}$$

Now, since the ψ_R constitute a complete set, we have

$$h = \sum_{R=0}^{\infty} \psi_R(\psi_R, h) \tag{5.20}$$

which is simply the Fourier expansion of h in terms of the ψ_R. Then

$$\begin{aligned} \sum_{R=N+1}^{\infty} \psi_R(\psi_R, h) &= \sum_{R=0}^{\infty} \psi_R(\psi_R, h) - \sum_{R=0}^{N} \psi_R(\psi_R, h) \\ &= h - \sum_{R=0}^{N} \psi_R(\psi_R, h) \end{aligned} \tag{5.21}$$

and substituting into Eq. (5.19):

$$L_N h = \sum_{R=0}^{N} (\nu_N + \lambda_R)\, \psi_R(\psi_R, h) - \nu_N h \tag{5.22}$$

In particular if $N = 4$, $\lambda_R = 0$ for $0 < R < N$ and consequently Eq. (5.22) reduces to Eq. (5.4) (with $\nu_4 = \nu$); by taking N larger and larger we include more and more details of the spectrum of L into the model. If we take $N = 9$ by including 5 more eigenfunctions, we can obtain the linearized version of the ES model which was mentioned above.

The above procedure applies only to the case of Maxwell's molecules. However, a slight generalization of the expansion (5.18) is capable of producing collision models in correspondence with any kind of linearized collision operator. In fact, nothing prevents us to expand h into a series of the ψ_R (eigenfunctions for the Maxwell collision operator) even if we are considering a different model; in this case we get:

$$Lh = \sum_{Q=0}^{\infty} (\psi_Q, h)\, L\psi_Q \tag{5.23}$$

and, by expanding again the result in terms of the ψ_R :

$$Lh = \sum_{R,Q=0}^{\infty} \beta_{QR} (\psi_Q, h)\, \psi_R \tag{5.24}$$

where

$$\beta_{QR} = (\psi_R, L\psi_Q) = \beta_{RQ} \tag{5.25}$$

Equation (5.24) generalizes Eq. (5.18) and reduces to the latter when $\beta_{QR} = \lambda_R\, \delta_{RG}$. If we now introduce the assumption $\beta_{QR} = -\nu_N\, \delta_{QR}$ for $Q,R > N$, we obtain the model

$$L_N h = \sum_{R,Q=0}^{N} (\nu_N \delta_{RQ} + \beta_{QR}) \psi_R (\psi_Q, h) - \nu_N h \tag{5.26}$$

which generalizes Eq. (5.22) for linearized collision operators other than Maxwell's. Taking $N = 4$ gives again the BGK model. It is clear now how the procedure can be extended to the nonlinear collision operator: it suffices to expand f into a series of the ψ's corresponding to the local Maxwellian Φ :

$$Q(f, f) = \sum_{P,Q,R=0}^{\infty} (\psi_P, f)(\psi_Q, f)\, Q(\Phi\psi_P, \Phi\psi_Q) \tag{5.27}$$

where now the scalar products are *not* weighted with a Maxwellian and expand the result in terms of the same ψ's:

$$Q(f, f) = \Phi \sum_{P,Q,R=0}^{\infty} \beta_{PQR} (\psi_P, f)(\psi_Q, f)\, \psi_R \tag{5.28}$$

where

$$\beta_{PQR} = (\psi_R, Q(\Phi\psi_P, \Phi\psi_Q)) \tag{5.29}$$

If now we assume

$$\beta_{PQR} = -\frac{1}{2}(\delta_{Po}\,\delta_{QR} + \delta_{Qo}\,\delta_{PR})\, \nu_N (\psi_0, f)^{-1} \qquad (P,Q,R > N) \tag{5.30}$$

then $Q(f, f)$ is substituted by a nonlinear model $J_N(f)$, given by:

$$J_N(f) = \Phi \sum_{P,Q,R=0}^{N} \beta_{PQR} (\psi_P, f)(\psi_Q, f)\, \psi_R + \nu_N \Big(\sum_{Q=0}^{N} \Phi(\psi_Q, f)\psi_Q - f \Big) \tag{5.31}$$

We note that, since Φ is the local Maxwellian, $(\psi_Q, f) = (\psi_Q, \Phi) = 0$ for $Q = 1,2,3,4$; the latter equality follows from the orthogonality properties of the ψ's. Accordingly, if we take $N = 4$ and take into account that $\beta_{PQR} = 0$ for $R < 4$, we re-obtain the nonlinear BGK model.

The linearized models for non-Maxwell molecules and the nonlinear models discussed above were proposed by Sirovich (25).

6. Models with Velocity-Dependent Collision Frequency

A characteristic feature of the linearized collision models which have been surveyed in sect. 5 is that they are based on bounded operators with a purely discrete spectrum (with an infinitely many times degenerate eigenvalue). This follows from the fact that the above mentioned collision operators can be written

$$L_N = K_N - \nu_N I$$

where ν_N is a constant, I the identity operator and K_N maps any function onto the finite-dimensional subspace spanned by the ψ_R $(R < N)$. In particular, by taking the projections of the eigenvalue equation for L_N onto this space and its orthogonal complement, one can show that the eigenfunctions are linear combinations of the ψ_R and hence polynomials (in particular the eigenfunctions are obviously the ψ_R themselves in the case of the model defined by Eq. (5.22)). It is known, however, that the collision operator for hard spheres or hard potentials with angular cut-off (26; see alco 2,5), is unbounded and displays a continuous spectrum (similar results should also hold for the case of a radial cut-off (27, see also 2,5) but have never been proved rigorously): if these features of the operator have any influence on the solution of particular problems, this influence is lost when we adopt one of the models proposed in sect. 5. It is therefore convenient to introduce and investigate models which retain the above mentioned features of the linearized collision operator; this can be done in many different ways.

Conceptually, the simplest procedure is based upon exploiting the fact that we either know (rigid spheres and angular cut-off) or conjecture (radial cut-off) that the operator $K^* = \nu^{-1/2} K \nu^{-1/2}$ is self-adjoint and completely continuous in $\mathcal{H}$ (see Refs. 2, 5, 26); accordingly, the kernel of K^* can be expanded into a series of its square summable eigenfunctions $\Phi_R \nu^{1/2}$ (such that $K\phi_R = \lambda_R \nu\phi_R$). In other words, we can write

$$Kh = \nu(c) \sum_{R=0}^{\infty} \lambda_R \phi_R (\nu\phi_R, h) \tag{6.1}$$

If one truncates this series, he gets the model

$$K_N h = \nu(c) \sum_{R=0}^{N} \lambda_R \phi_R (r\phi_R, h) \tag{6.2}$$

Since the first five ϕ_R are the collision invariants and $\lambda_R = 1$ $(0 \leq R \leq 4)$, if we take $N = 4$, we have

$$(6.3) \qquad L_4 h = \nu(c) \sum_{\alpha=0}^{4} \psi_\alpha(\psi_\alpha, \nu h) - \nu(c)\, h$$

where the collision invariants are orthonormalized according to

$$(6.4) \qquad (\psi_\alpha, \nu\psi_\beta) = \delta_{\alpha\beta}$$

Equation (5.3) is nothing else than the linearized version of the nonlinear model with velocity dependent collision frequency which was briefly discussed in sect. 5.

If we want to obtain models corresponding to $N \geq 5$, we are faced with the trouble that we do not have analytical expressions for the ϕ_R $(R \geq 5)$ it is true that we can compute them numerically, but this is obviously a noticeable complication. Besides, the procedure by which we have derived Eqs. (6.2) and (6.3) shows that $\nu(c)$ is rigidly fixed by the original molecular model; hence we do not have any parameters to be adjusted in order to reproduce the correct continuum limit (see sect. 5)

We can introduce an adjustable parameter by a slight modification of the above procedure: instead of simply truncating the series in Eq. (6.1) (i.e., putting $\lambda_R = 0$ for $R > N$), we can put $\lambda_R = k$ for $R > N$. The result is exactly the same model as in Eq. (6.2), except for the fact that $\nu(c)$ is substituted by $(1-k)\,\nu(c)$ and λ_R by $(\lambda_R - k)/(1-k)$; in particular, for $N = 4$, the only change is an adjustable factor in $\nu(c)$ which gives the model the same flexibility as the BGK model.

A larger flexibility can be obtained by using higher order models; here we meet the above mentioned trouble that we do not know the ϕ_R $(R \geq 5)$. This difficulty can be avoided by a procedure analogous to the non-Maxwell modeling considered in sect. 5; i.e., we take a complete set of orthogonal functions and expand everything in terms of these functions, as we did in sect. 5. Of course, if we want to retain the features of the models discussed in this section, we have to make a proper choice of the set; the simplest one is to take polynomials orthogonalized with respect to the weight $f_0\ \nu(c)$, f_0 being the basic Maxwellian. The simplest model takes again the form shown in Eq. (6.3) but now we can extend it to arbitrarily high orders without troubles (the only things to be computed numerically are coefficients, not functions); the procedure is obvious and will not be described in detail.

Another procedure for constructing models can be based upon the requirement that the model is able, in the continuum limit, to reproduce the behavior not only of some coefficients but also of the distribution function; such models can be constructed (26) by insisting that the equation

$$Lh = \xi_i g - P(\xi_i g) \tag{6.5}$$

(where P is the projector onto the space spanned by the five collision invariants) has the same solution of the corresponding equation with L_N in place of L for a certain class of source functions g. To be precise one starts with the five collision invariants for g and then constructs step by step the set of functions h, using the results of each step as sources for the next step and orthogonalizing the set when necessary; again the trouble is that, except for Maxwell molecules, we do not have analytical tools for solving Eq. (6.5) and consequently constructing the above mentioned models in an explicit fashion. Sometimes, however, the procedure can be worthwhile because it is capable of giving very accurate results for non-Maxwell molecules; this is particularly true when the final results can be expressed in terms of quadratures to be performed numerically.

CHAPTER III

THE KINETIC BOUNDARY LAYERS

7. Introductory Remarks

As we mentioned in sect. 1, when the Knudsen number is small, the description based on the Boltzmann equation can be shown to be equivalent to a macroscopic description based on the Navier-Stokes equations. This result is achieved by means of the Hilbert and Chapman-Enskog methods, i.e., methods based on expansions in powers of the mean free path. These theories are not complete, however, and, in order to complete them, one has to derive boundary conditions for the macroscopic differential equations resulting from the theory. In order to obtain these boundary conditions, one has to go through regions having the thickness of a mean free path in the neighborhood of solid walls; in these regions, called the Knudsen layers or the kinetic boundary layers, the Hilbert and Chapman-Enskog expansions are invalid. The same problem arises (29) in connection with the initial layer, i.e., a time interval of the order of the mean free time after $t = 0$, but the problem of boundary layers is much more important and difficult to handle. In fact, the influence of the boundary layers is already felt at the first order in Kn, i.e., at the Navier-Stokes level. To this order one can show that one has to match the Navier-Stokes equations with the following "extrapolated" boundary conditions:

$$\mathbf{v} - \zeta\mathbf{n}\cdot\frac{\partial}{\partial\mathbf{x}}(\mathbf{n}\times(\mathbf{v}\times\mathbf{n})) - \omega\left(\frac{2R}{T}\right)^{1/2}\mathbf{n}\times\left(\frac{\partial T}{\partial\mathbf{x}}\right) = \mathbf{u}_0$$

$$T - \tau\mathbf{n}\cdot\frac{\partial T}{\partial\mathbf{x}} - \chi(2RT)^{-1/2}\left(\mathbf{n}\times\left(\mathbf{n}\times\frac{\partial}{\partial\mathbf{x}}\right)\right)(\mathbf{n}\times(\mathbf{n}\times\mathbf{v})) = T_0 \qquad (7.1)$$

where $\mathbf{u}_0$, T_0 are the velocity and temperature of the boundary, $\zeta, \omega, \tau, \chi$ are coefficients of the order of mean free path. In particular, ζ measures the tendency of the gas to slip over a solid wall in presence of velocity gradients and is called the slip coefficient; τ measures the tendency of the gas to have a temperature different from the wall temperature in presence of temperature gradients and is called the temperature jump coefficient; ω and χ measure cross effects. We note that when the mean free path is not only small but completely negligible, then the boundary conditions reduce to $v = u_0$, $T = T_0$ i.e., the gas does not slip and accommodates completely to the wall temperature.

The slip coefficient ζ has been studied by many authors and can be evaluated as a by-product of the so-called Kramers problem (30). This problem consists in finding the molecule distribution function of a gas in the following situation: the gas fills the half space $x > 0$ bounded by a physical wall in the plane $x = 0$, and is non-uniform because of a gradient along the x-axis of the z-component of the mass velocity; this gradient tends to a constant when $x \to \infty$.

It is seen that this problem can be considered as the limiting case of plane Couette flow, when one of the plates is pushed to infinity. More in general, the Kramers problem can be interpreted as a connection problem through the kinetic boundary layer; in this case "infinity" is simply the region where the Hilbert solution holds and the velocity gradient "at infinity" can be regarded as constant because it does not vary sensibly on the scale of the mean free path (this is rigorously correct up to terms of order Kn^2).

Both these interpretations of the Kramers problem suggest that a convenient linearization is about a Maxwellian endowed with a translational velocity $k\,x$ in the z-direction. Because of the non-uniformity of this Maxwellian distribution, linearization gives an inhomogeneous linearized Boltzmann equation:

$$2\,k\,c_1 c_3 + c_1 \frac{\partial h}{\partial x} = Lh \tag{7.2}$$

where $c = (c_1, c_2, c_3) = (\xi_1, \xi_2, \xi_3 - kx)$. Equation (7.2) can be reduced to the homogeneous Boltzmann equation by subtracting a particular solution. One particular solution independent of x, is suggested by the Hilbert theory; this solution, $L^{-1}(2kc_1\,c_3)$ is given by $-2kc_1\,c_3\,\theta$ for the BGK model (θ being the mean free time equal to ν^{-1}). Therefore, if we adopt the BGK model, we have:

$$h = -2k\,c_1\,c_3\,\theta + 2c_3 Y(x, c_1) \tag{7.3}$$

where $Y(x, \xi)$ satisfies Eq. (8.1) below, if x is measured in θ units. We write ξ in place of c_1, the x-component of c, since no confusion arises. The mass velocity is given by

$$v_3 = kx + \pi^{-1/2} \int_{-\infty}^{\infty} e^{-\xi^2} Y(x, \xi) \, d\xi \tag{7.4}$$

the first term being the contribution from the Maxwellian.

8. Elementary Solutions of the Shear Flow Equation

The equation to be solved can be written as follows

$$\xi \frac{\partial Y}{\partial x} + Y(x, \xi) = \pi^{-1/2} \int_{-\infty}^{\infty} e^{-\xi_1^2} Y(x, \xi_1)\, d\xi_1 \tag{8.1}$$

and can be shown to describe, in general, shear flows when the solution depends upon a single space variable.

Let us begin with separating the variables. Putting

$$Y(x, \xi) = g(\xi)\, X(k) \tag{8.2}$$

it is easily seen that, either $Y = A_0$ (arbitrary constant) or

$$Y_u(x, \xi) = e^{-x/u}\, g_u(\xi) \tag{8.3}$$

where $g_u(\xi)$ satisfies

$$\left(-\frac{\xi}{u} + 1\right) g_u(\xi) = \pi^{-1/2} \int_{-\infty}^{\infty} g_u(\xi_1)\, e^{-\xi_1^2}\, d\xi_1 \tag{8.4}$$

and u, the separation parameter, has been used to label the elementary solutions.

Though, a priori, u may assume any complex value, it is easily seen that u is a real number. This follows from a direct reasoning (30). Thus the values of u must be real. This requires some care, because one cannot divide by $u-\xi$ in Eq. (8.4). This difficulty is overcome by letting $g_u(\xi)$ be a generalized function. Then, if we disregard a multiplicative constant (i.e., normalizing g_u in such a way that the right hand side of Eq. (8.4) is equal to 1), $g_u(\xi)$ will be a generalized function of the type

$$g_u(\xi) = P\, \frac{u}{u-\xi} + p(u)\, \delta(u-\xi) \tag{8.5}$$

where p(u) is an arbitrary constant depending, however, upon u. In order that Eq.

(8.4) be satisfied by Eq. (8.5), the normalization condition for $g_u(\xi)$ must be satisfied, i.e., the right hand side of Eq. (8.4) must be equal to 1. This condition can be satisfied for any real u and serves for determining p(u)

$$p(u) = e^{u^2} \, P \int_{-\infty}^{\infty} \frac{\xi e^{-\xi^2}}{\xi - u} \, d\xi \tag{8.6}$$

$$= \pi^{1/2} \left(e^{u^2} - 2u \int_0^u e^{t^2} \, dt\right)$$

The first of these expressions follows directly, the second by manipulation of the first; p(u) can be expressed in terms of tabulated functions (30).

The generalized eigenfunctions $g_u(\xi)$ have many properties of orthogonality and completeness. Some of the orthogonality properties in the full range $(-\infty < \xi < \infty)$ can easily be proved by means of the usual procedures for proving orthogonality of eigenfunctions. Other properties of orthogonality in partial ranges (notably $0 < \xi < \infty$) and completeness are far from trivial to prove, since they require solving singular integral equations. However, standard techniques are available for treating such problems (31) and the following results can be obtained (30, 32).

THEOREM I: The generalized functions $g_u(\xi)$ $(-\infty < u < +\infty)$ and $g_\infty = 1$, complemented with $g^* = \xi$, form a complete set for the functions $g(\xi)$ defined on the real axis, satisfying a Holder condition in any open interval of the real axis and such that

$$\int_{-\infty}^{\infty} \xi^2 \, e^{-\xi^2} g(\xi) \, d\xi < \infty \tag{8.7}$$

Also, the coefficients of the generalized expansion:

$$g(\xi) = A_0 + A_1 \xi + \int_{-\infty}^{\infty} A(u) \, g_u(\xi) \, du \tag{8.8}$$

are uniquely and explicitely determined by

$$A_0 = 2\pi^{-1/2} \int_{-\infty}^{\infty} \xi^2 e^{-\xi^2} g(\xi)\, d\xi \tag{8.9}$$

$$A_1 = 2\pi^{-1/2} \int_{-\infty}^{\infty} \xi e^{-\xi^2} g(\xi)\, d\xi \tag{8.10}$$

$$A(u) = (C(u))^{-1} \int_{-\infty}^{\infty} \xi e^{-\xi^2} g(\xi) g_u(\xi)\, d\xi \tag{8.11}$$

where

$$C(u) = u\, e^{-u^2} \{ (p(u))^2 + \pi^2 u^2 \} \tag{8.12}$$

THEOREM II: The generalized eigenfunctions $g_u(\xi)\,(0 < u < \infty)$ and $g_\infty = 1$ form a complete set for the functions $g(\xi)$ defined on the positive real semiaxis, satisfying a Holder condition in any open interval of this semiaxis, bounded by $A\,|\xi|^{-\gamma}$ with $\gamma < 2$ in the neighborhood of $\xi = 0$ and integrable with respect to the weight $\xi^2 e^{-\xi^2}$ Also, the coefficients of the generalized expansion

$$g(\xi) = A_0 + \int_0^{\infty} A(u)\, g_u(\xi)\, du \tag{8.13}$$

are uniquely and explicitly determined by

$$A_0 = 2\pi^{-1/2} \int_0^{\infty} P(\xi) \xi e^{-\xi^2} g(\xi)\, d\xi \tag{8.14}$$

$$A(u) = (C(u)P(u))^{-1} \int_0^{\infty} \xi e^{-\xi^2} P(\xi)\, g_u(\xi)\, g(\xi)\, d\xi \tag{8.15}$$

Here we have put

(8.16) $$P(\xi) = \xi \exp \{-\frac{1}{\pi}\int_0^\infty \tan^{-1}(\frac{\pi t}{p(t)}) \frac{dt}{t+\xi}\} \quad (\xi > 0)$$

where the arc-tangent varies from $-\pi$ to 0 when t varies from 0 to ∞. Theorem I shows that the generalized eigenfunctions are orthogonal with respect to the weight $\xi e^{-\xi^2}$ on $(-\infty, +\infty)$. Since, however, this weight is not everywhere positive, some irregularities are noticed. Firstly, we note that the expressions for A_0 and A_1 (8.9) and (8.10)) are exactly interchanged with respect to what one would expect by analogy with familiar expansions. Secondly, the appearance of $g_* = \xi$ is completely unexpected, because g_* does not solve Eq. (8.4) for any value of $u(f_\infty$ does for $u = \infty)$. There is, however, a particular solution of Eq. (8.1), $Y_* = \xi - x$, which does not have separated variables (in the usual sense) and yet cannot be represented as a super-position of the elementary solutions $Y_u = e^{-x/u} g_u(\xi)$ and $Y_\infty = 1$. The latter statement follows from theorem I itself, because if we could write

(8.17) $$\xi - x = A_0 + \int_{-\infty}^{\infty} A(u)\, e^{-x/u} g_u(\xi)\, du$$

for some x (we can take $x = 0$, without loss of generality) and $-\infty < \xi < \infty$, then

(8.18) $$0 = A_0 - \xi + \int A(u) g_u(\xi)\, du \qquad (-\infty < \xi < \infty)$$

for suitable A_0 and $A(u)$. But this implies that the function which is identically zero on the real axis has two different expansions of the form (8.8): the trivial one with coefficients $A_0 = A_1 = A(u) = 0$ and another given by Eq. (8.18). This is in contrast with the uniqueness of the expansion (8.8), asserted by theorem I; hence Eq. (8.17) is wrong and $\xi - x$ cannot be represented as a superposition of separated variables solutions.

Accordingly the elementary solutions $Y_u (-\infty < u < \infty)$ and Y_∞ are not sufficient to construct the general solutions. We want to prove now that, if we add $Y_* = \xi - x$ to the set, we obtain a complete set, i.e., the general solution of Eq.

(8.1) is given by

$$Y(x, \xi) = A_0 + A_1(x - \xi) + \int_{-\infty}^{\infty} A(u)e^{-x/u}\, g_u(\xi)du \tag{8.19}$$

In fact, given any solution $Y(x, \xi)$ of Eq. (8.1), theorem I proves that, for any fixed x, we can find $A_0, A_1, A(u)$ such that Eq. (8.19) holds. We have now to show that A_0, A_1, and $A(u)$ do not depend on x and our result will be proved. If we substitute Eq. (8.19) into Eq. (8.1) and use the fact that Y_0, Y_*, Y_u satisfy Eq. (8.1), we obtain

$$\frac{\partial A_0}{\partial x} + \frac{\partial A_1}{\partial x}(x - \xi) + \int_{-\infty}^{\infty} \frac{\partial A}{\partial x} e^{-x/u}\, g_u(\xi)\, du = 0 \tag{8.20}$$

But this equation gives, for any fixed x, the expansion of the function which is identically zero on the real axis by the uniqueness of the expansion, it follows that

$$\frac{\partial A_0}{\partial x} = \frac{\partial A_1}{\partial x} = \frac{\partial A}{\partial x} = 0,$$

as was to be shown.

Theorem I, therefore, insures that Eq. (8.19) gives the most general solution of Eq. (8.1). Theorem II is equally or, perhaps, more important, because it allows us to solve boundary value problems. This theorem shows that the generalized eigenfunctions are orthogonal on $(0, +\infty)$ with respect to the weight $\xi e^{-\xi^2}\, P(\xi)$. This orthogonality property is more classical than the full range orthogonality, because the weight function is positive. The only trouble is now the complicate expression of $P(\xi)$; it is to be noted that $P(\xi)$, though far from being an elementary function, satisfies two important identities which make the manipulation of integrals involving $P(\xi)$ much easier than would be expected. These identities are the following (30, 32):

$$2\pi^{-1/2} \int_0^{\infty} \frac{te^{-t^2}\, P(t)}{t + u}\, dt = (P(u))^{-1} \tag{8.21}$$

$$u - \frac{1}{\pi} \int_0^\infty \tan^{-1} \left(\frac{\pi t}{p(t)} \right) dt - \frac{\pi^{1/2}}{2} \int_0^\infty \frac{t e^{t^2} (P(t))^{-1}}{(p(t))^2 + \pi^2 t^2} \frac{dt}{t+u} = P(u)$$

(8.22)

Also:

(8.23) $$P(0) = 2^{-1/2}$$

9. Application of the General Method to the Kramers Problem

In this section we shall apply the above results to the Kramers problem described in sect. 7.

Concerning the boundary conditions, we shall assume that the molecules are re-emitted from the wall according to a Maxwellian distribution completely accommodated to the state of the wall. Therefore the boundary condition for h reads as follows:

$$h(0, c) = 0 \qquad (c_1 > 0) \tag{9.1}$$

and this, in terms of Y, becomes

$$Y(0, \xi) = k\theta\xi \tag{9.2}$$

where k is the limiting value of the velocity gradient. Besides Y must satisfy the condition of boundedness at infinity.

According to the discussion in sect. 8, the general solution of Eq. (8.1) which also satisfies the condition of boundedness at infinity is given by

$$Y(x, \xi) = A_0 + \int_0^\infty A(u)e^{-x/u}\, g_u(\xi)du \tag{9.3}$$

and the condition to be satisfied at the plate gives:

$$k\theta\xi = A_0 + \int_0^\infty A(u)g_u(\xi)du \tag{9.4}$$

But solving this equation means expanding $k\theta\xi$ according to theorem II of sect. 8; therefore, A_0 and A(u) are given by Eqs. (8.14) and (8.15) if $k\theta\xi$ is substituted for $g(\xi)$. The result is as follows:

$$A_0 = -k\theta\pi^{-1/2} \int_0^\infty \tan^{-1}\left(\frac{\pi\xi}{p(\xi)}\right) d\xi$$

$$(9.5) \qquad = k\theta\pi^{1/2} \int \frac{\xi e^{\xi^2}}{(p(\xi))^2 + \pi^2 \xi^2} d\xi$$

$$(9.6) \qquad A(u) = -k\theta\pi^{1/2} e^{u^2} (P(u))^{-1} \{(p(u))^2 + \pi^2 u^2\}^{-1}$$

where use has been made of Eq. (8.21) which yields Eq. (9.6) directly and the following identity by asymptotically expanding for large values of u comparing with Eq. (8.22) or (8.16).

$$(9.7) \qquad 2\pi^{-1/2} \int_0^\infty \xi^2 e^{-\xi^2} P(\xi) d\xi = \frac{1}{\pi} \int_0^\infty \tan^{-1} \left(\frac{\pi t}{p(t)}\right) dt$$

Eq. (9.7) yields for the first expression of A_0, while the second one is obtained by partial integration.

Substituting Eqs. (9.5) and (9.6) into Eq. (9.3) gives the solution of the Kramers problem. The mass velocity is readily obtained from Eqs. (9.3) and (7.4):

$$(9.8) \qquad v_3(x) = kx + A_0 + \int_0^\infty A(u) e^{-x/u} du$$

where A_0 and A(u) are given by Eqs. (9.5) and (9.6). From Eq. (9.8) we recognize that A_0 is the macroscopic slip of the gas on the plate (see sect. 7): it has the form ζk, where ζ is the slip coefficient

$$(9.9) \qquad \zeta = \theta\pi^{1/2} \int \frac{\xi e^{\xi^2} d\xi}{(p(\xi))^2 + \pi^2 \xi^2} = 2\ell \int \frac{\xi e^{\xi^2} d\xi}{(p(\xi))^2 + \pi^2 \xi^2}$$

Here ℓ is the mean free path related to θ by $\theta = 2\pi^{-1/2} \ell (2RT_0 = 1)$. The integral appearing in Eq. (9.9) has been evaluated numerically with the following result (3,4)

$$(9.10) \qquad \zeta = (1.01615)\theta = (1.1466)\ell$$

A direct evaluation of the microscopic slip, i.e., the velocity of the gas at the wall, results without any numerical calculation. As a matter of fact, we have:

$$v_3(0) = k(\zeta - \ell \int_0^\infty \frac{e^{\xi^2}(P(\xi))^{-1}}{(p(\xi))^2 + \pi^2\xi^2} d\xi) = (\frac{2}{\pi})^{1/2} k\ell \qquad (9.11)$$

where the last result is obtained by letting $u \to 0$ in Eq. (8.22) and taking into account Eqs. (8.23) and (9.9).

Analogously, we can evaluate the distribution function of the molecules arriving at the plate. We obtain

$$Y(0, \xi) = 2\pi^{-1/2} k\ell\xi + 2\pi^{-1/2} k\ell\, P(-\xi) \qquad (\xi < 0) \qquad (9.12)$$

where Eq. (8.22) has been used. Then $h(0, c)$ (the perturbation of the Maxwellian distribution at the plate) is given by

$$h(0, c) = 4\pi^{-1/2} k\ell c_3\, P(|c_1|) \qquad (c_1 < 0) \qquad (9.13)$$

and the function $P(\xi)$ $(\xi > 0)$ receives a physical interpretation in terms of the distribution function of the molecules arriving at the wall. From Eqs. (8.22) it is easily inferred that

$$|c_1| + 0.7071 < P(|c_1| < |c_1| + 1.01615 \qquad (9.14)$$

Hence the distribution function of the arriving molecules is rather close to a Hilbert distribution; in fact, a Hilbert expansion would predict Eq. (9.13) with $P(|c_1|)$ linear in $|c_1|$ (this is the distribution holding outside the kinetic layer (see Eqs. (7.3) and (9.3). The fact that the distribution of the molecules arriving at the plate is close to the one prevailing outside the kinetic boundary layer is not surprising; in fact, each molecule has the velocity acquired after its last collision, which, in average, happened a mean free path from the wall, i.e., in a region where the distribution function is of the Hilbert type. It is interesting to note that Maxwell

(33) assumed that the distribution function of the arriving molecules was exactly the one prevailing far from the wall; by using this assumption and conservation of momentum, he was able to evaluate the slip coefficient, without solving the Kramers problem. He found $\zeta = \ell$ (with an error of 15%) and

$$h(0, c) = 4\pi^{-1/2} k \ell c_3 (|c_1| + 0.8863) \qquad (c_1 < 0) \tag{9.15}$$

i.e., a good approximation to the correct result (see Eqs. (9.13) and (9.14)).

10. Further Results on the Knudsen Layer

The method of elementary solutions, i.e. separated variable solutions, developed in Sect. 8 and applied to the Kramers problem in Sect. 9, can be extended to more general problems and models (2, 5, 32) and even to a generic linearized Boltzmann equation (35). It turns out, however, that, in general, explicit solutions cannot be found. Only recently a rather cumbersome analytic method for evaluating the temperature jump coefficient τ has been found; both numerical and variational calculations of τ, however, are available since several years (see Chapters IV and V).

Other interesting calculations are concerned with the evaluation of the slip coefficient with models different from the BGK. We quote the fact that the slip coefficient ζ for the ES model is exactly the same as the BGK one, while for the modified BGK model with velocity dependent collision frequency, ζ can also be expressed in analytical form and the resulting quadrature performed with great accuracy by numerical means (37). It has been found that for a collision frequency $\nu(\xi)$ increasing linearly as $\xi \to \infty$ the value of ζ is somewhat lower (3 %) than for the BGK model (for a fixed value of the viscosity coefficient).

G. Tironi and the author of the present notes tried (38) to extend the validity of the Navier-Stokes equations beside their strict limit of validity by using the concept of Knudsen layers, even when the latter are so thick as to occupy a significant part of the flow region. The idea of the method goes back to Maxwell (33) and Langmuir (20); but Maxwell's reasoning is restricted to walls of negligible curvature while Langmuir's method suffers the ambiguity of containing the unknown probability density of the number of collisions which took place at a certain distance from the wall. Cercignani and Tironi restricted themselves to the steady linearized BGK model, but the main reasonings can be repeated for more general models or the full Boltzmann equation; using the BGK model allows a quick comparison with available accurate solutions of both numerical and variational nature.

The method consists in using the Navier-Stokes equations plus suitable boundary conditions. The latter are, in general, integrodifferential relations and are obtained through an approximated analysis of the kinetic layers.

Recently, Loyalka (40) pointed out some numerical errors in the applications of Cercignani and Tironi's method (38) and suggested an improvement of the method. Both contributions of Loyalka are very important because they bring the method into a very close agreement with exact solutions as far as the evaluation of global quantities is concerned.

CHAPTER IV

APPROXIMATE METHODS OF SOLUTION

11. General Aspects

In sect. 4 we have mentioned some methods of solution for the Boltzmann equation based on perturbation methods, i.e., the Hilbert and Chapman-Enskog expansions, the Knudsen iteration, the method of successive collisions and the linearization of the Boltzmann equation. The latter procedure is usually coupled with the use of kinetic models. These models, however, mentioned in sect. 5, have been shown to be capable of arbitrarily approximating not only the Boltzmann linearized equation but also its solution. This fact makes the models very useful, particularly in those cases when their solution is explicit (in terms of quadratures or functions, whose qualitative behavior can be studied by analytical means).

A general method which allows the analytical manipulation of the linearized models is the method of separation of variables presented in chapter III. This method can be extended to the case of time dependent problems for the BGK model with constant or velocity dependent collision frequency (2, 21, 41, 42, 5). The extension is based upon taking the Laplace transform with respect to time of the linearized Boltzmann equation.

Then, if h depends on a single space variable x and L is replaced by a suitable collision model, one can separate the variables as done in sect. 8. The main difference is that now a discrete spectrum can appear besides the continuum spectrum, if the collision frequency is constant (41); but, if the collision frequency is velocity dependent, the continuous spectrum, instead of filling a line of the complex plane, fills a two-dimensional region (42). This implies the use of a more general theory than the theory of singular integral equations in order to solve boundary value problems; the construction of such a theory has been, however, carried out without difficulty (42, 5).

It is to be noted that these analytical procedures are rather complicated even for relatively simple problems; this suggests that for more complicated problems, of both linear or nonlinear nature, one should look for less sophisticated procedures, yielding approximate but essentially correct results. Such procedures can be easily constructed for linearized problems or in the limit of either large or small Knudsen numbers; the intermediate range of Knudsen numbers (transition region) in nonlinear situations is, at present, a matter of interpolation procedures of more or less sophisticated nature. Besides, it is to be noted that a good procedure does not necessarily mean a good method of solution, since in many cases the procedure consists in deducing a system of nonlinear partial differential equations. The latter have to be solved in correspondence with particular problems, and, in general they are tougher than the Navier-Stokes equations for the same problem. As a consequence, one has to resort to numerical procedures to solve them. The approximation procedures can be grouped under two general headings: moment methods and integral equation method. In the former case one constructs certain partial differential equations, as mentioned above, in the latter one tries to obtain either expansions valid for large Knudsen numbers or numerical solutions. In connection with both methods one can simplify the calculations by the use of models (sometimes in an essential manner), but one has to remember that the accuracy of kinetic models in nonlinear problems is less obvious than in the linearized ones. Finally, in connection with both methods one can apply variational procedures; again, the latter are more significant and, probably, much more accurate for linearized problems.

In this chapter we shall briefly review this matter, with a certain emphasis on problems which have been solved accurately and methods which seem to provide a systematic approach to the transition regime of rarefaction.

12. Moment Methods

If one multiplies both sides of the Boltzmann equation by functions $\phi_i(\xi)$ (i = 1, . . . , N,. set and integrates over the molecular velocity, he obtains infinitely many relations to be satisfied by the distribution function

$$\frac{\partial}{\partial t} \int \phi_i(\xi) f(x, \xi, t) d\xi + \frac{\partial}{\partial x} \cdot \int \phi_i(\xi) f(x, \xi, t) \xi \, d\xi = \int \phi_i(\xi) Q(f, f) d\xi \qquad (i = 1, \ldots, N, \ldots) \tag{12.1}$$

This system of infinitely many relations (Maxwell's transfer equations) is equivalent to the Boltzmann equation, because of the completeness of the set (ϕ_i). The common idea of the so-called moment methods is to satisfy only a finite number of transfer equations or moment equations.

This leaves the distribution function f largely undetermined, since only the infinite set (12.1) (with proper initial and boundary conditions) can determine f. This means that we can choose, to a certain extent, f arbitrarily and then let the moment equations determine the details which we have not specified. The different "moment methods" differ in the choice of the set ϕ_i and the arbitrary input for f. Their common feature is to choose f in such a way that f is a given function of ξ containing N undetermined parameters depending upon x and t, M_i (i = 1,. . . ,N); this means that if we take N moment equations we obtain N partial differential equations for the unknowns M_i (x, t). In spite of the large amount of arbitrariness, it is hoped that any systematic procedure yields, for sufficiently large N, results essentially independent of the arbitrary choices. On practical grounds, another hope is that, for sufficiently small N and a judicious choice of the arbitrary elements, one can obtain accurate results. We shall not attempt to describe all the possible choices and not even all those which have been actually proposed, but shall limit ourselves to the simplest and most used ones.

The simplest choice (9) is to assume f to be a Maxwellian f_0 times a polynomial:

$$f = f_0 \sum_{k=0}^{N-1} Q_k(x, t) \, H_k(\xi) \tag{12.2}$$

where, for convenience, the polynomial is expressed in terms of the three--dimensional Hermite polynomials $H_k(\xi)$ orthogonal with respect to the weight f_0 It is also convenient to choose f_0 to be the local Maxwellian (this implies that $Q_0 = 1$, $Q_1 = Q_2 = Q_3 = Q_4 = 0$). There are N arbitrary quantities which can be identified with the basic moments (ρ, v_i, T, p_{ij}, q_i and higher order moments) and can be determined by taking $\phi_i = H_i$ eq. (2.1) (i = 0,...N−1). A reasonable choice is N = 13; in such a case the unknowns are $\rho, v_i, T, p_{ij} - p^{\delta}_{ij}, q_i$ and the equations are known under the name of Grad's thirteen moments equations.

There is an obvious disadvantage in Grad's choice, i.e., the fact that the distribution function is assumed to be continuous in the velocity variables and this is not true at the boundaries.

In order to avoid this inconvenience, one can take for f a distribution function which already takes into account the discontinuities in velocity space, e.g., a function which piecewise reduces to eq. (12.2). Then one can choose between piecewise continuous functions (43) or continuous functions (44) for the set $\{ \phi_i \}$. The second methods seems to be more satisfactory and easier to handle: in its crudest version it reduces to assuming that the distribution function is piecewise Maxwellian, the discontinuities being located exactly where predicted by the free molecular solution.

It is obvious that these procedures can be used to obtain reasonably good results by judicious choices of the arbitrary elements, but is to be remembered that they eventually become very complicated and generate equations which can be solved only numerically.

13. The Integral Equation Approach

The Boltzmann equation can be transformed from an integrodifferential to a purely integral form. The transformation can be achieved in many ways. The basic idea is to add a term $\mu(\xi)\,h$ ($\mu(\xi) > 0$, a function of the molecular velocity) to both sides of the Boltzmann equation and then construct the inverse of the differential operator appearing in the left hand side. This can be easily achieved by integrating along the characteristics of the latter operator. The net result is

$$f = \bar{f}_0 + N(f) \tag{13.1}$$

where $\bar{f}_0$ is a "source term" (in general depending on f) and N a nonlinear operator acting upon f; the separation between f_0 and N(f) is purely formal, as always happens in nonlinear problems. One can use eq. (13.1) as a tool for an iteration method: we start with a guess $f^{(0)}$ for f and then evaluate $f^{(1)}$, $f^{(2)}$, $f^{(3)}$ etc. by means of the iteration scheme:

$$f^{(n)} = \bar{f}_0^{(n-1)} + N(f^{(n-1)}) \tag{13.2}$$

Since there are different integral equations, there will be different iteration methods. Formally, the simplest one corresponds to the integral equation obtained by a simple integration along the characteristics. This iteration procedure is called the Knudsen iteration, but is not frequently used, since it presents serious troubles, especially in one dimensional plane geometries.

Better results are obtained by taking $\mu(\xi) > 0$ in the general procedure sketched above. In this case, the iteration method (integral iteration, or successive collisions method (45)) gives satisfactory results for sufficiently large Knudsen numbers.

A more far-reaching advantage of the integral equation approach can be obtained in connection with the use of models. In fact, all the collision models (linear or not) can be split into two parts, one of which (say, $\nu\Phi$) depends only on a finite number of moments of the distribution function (five, if Φ is the local Maxwellian, as is the case for the BGK model), while the second one (say, νf) can be written as the distribution function times a smooth function of f (a constant in the linearized case). If we choose $\mu = \nu$ in building the integral equation, eq. (13.1), then N(f) and $\bar{f}_0$ depend only on a finite number of moments (plus the initial and boundary conditions). This means that f is known when these moments are known. Since, however, these moments are defined in terms of f one can use eq. (13.1) to construct integral equations for the basic moments. These equations can be very complicated but have the essential advantage that the independent variables are only four $(\mathbf{x}, t)$ instead of seven $(\mathbf{x}, \boldsymbol{\xi}, t)$ a very important feature for numerical computations. In fact, in the case of one-dimensional steady problems one is reduced to a system of few integral equations with one independent variable, i.e., something feasible on a high speed computer. In the linearized case, one can even achieve reasonably accurate results with limited amounts of computation time. The solutions obtained by these accurate numerical procedures are very important for two reasons:

1) By comparison with experimental data, they can show the essential accuracy of certain models in describing certain experimental situations.

2) They can be used to test the accuracy of any approximate method of solution.

We list some of the problems solved with this approach. In the nonlinear case:

1) The shock wave structure (46);

2) Heat transfer between parallel plates, with the BGK model (47) and with the ES model (23);

3) Heat transfer between concentric cylinders (48).

In the linearized case we have the following list:

1) Plane Couette flow with the BGK model (49) and with the ES model (50).

2) Plane Poiseuille flow with the BGK model (51) and with the ES model (50).

3) Cylindrical Poiseuille flow with the BGK model (52) and with the ES model (53, 54).

4) Poiseuille flow in annular tubes (55).

5) Rotating cylinder (56).

6) Cylindrical Couette flow (57).

7) Heat transfer between parallel plates (58).

14. The Variational Principle

As we have seen, the most efficient procedures for solving the Boltzmann equation, in its full or linearized form, are based on moment methods or on the use of kinetic models. When going from the Boltzmann equation to moment methods or kinetic models, one gives up any intent of accurately investigating the distribution function and restricts himself to the study of the space variation of some moments of outstanding physical significance, such as density, mass velocity, temperature and heat flux. However, it is to be noted that even such restricted knowledge is not required for the purpose of comparing with experimental results. As a matter of fact, the typical output of an experimental investigation of Poiseuille flow is a plot of the flow rate versus the Knudsen number, while the actual velocity profile is usually the object of purely theoretical considerations. Analogously, the outstanding quality is the stress constant in Couette flow, the heat flux constant in heat transfer problems, the drag on the body in the flow past a body. From the point of view of evaluating these overall quantities, any computation of the flow fields appears to be a waste of time. Of course, the knowledge of flow fields is always interesting and illuminating, but frequently it happens that we have such a clear qualitative insight of the space behavior of the unknowns, that we can imagine, for them, simple approximated analytical expressions containing a small number of adjustable constants. It appears, therefore, that a method which would succeed in giving both a precise rule for determining the above mentioned constants and a highly accurate evaluation of the overall quantities of outstanding interest should turn out to be mostly useful. The features which we have just mentioned are typical of variational procedures; it is, therefore, in this direction that one has to look for the desired method. The procedure is well established and useful only in the case of the linearized Boltzmann equation; accordingly, we shall restrict ourselves to this case.

A nontrivial variational procedure for a linear equation is usually based on the self-adjointness of the linear operator $\mathcal{L}$ which appears in the equation

$$\mathcal{L}h = S \tag{14.1}$$

where S is a source term. Once $\mathcal{L}$ has been shown to be self-adjoint with respect to a certain scalar product $((\ ,\ ))$, then the functional

$$J(\tilde{h}) = ((\tilde{h}, \mathcal{L} h)) - 2((S, \tilde{h})) \tag{14.2}$$

is easily shown to satisfy

$$\delta J = 0 \tag{14.3}$$

for infinitesimal departures $\delta h = \tilde{h} - h$ from the solution of eq. (14.1). If additional conditions are satisfied then $J(\tilde{h})$ has a relative or absolute maximum for $\tilde{h} = h$. In the case of the steady linearized Boltzmann equation, S is a surface term (related, in general, to the inhomogeneous part of the boundary conditions), while

$$\mathcal{L}_0 = L - \xi \frac{\partial}{\partial x} \tag{14.4}$$

with homogeneous boundary conditions of the type discussed in Chapter I, sect. 3. Although L is self-adjoint in $\mathcal{H}$ and hence also in the space of square summable function of ξ and x, $\xi \cdot \partial/\partial x$ is not self-adjoint in such a space or in any space with positive definite metric. If we introduce, however, the reflection operator P defined by

$$P(f(x, \xi)) = f(x, -\xi) \tag{14.5}$$

it is possible to show (59, 5) that because of the properties of the kernel $R(\xi' \to \xi)$ appearing in the boundary conditions, Eq. (3.2), the operator $\mathcal{L} = P\mathcal{L}_0$ is self-adjoint, i.e. $((g, \mathcal{L} h))$ $((\mathcal{L} g, h))$, where

$$((h, g)) = \int\int h(x, \xi)\, g(x, \xi)\, f_0(\xi) dx d\xi \tag{14.6}$$

Then, let us consider the functional

$$(14.7) \quad J(\tilde{h}) = ((\tilde{h}, \mathcal{L}\tilde{h} - 2Pg_0)) + ((\tilde{h}^+ - A\tilde{h}^- - 2h_0, P\tilde{h}^-))_B$$

A, h^+, h^- are notations explained by the following rewriting of Eq. (3.2):

$$(14.8) \quad f^+ = f_w \, A(f^-/f_w)$$

where f_w is the wall Maxwellian, and

$$(14.9) \quad ((g, h))_B = \int dS \int d\xi \; g \, h \, f_0 \; |\xi \cdot n|$$

It is easy to show (59, 5) that $\delta J = 0$ for arbitrary (infinitesimal) $\delta \tilde{h}$ if and only if $\tilde{h} = h$, where h satisfies

$$(14.10) \quad \mathcal{L}_0 h = g_0$$

with the boundary conditions

$$(14.11) \quad h^+ = h_0 + Ah^-$$

Eq. (14.10) is the inhomogeneous linearized Boltzmann equation and Eq. (14.11) the general form assumed by the linearized boundary conditions (2, 5).

Accordingly, we can make a guess containing a certain number of parameters for $\tilde{h}$ and then determine them, according to the variational principle, by making $J(\tilde{h})$ stationary. If the boundary conditions have an inhomogeneous term of the form $2\mathbf{U} \cdot \xi$, where $\mathbf{U}$ is the boundary velocity or $\Delta T(\xi^2 - 5/2)$ where ΔT is the deviation of the boundary from the average temperature, the value of $J(\tilde{h})$ for $\tilde{h} = h$, takes on the form of a surface integral of $p_{ji} \, n_i \, U_j$ which give the drag, or $q_i \, n_i$ which gives the heat transfer through the body surface. This means that, by means of the variational principle, if one approximates h with an average error of 10 %, he can approximate the drag or the heat transfer with an error of order 1 % , because terms in the δh cancel according to Eq. (14.3) and the error is of order $|\delta h|^2$. In the case of Poiseuille flow, the boundary conditions are homogeneous and the source term is $g_0 = k\,\xi_3$; this means that $J(h)$ is proportional to the volume flow rate, which, as a consequence, can be evaluated accurately. The usefulness of the variational principle as explained above is somewhat limited by two circumstances: we have to make guesses on the distribution function rather than on moments of the distribution function itself; although this can be done (53), we must expect, in general, that we obtain very complicated trial functions. The advantages of the variational approach are much enhanced if we use it in connection with two other tools: kinetic models and integral equation approach. In fact, if we use kinetic models in the integral form, a guess on a finite number of moments implies a guess on the distribution function which automatically satisfies the boundary conditions (since the latter are built in the equations). If we denote the basic moments by ρ_j ($j = 1, \ldots, N$) and ρ is a column vector which summarized the ρ_j, then ρ satisfies (50)

$$\rho = \rho^0 + A\,\rho \qquad (14.12)$$

where A is a matrix integral operator. In the case of the models defined by eq. (4.14), one can show that

$$\overset{+}{A}\,\alpha\,\sigma = \sigma\,\alpha\,A \qquad (14.13)$$

where $\alpha = ((\nu_N\,\delta_{RQ} + B_{QR}))$, $\overset{+}{A}$ is the transposed of the operator A with respect to the scalar product

$$(\rho, \psi)_V \qquad \rho(x)\cdot\psi(x)\,dx \qquad (14.14)$$

and σ is a diagonal matrix [illegible] components $(-1)^p\,\delta_{jk}$, $(-1)^p$ meaning ± 1 according to the parity of ρ_j Then the functional

$$J(\tilde{\rho}) = (\alpha\,\sigma\,\tilde{\rho},\ \tilde{\rho} - A\tilde{\rho} - 2\rho^{(0)})_V \qquad (14.15)$$

is stationary for $\tilde{\rho} = \rho$ satisfies Eq. (14.12). Also, if the symmetry of the problem is such that only ρ_j's with a definite parity are different from zero, then σ coincides with either $+I$ or $-I$, I being the unit matrix, and one can easily show that the stationarity principle becomes a minimum or a maximum principle. Finally, one can relate the value attained by J for $\tilde{\rho} = \rho$ to the global quantities of basic interest (drag, flow rate, heat flux) and hence retain the feature of having a very precise procedure for evaluating these quantities.

CHAPTER V

SPECIFIC PROBLEMS

15. Application of the Method of Elementary Solutions to the Flow Between Parallel Plates

As we saw in Chapter III, the half-space problems connected with Eq. (8.1) can be solved by analytical means. This is not true for flow between parallel plates, as Couette flow and Poiseuille flow. However, the method of elementary solutions can be used to obtain series solutions and gain insight about the qualitative behavior of the solution. Let us rewrite the general solution, Eq. (8.19), of Eq. (8.1) as follows

$$Y(x, \xi) = A_0 + A_1(x - \xi) + \int_{-\infty}^{\infty} A(u) \exp\left(-\frac{x}{u} - \frac{\delta}{2|u|}\right) g_u(\xi)\, du \tag{15.1}$$

where δ is the distance between the plates in θ units. A(u) has been redefined by inserting a factor $\exp(-\delta/2|u|)$ for convenience and the plates are assumed to be located at $x = \pm\ \delta/2$.

The general form of Eq. (15.1) shows that any solution is split into two parts, one of which

$$A_0 + A_1(x - \xi) \tag{15.2}$$

gives just the Navier-Stokes solution for the considered geometry (constant stress, straight profile of velocity) and the other part

$$\int_{-\infty}^{\infty} A(u)\ f_u(\xi)\ \exp\left(-\frac{\delta}{2|u|} - \frac{x}{u}\right) du \tag{15.3}$$

is expected to be important only near boundaries, because of the fast decay of the exponential factor. Therefore for sufficiently large δ the picture is the following: a core, where a continuum description (based on the Navier-Stokes equations) prevails, surrounded by kinetic boundary layers, produced by the interaction of the molecules with the walls. As δ becomes smaller, however, the exponentials are never negligible, i.e., the kinetic layers merge with the core to form a flow field which cannot be described in simple terms. Finally, when δ is negligibly small, $Y(x, \xi)$ does not depend sensibly on **x**, and the molecules retain the distribution they had just after their last interaction with a boundary. In the case of Couette flow (i.e., when there are two plates at $x = \pm\, \delta/2$ moving with velocities $\pm U/2$ in the z-direction) the situation is well described by the above short discussion, although it is possible to obtain a more detailed picture by finding approximate expressions for A_1 and $A(u)$ (A_0 is zero and $A(u)$ is odd in u because of the antisymmetry inherent with the problem). Accordingly, we shall consider in more detail the case of Poiseuille flow between parallel plates, which lends itself to more interesting considerations.

Plane Poiseuille flow is the flow of a fluid between two parallel plates induced by a pressure gradient parallel to the plates. In the continuum case no distinction is made between a pressure gradient arising from a density gradient and one arising from a temperature gradient. This distinction, on the contrary, is to be taken into account when a kinetic theory description is considered. We shall restrict ourselves to the former case (for the case of a temperature gradient see ref. 61).

The basic linearized Boltzmann equation for Poiseuille flow in a channel of arbitrary cross-section (including the slab as particular case) will now be derived. We assume that the walls re-emit the molecules with a Maxwellian distribution f_0 with constant temperature and an unknown density $\rho = \rho\,(z)$ (z being the coordinate parallel to the flow). If the length of the channel is much larger than any other typical length (mean free path, distance between the walls), then we can linearize about the above mentioned Maxwellian f_0; in fact $\rho(z)$ is slowly varying and f_0 would be the solution in the case of a rigorously constant ρ. Accordingly we have

$$\xi_1 \frac{\partial h}{\partial x} + \xi_2 \frac{\partial h}{\partial y} + \xi_3 \frac{\partial h}{\partial z} + \frac{1}{\rho}\frac{\partial \rho}{\partial z}\, \xi_3 = Lh \tag{15.4}$$

Because of the assumption of a slowly varying ρ (long tube), we can regard $(1/\rho)(\partial\rho/\partial z)$ as constant (i.e., we disregard higher order derivatives of ρ as well as powers of first order derivatives). If $(1/\rho)(\partial\rho/\partial z)$ is regarded as a constant, it follows that $\partial h/\partial z = 0$ since z does not appear explicitly in the equation nor in the boundary conditions. The latter can be written

$$h(x, y, z, \boldsymbol{\xi}) = 0 \qquad ((x, y) \in \partial\Sigma\,;\; xn_1 + yn_2 > 0) \tag{15.5}$$

where $\partial\Sigma$ is the contour of the cross section and $\mathbf{n} = (n_1, n_2)$ the normal pointing into the channel. Therefore, we can write

$$\xi_1 \frac{\partial h}{\partial x} + \xi_2 \frac{\partial h}{\partial y} + k\xi_3 = Lh \tag{15.6}$$

where $k = (1/\rho)(\partial\rho/\partial z)$. Equation (15.6) governs linearized Poiseuille flow in a very long tube of arbitrary cross section. If we specialize to the case of a slab and use the BGK model, we have:

$$h = 2\xi_3\, Z(x, \xi_1) \tag{15.7}$$

where $Z(x, \xi)$ satisfies

$$\xi \frac{\partial Z}{\partial x} + \frac{1}{2} k = \pi^{-1/2} \int_{-\infty}^{\infty} e^{-\xi_1^2} Z(x, \xi_1)\, d\xi_1 - Z(x, \xi) \tag{15.8}$$

$$Z\left(-\frac{\delta}{2} \operatorname{sgn} \xi, \xi\right) = 0 \tag{15.9}$$

provided x is measured in θ units.

Equation (15.8) differs from Eq. (15.1) because of the inhomogeneous term $k/2$. If we find a particular solution of Eq. (15.8) then we can add it to the general solution of Eq. (15.1) n order to have the general solution of Eq. (15.8). By differentiation of the latter equation, we deduce that $\partial Z/\partial x$ satisfies Eq. (15.1); since the general solution of the latter contains exponentials (which reproduce themselves by integration and differentiation) and a linear function of x, we try a particular solution of Eq. (15.8) in the form of a quadratic function of x (with coefficients depending upon ξ). It is verified that solutions of this form exist and one of them is

$$(15.10) \qquad Z_0(x, \xi) = \frac{1}{2} k \left(x^2 - \frac{\delta^2}{4} - 2x\xi - (1 - 2\xi^2)\right)$$

Therefore,

$$(15.11) \qquad Z(x, \xi) = Z_0(x, \xi) + Y(x, \xi)$$

where $Y(x, \xi)$ is given by Eq. (8.1). Equation (15.5) gives the following boundary condition for $Y(x, \xi)$:

$$(15.12) \qquad Y\left(-\frac{\delta}{2} \operatorname{sgn} \xi, \xi\right) = -\left(|\xi| - (1 - 2\xi^2)\frac{1}{\delta}\right)\frac{k\delta}{2}$$

Since the symmetry inherent to our problem implies that

$$(15.13) \qquad Y(x, \xi) = Y(-x, -\xi)$$

$A_1 = 0$ and $A(u) = A(-u)$ If we take this into account, Eq. (15.12) becomes

$$(15.14) \quad A_0 + \int_0^\infty A(u) g_u(\xi) du = -\frac{k\delta}{2}\left(\xi - (1 - 2\xi^2)\frac{1}{\delta}\right) - \int_0^\infty \frac{uA(u)}{u + \xi} e^{-\frac{\delta}{u}} du \qquad (\xi > 0)$$

and the equation for $\xi < 0$ is not required because $A(u) = A(-u)$. If we call the right hand of this equation with $g(\xi)$, Eq. (15.14) becomes Eq. (8.13) and we can apply Eqs. (8.14) and (8.15), thus obtaining

$$(15.15) \quad A_0 = -\left\{\sigma + \frac{1}{\delta}\left(\frac{1}{2} + \sigma^2\right) - \int_0^\infty u e^{-\frac{\delta}{u}} (P(u))^{-1} A(u) du\right\} \frac{k\delta}{2}$$

$$A(u) = \frac{k}{2}\pi^{1/2}\left(u + \frac{\delta}{2} + \sigma\right)e^{u^2}(P(u))^{-1}\{(p(u))^2 + \pi^2 u^2\}^{-1}$$
$$+ \frac{\pi^{1/2}}{2} e^{u^2}(P(u))^{-1}\{(p(u))^2 + \pi^2u^2)^{-1}\int \frac{\xi(P(\xi))^{-1}}{u+\xi} e^{-\frac{\delta}{\xi}} A(\xi)d\xi \qquad (15.16)$$
$$(\xi > 0)$$

where permissible inversions of the order of integration have been performed and Eq. (8.21) used. Here $\sigma = \zeta/\theta$ (see Eqs. (9.9) and (9.10)). Thus the problem has been reduced to the task of solving an integral equation in the unknown A(u), Eq. (15.16). This equation is a classical Fredholm equation of the second kind with symmetrizable kernel. The corresponding Neumann-Liouville series can be shown to converge for any given positive value of δ (62).

It is also obvious that the larger δ, the more rapid is the convergence. This allows the ascertainment of certain results in the near-continuum regime. In particular, if terms of order $\exp(-3(\delta/2)^{3/2})$ are negligible, only the zero order term of the series need be retained:

$$A(u) = \frac{k}{2}\pi^{1/2}\left(\left(\frac{\delta}{2} + \sigma\right) + u\right)e^{u^2}(P(u))^{-1}\{(p(u))^2 + \pi^2u^2\}^{-1} \qquad (15.17)$$

Within the same limits of accuracy, A_0 is given by

$$A_0 = -\frac{k\delta}{2}\left(\sigma + \frac{1}{\delta}\left(\frac{1}{2} + \sigma^2\right)\right) \qquad (15.18)$$

We note that this zero-order approximation is by far more accurate than a continuum treatment (even if slip boundary conditions are used in the latter). In fact, even in the zero-order approximation

1) kinetic boundary layers are present near the walls;

2) in the main body of the flow the mass velocity satisfies the Navier-Stokes momentum equation; however, the corresponding extrapolated boundary conditions at the walls are not the usual slip boundary conditions, but show the presence of a second order slip:

(15.19) $$v_3(\pm\frac{\delta}{2}) = \pm\,\sigma(\frac{\partial v_3}{\partial x})_{x=\pm\frac{\delta}{2}} - \frac{1}{2}(\frac{1}{2} + \sigma^2)\,(\frac{\partial^2 v_3}{\partial x^2})_{x=\pm\frac{\delta}{2}}$$

In order to obtain these results, we observe that the mass velocity is given by

(15.20) $$\begin{aligned} v_3(x) &= \pi^{-1/2} \int_{-\infty}^{\infty} Z(x,\xi)\, e^{-\xi^2}\, d\xi \\ &= \frac{k}{2}(x^2 - \frac{\delta^2}{4}) + A_0 + \int_{-\infty}^{\infty} A(u) \exp(-\frac{x}{u} - \frac{\delta}{2|u|})du \end{aligned}$$

where Eqs. (15.10) and (15.11) have been taken into account. Equation (15.20) is exact; if terms of order $\exp(-3(\delta/2)^{3/2})$ can be neglected, A_0 and $A(u)$ are given by Eqs. (25.18) and (15.17). In particular, the integral term in Eq. (15.20) describes the space transients in the kinetic boundary layers; in the main body of the flow the integral term is negligible and we have

(15.21) $$v_3(x) = \frac{k}{2}(x^2 - \frac{\delta^2}{4} - \sigma\delta - (\frac{1}{2} + \sigma^2))$$

It is easily checked that this expression solves the Navier-Stokes momentum equation for plane Poiseuille flow and satisfies the boundary conditions (15.19).

We can also easily write down the distribution function in the main body of the flow. As a matter of fact, $Y(x, \xi)$ reduces here to A_0, so that Eqs. (25.10) and (25.11) give

(15.22) $$Z(x,\xi) = \frac{k}{2}(x^2 - \frac{\delta^2}{4} - 2x\xi - (1-2\xi^2) - \sigma\delta - (\frac{1}{2} + \sigma^2))$$

By taking into account Eq. (21.21), Eq. (21.22) can be rewritten as follows

(15.23) $$Z(x,\xi) = v_3(x) - \theta \frac{\partial v_3}{\partial x}\xi + (\xi^2 - \frac{1}{2})\theta^2 \frac{\partial^2 v_3}{\partial x^2}$$

where general units for x have been restored (i.e., we have written x/θ in place of x). Equation (21.23) clearly shows that, in the main body of the flow, the distribution function is of the Hilbert-Enskog type (power series in θ), as was to be expected. However, it is not the usual distribution corresponding to the Navier-Stokes level of description. As a matter of fact, Eq. (15.23) gives a Burnett distribution function (2,3, 5,7) and this explains the appearance of a second order slip from a formal point of view. From an intuitive standpoint, the second order slip can be attributed to the fact that molecules with non-zero velocity in the z-direction move into a region with different density before having any collisions and there is a net transport of mass because of the density gradient; i.e., molecules move preferentially toward smaller densities even before suffering any collision, and, therefore, at a mean free path from the wall an effect of additional macroscopic slip appears.

The presence of an additional slip means that, for a given pressure gradient and a given geometry, more molecules pass through a cross section than predicted by Navier-Stokes equations with first-order slip. This is easily shown for sufficiently large δ by using Eq. (15.20), which gives for the flow rate

$$F = \int_{-d/2}^{d/2} \rho v_3(x)dx = -\frac{1}{2}\frac{d\rho}{dx}\, d^2\left(\frac{1}{6}\delta + \sigma + \frac{2\sigma^2 - 1}{\delta}\right) \qquad (15.24)$$

$$(\delta >> 1)$$

provided terms of higher order in $1/\delta$ are neglected (Eqs. (21.17) and (25.18) have been used). In Eq. (15.24) x and z are in general units and $d = \delta\theta$ is the distance between the plates in the same units. Therefore, for given geometry and pressure gradient, the non-dimensional flow rate is

$$Q(\delta) = \frac{1}{6}\delta + \sigma + \frac{2\sigma^2 - 1}{\delta} \qquad (\delta >> 1) \qquad (15.25)$$

The last term is the correction to the Navier-Stokes result; it arises in part from the second order slip and in part from the kinetic boundary layers. In fact, the gas near the walls moves more slowly than predicted by an extrapolation of Eq. (15.21); this brings in a correction by $Q(\delta)$ of the same order of the second order slip. This correction reduces the effect of the second order slip but does not eliminate it completely at least for sufficiently large values of δ . However, it is clear that the effect is not completely cancelled even for smaller values of δ, because the molecules with velocity almost parallel to the wall give a sensible contribution to the motion by travelling downstream for a mean free path. This qualitative reasoning is confirmed by a study of the nearly-free regime ($\delta \to 0$). This study can be based either on the iteration procedures described in sect. 13 (63) or on a different use of the method of elementary solutions (62). In both cases the conclusion is

$$Q(\delta) \cong - \pi^{-1/2} \log \delta \qquad (\delta \to 0) \tag{15.26}$$

This means that higher order contributions from kinetic layers destroy the $1/\delta$ term in Eq. (15.25) but leave a weaker divergence for $\delta \to 0$ (essentially related to the above mentioned molecules travelling parallel to the plates). The behavior for large values of δ (Eq. (15.25)) and for small values of δ imply the existence of at least one minimum in the flow rate. This minimum was experimentally found a long time ago by Knudsen (64) for circular tubes and then by different authors for long tubes of various cross sections. The above discussion gives a qualitative explanation of the presence of the minimum, although its precise location for slabs and more complicated geometries must be found by approximate techniques (see sect. 16).

16. Examples of Application of the Integral Equation Approach and the Variational Method

In this section we shall consider very simple problems as examples of application of both the integral equation approach and the variational procedure. The first example is Kramer's problem with BGK model (Chapter III, sect. 9); as we know, this problem can be solved exactly, but we shall ignore this at present. If we follow the procedure indicated in sect. 13, we can express the perturbation of the distribution function in terms of a single moment, the z component of the mass velocity $v_3(x)$; in fact, density, temperature and the remaining components of v remain unperturbed in a linearized treatment. Accordingly, we can construct an integral equation for v_3 by taking the corresponding moment of the expression of h in terms of v_3. If we put $(\phi(x) = (v_3(x)-kx)/(k\theta)$ i.e., if we subtract the asymptotic behavior and make the unknown nondimensional, we obtain the following integral equation:

$$\pi^{1/2}\ \phi(x) = T_1(x) + \int T_{-1}(|x-y|\phi(y)\ dy \qquad (x > 0) \qquad (16.1)$$

where we have put

$$T_n(x) = \int_0^\infty t^n \exp(-t^2 - \frac{x}{t})\ dt \qquad (16.2)$$

θ has been taken to be unity, while the velocity unit is, as usual, (2RT) (2RT) . Equation (16.2) defines a family of transcedental functions, which are always met when using the integral equation approach for solving models with constant collision frequency. The T_n functions have some notable properties:

$$\frac{dT_n}{dx} = -T_{n-1} \qquad (16.3)$$

$$T_n(x) = \frac{n-1}{2}T_{n-2} + \frac{x}{2}T_{n-3} \qquad (16.4)$$

$$(16.5) \qquad T_n(0) = \frac{1}{2}\Gamma\left(\frac{n+1}{2}\right)$$

One can derive expansions valid for small x and expansions for large x and compute the T_n to any desired accuracy. Let us put

$$(16.6) \qquad \mu = \lim_{x\to\infty}\phi(x) = \pi^{1/2}\,\frac{\zeta}{2\ell}$$

where ζ is the slip coefficient and ℓ the mean free path; then, if we define

$$(16.7) \qquad \psi(x) = \phi(x) - \mu$$

we have

$$(16.8) \qquad \pi^{1/2}\;\psi(x) - \int_0^\infty T_{-1}(|x-y|)\,\psi(y)dy = T_1(x) - \mu T_0(x) \qquad (x \geq 0)$$

where Eq. (16.3) has been used.

Since $\psi(x)$ is integrable on $(0,\infty)$, we can integrate Eq. (26.8) from x to ∞ twice and obtain

$$(16.9) \qquad \int T_1(|x-y|)\,\psi(y)dy = \mu T_2(x) - T_3(x) \qquad (x \geq 0)$$

where Eqs. (16.3) and (16.5) have been used.

For this problem, the basic functional is

$$(16.10) \qquad J(\tilde\phi) = \pi^{1/2}\int_0^\infty dx\,\{\,(\tilde\phi(x))^2 - \int_0^\infty T_{-1}(|x-y|)\tilde\phi(x)\tilde\phi(y)dx - 2\tilde\phi(x)T_1(x)\,\}$$

This functional attains a minimum when $\tilde{\phi} = \phi$, ϕ being the solution of Eq. (16.1). Equation (16.9) with $x = 0$ gives

$$\int T_1(y)\,\psi(y)\,dy = \frac{1}{4}\,\mu\pi^{1/2} - \frac{1}{2} \qquad (16.11)$$

and using Eq. (16.6) we have:

$$\zeta = \ell\left(\frac{2}{\pi} + \frac{4}{\pi}\int_0^\infty \phi(y)T_1(y)\,dy\right) \qquad (16.12)$$

We now note that for $\tilde{\phi} = \phi$ we have

$$J(\phi) = \min(J(\tilde{\phi})) = -\int_0^\infty \phi(y)T_1(y)\,dy \qquad (16.13)$$

and Eq. (16.12) becomes

$$\zeta = \ell\left\{\frac{2}{\pi} - \frac{4}{\pi}\min(J(\tilde{\phi}))\right\} \qquad (16.14)$$

In such a way we have found a direct connection between the slip coefficient and the minimum value attained by the functional $J(\tilde{\phi})$; this means that even a poor estimate for ϕ can give an accurate value for ζ. Accordingly, we make the simplest choice for the trial function $\phi = c$ (constant). We find by straightforward computation

$$J(c) = \frac{1}{2}c^2 - \frac{1}{2}\pi^{1/2}c \qquad (16.15)$$

The minimum value is attained when

$$c = \frac{1}{2}\pi^{1/2} \qquad (16.16)$$

corresponding to

(16.17) $$J(\frac{1}{2}\pi^{1/2}) = \frac{1}{\delta}\pi$$

Equation (16.15) then gives for ζ

(16.18) $$\zeta = (\frac{1}{2} + \frac{2}{\pi})\ell = (1.1366)\ell$$

to be compared with the exact value $\zeta = (1.1466)\ell$ given in chapter III. We see that even a very simple choice for $\tilde{\phi}$, a choice which we know to be very inadequate to describe the kinetic boundary layer, yields a rather accurate estimate of the slip coefficient ζ.

As the next example, we consider Couette flow between two parallel plates (described in sect. 15). If $\phi(x)$ now denotes the ratio of the mass velocity to U/2 ($\pm$ U/2 being the velocities of the plates), we obtain the following integral equation (49):

(16.19) $$\pi^{1/2}\phi(x) = T_0(\frac{\delta}{2} - x) - T_0(\frac{\delta}{2} + x) + \int_{-\delta/2}^{\delta/2} T_{-1}(|x-y|)\phi(y)dy$$

where δ is the distance between the plates in θ units. Analogously we can find

(16.20) $$\pi_{xz} = T_1(\frac{\delta}{2} - x) + T_1(\frac{\delta}{2} + x) - \int_{-\delta/2}^{\delta/2} \mathrm{sgn}(x-y)T_0(|x-y|)\phi(y)dy$$

where π_{xz} is the ratio of the stress tensor p_{xz} to its free molecular value $(-\rho U\pi^{-1/2}/2)$. Note that π_{xz} is constant, in spite of the fact that Eq. (16.20) shows an x-dependence; the constancy of π_{xz} is obvious from conservation of momentum and can be recovered from Eq. (16.20) by differentiating and comparing with Eq. (16.19). Therefore we can take Eq. (16.20) at any point for evaluating π_{xz} ; the simplest choice is to take the arithmetical mean of the expression given by Eq. (16.20) for $x = \pm\,\delta/2$. We obtain

$$\pi_{xz} = \frac{1}{2} + T_1(\delta) + \frac{1}{2}\int_{-\delta/2}^{\delta/2} \left(T_0\left(\frac{\delta}{2} + y\right) - T_0\left(\frac{\delta}{2} - y\right)\right)\phi(y)\,dy \qquad (16.21)$$

Equation (16.19) was solved numerically by Willis (49) for δ ranging from 0 to 20. If we want to use the variational procedure, we construct the functional

$$J(\phi) = \int_{-\delta/2}^{\delta/2} (\phi(x))^2\,dx - \pi^{-1/2} \int_{-\delta/2}^{\delta/2} T_{-1}(|x-y|)\,\bar{\phi}(x)\bar{\phi}(y)\,dx\,dy$$

$$+\ 2\pi^{-1/2} \int_{\delta/2}^{\delta/2} \left(T_0\left(\frac{\delta}{2} + x\right) - T_0\left(\frac{\delta}{2} - x\right)\right)\bar{\phi}(x)\,dx \qquad (16.22)$$

which attains its minimum value

$$J(\phi) = \min J(\tilde{\phi}) = \pi^{-1/2} \int \left(T_0\left(\frac{\delta}{2} + x\right) - T_0\left(\frac{\delta}{2} - x\right)\right)\phi(x)\,dx \qquad (16.23)$$

when $\bar{\phi} = \phi(x)$ solves Eq. (16.19). Comparing Eq. (16.21) and (16.23), we deduce

$$\pi_{xz} = \frac{1}{2} + T_1(\delta) + \frac{\pi^{1/2}}{2} \min J(\bar{\phi}) \qquad (16.24)$$

In such a way, the stress has a direct connection with the minimum value attained by $J(\bar{\phi})$. As a trial function, we can take the continuum solution, i.e.

$$\bar{\phi}(x) = A\,x \qquad (16.25)$$

where A is an indeterminate constant. After some easy manipulations based on the properties of the T_n functions, we find (60) an expression for $J(Ax)$ in terms of $T_n(\delta)$ $(1 < n < 3)$. The minimum condition is easily found, the corresponding $A = A_0(\delta)$ and $J(A_0x)$ being again expressed in terms of the $T_n(\delta)$. Inserting $J(A_0x)$ into Eq. (16.24) we obtain π_{xz} ; the resulting values are tabulated versus δ in Table I, third column. The second column of the same table gives the results obtained by Willis by means of a numerical solution of Eq. (16.19). It will be noted that the agreement is excellent. But one can see something more, i.e., that the results obtained by the variational procedure are more accurate than Willis'. As a matter of fact, the variational procedure gives for π_{xz} a value approximated from above, and it is noted that values in the third column are never larger than the corresponding ones in the second column. The fourth column gives the results obtained by using a cubic trial function:

$$\tilde{\phi}(x) = A x + B x^3 \qquad (16.26)$$

In this case, the algebra is more formidable but still straightforward. It is to be noted that the resulting values for π_{xz} are only slightly different from the previous one: this is another test of the accuracy of the method.

Another example is offered by Poiseuille flow. The starting point is Eq. (15.6). If we adopt the BGK model, we can obtain an integral equation for the mass velocity $v_3(\mathbf{x})$ where $\mathbf{x}$ is the two-dimensional vector describing the cross section of the channel. If we put $v_3(\mathbf{x}) = 1/2\, k\theta(1-\phi(\mathbf{x}))$, where $k = p^{-1}\, \partial p/\partial z$, the integral equation to be solved can be written as follows

$$\phi(\mathbf{x}) = 1 + \pi^{-1} \iint T_0(|\mathbf{x}-\mathbf{y}|)\, |\mathbf{x}-\mathbf{y}|^{-1}\, \phi(\mathbf{y})\, d\mathbf{y} \qquad (16.27)$$

where $\mathbf{x}$ is measured in θ units and $\Sigma(\mathbf{x})$ is the part of the cross-section whose points can be reached from $\mathbf{x}$ by straight lines without intersecting boundaries ($\Sigma(\mathbf{x})$ is the whole cross section Σ when the boundary curvature has constant sign). The functional to be considered is now

TABLE I

δ	Willis' result (ref. 49)	Linear trial function	Cubic trial function
0.01	0.9913	0.9914	0.9914
0.10	0.9258	0.9258	0.9258
1.00	0.6008	0.6008	0.6008
1.25	0.5517	0.5512	0.5511
1.50	0.5099	0.5097	0.5096
1.75	0.4745	0.4743	0.4742
2.00	0.4440	0.4438	0.4437
2.50	0.3938	0.3935	0.3933
3.00	0.3539	0.3537	0.3535
4.00	0.2946	0.2945	0.2943
5.00	0.2526	0.2524	0.2523
7.00	0.1964	0.1964	0.1963
10.00	0.1474	0.1474	0.1473
20.00	0.0807	0.0805	0.0805

$$J(\bar{\phi}) = \int_{\Sigma} (\bar{\phi}(x))^2\, dx - \int_{\Sigma} \bar{\phi}(x) \int_{\Sigma(x)} T_0(|x-y|)\, |x-y|^{-1}\, \bar{\phi}(y)dy \tag{16.28}$$
$$- 2 \int \bar{\phi}(x)dx$$

The value of this functional attains a minumum when $\bar{\phi} = \phi$, where ϕ satisfies Eq. (22.27), and the minimum value is

$$J(\bar{\phi}) = - \int_{\Sigma} \bar{\phi}(x)dx \tag{16.29}$$

a quantity obviously related to the flow rate. Actual calculations have been performed for the case of a slab (60) and a cylinder (65) by assuming a parabolic profile (i.e., a continuum-like solution). In the plane case one can obtain the flow rate in terms of T_n functions, while in the cylindrical case one has also to perform numerical quadratures involving these functions.

In both cases one can evaluate the flow rate with great accuracy and compare the results with the numerical solutions by Cercignani and Daneri (51) and Cercignani and Sernagiotto (52). As shown by Tables II and III the agreement is very good The results compare also very well with experimental data and, in particular, the flow rate exhibits a minimum for $\delta \simeq 1.1$ in the plane case (δ = distance of the plates in θ units) and for $\delta \simeq 0.3$ in the cylindrical case (δ = radius of the cylinder in θ units). The location of the minimum is in an excellent agreement with experimental data, while the values of the flow rate show deviations of order 3% from experimental data in the cylindrical case. It is to be noted that the same problem can be treated with the ES model (53) to show that, once one knows the BGK solution (Pr = 1) one can obtain the ES for any Prandtl number; in particular, the non-dimensional flow rate in the cylindrical case is given by

$$Q(\delta, Pr) = Q(\delta Pr, 1) + (1 - Pr)\frac{\delta}{4} \tag{16.30}$$

TABLE II

δ	Results from ref. 51 from above	from below	Variational method
0.01	3.0499	......	3.0489
0.10	2.0331	2.0326	2.0314
0.50	1.6025	1.6010	1.6017
0.70	1.5599	1.5578	1.5591
0.90	1.5427	1.5367	1.5416
1.10	1.5391	1.5352	1.5379
1.30	1.5441	1.5390	1.5427
1.50	1.5546	1.5484	1.5530
2.00	1.5963	1.5862	1.5942
2.50	1.6497	1.6418	1.6480
3.00	1.7117	1.7091	1.7092
4.00	1.8468	1.8432	1.8440
5.00	1.9928	1.9863	1.9883
7.00	2.2957	2.2851	2.2914
10.00	2.7669	2.7447	2.7638

TABLE III

δ	Numerical solution (ref. 52)	Variational method
0.01	1.4768	1.4801
0.10	1.4043	1.4039
0.20	1.3820	1.3815
0.40	1.3796	1.3788
0.60	1.3982	1.3971
0.80	1.4261	1.4247
1.00	1.4594	1.4576
1.20	1.4959	1.4937
1.40	1.5348	1.5321
1.60	1.5753	1.5722
2.00	1.6608	1.6559
3.00	1.8850	1.8772
4.00	2.1188	2.1079
5.00	2.3578	2.3438
7.00	2.8440	2.8245
10.00	3.5821	3.5573

If the value $Pr = 2/3$ is chosen, as appropriate for a monoatomic gas, we obtain results deviating from experimental data by only 1% (54).

Other problems which have been treated with the variational technique are heat transfer between parallel plates (58) and from a sphere (65), flow past a solid body (66–67) and heat transfer between concentric cylinders (58). Also, applications to models with velocity dependent collision frequency have been considered, especially in connection with the evaluation of the slip coefficient (28, 37). In particular, for the flow past a sphere at low Mach number, the numerical results are listed in Table IV, column II. R is the ratio of the sphere radius to $2\ell/\sqrt{\pi}$, where ℓ is the mean path given by

$$\ell = \frac{\mu}{p}\left(\frac{\pi R T_\infty}{2}\right)^{1/2} \tag{16.31}$$

In column I are listed the values for the drag obtained from Millikan's formula, based on a fit of his experimental data:

$$D = \frac{A + B}{2\pi^{-1/2} R + A + B\exp(-2\pi^{-1/2} CR)} \tag{16.32}$$

where:

$$A = 1.234, \quad B = 0.414, \quad C = 0.876 \tag{16.33}$$

As it is seen the variational solution disagrees from Millikan's formula, i.e., from his experimental data, by at most 2%. We conclude that our solution is in excellent agreement with the experimental data. The remaining discrepancy can be ascribed to one or more of the following cases:

1) experimental errors;

2) use of the BGK model with the consequence of an incorrect Prandtl number (the temperature is not constant);

3) errors due to the approximate nature of the trial function.

We note that the latter should be less than 1%, on the basis of an analogy with previously treated problems, and should go in the wrong direction, i.e., would

TABLE IV

R	Variational results	Millikan's formula
0.05	0.9778	0.9784
0.075	0.9651	0.9677
0.1	0.9529	0.0571
0.25	0.8864	0.8959
0.5	0.7900	0.8036
0.75	0.7088	0.7236
1	0.6404	0.6549
1.25	0.5824	0.5961
1.5	0.5332	0.5456
1.75	0.4910	0.5021
2	0.4546	0.4645
2.5	0.3951	0.4029
3	0.3488	0.3551
4	0.2818	0.2863
5	0.2360	0.2396
6	0.2029	0.2058
7	0.1779	0.1804
8	0.1583	0.1606
9	0.1426	0.1447
10	0.1297	0.1317

increase the disagreement from Millikan's formula to, perhaps, 3%. This disagreement could be possibly reduced within the range of experimental data by the use of a more accurate model, as in the case of Poiseuille flow (54).

REFERENCES

[1] BRUSH, S.G., "Kinetic Theory", vol. 2 – Pergamon Press, Oxford (1966).

[2] CERCIGNANI, C., "Mathematical Methods in Kinetic Theory", Plenum Press McMillan, New York (1969).

[3] GRAD, H. – "Principles of Kinetic Theory in "Handbuch der Physik", vol. XXII, Springer (1958).

[4] CERCIGNANI, C., "Transport Theory and Statistical Physics", 2,211 (197)

[5] CERCIGNANI, C., "Theory and Application of the Boltzmann Equation", Scottish Academic Press (1975).

[6] BOLTZMANN, L., Sitzungsberichte Akad. Wiss., Vienna p. II, 66, 275 (1872).

[7] CHAPMAN, S. and COWLING, T.G., "The Mathematical Theory of Nonuniform Gases", Cambridge University Press, Cambridge (1952).

[8] LANFORD, O.H., III: Private Communication.

[9] GRAD, H., "On the Kinetic Theory of Rarefied Gases", Comm. Pure and Appl. Mathematics 2, 331-407, 1949.

[10] CARLEMAN, T., "Problèmes Mathematiques dans la Theorie Mathematique des gaz" Almqvist and Wicksells, Uppsala (1953).

[11] HIRSCHFELDER, J.O., CURTISS, C.F., and BIRD, R.B., "Molecular Theory of Gases and Liquids", Wiley, New York (1954).

[12] DARROZES, J.S., and GUIRAUD, J.P., Compt., Rend. Ac. Sci. (Paris), A262, 1368 (1966).

[13] CERCIGNANI, C., "Boundary Value Problems in Linearized Kinetic Theory" in "Transport Theory", G. Birkhoff et al., eds. SIAM-AMS Proceedings, vol. I, p. 240, AMS, Providence (1968).

[14] KUSCER I., in "Transport Theory Conference" AEC Report ORO–3858–1, Blacksburgh, Virginia (1963).

[15] SHEN, S.F., "Entropie", 18, 138 (1967).

[16] CERCIGNANI, C. and LAMPIS, M., "Transport Theory and Statistical Physics", 1, 101 (1971).

[17] CERCIGNANI, C. – "Transport Theory and Statistical Physics", 2, 27 (1972).

[18] BHATNAGAR, P.L., GROSS, E.P., KROOK, M., "A Model for Collision Processes in Gases". Physical Review, 94, 511-525, 1964.

[19] WELANDER, P., "On the Temperature Jump in Rarefied Gas", Arkiv für Fysik, 7, 507-553, 1954.

[20] KROOK, M. "Continuum Equations in the Dynamics of Rarefied Gases", Jnl. Fluid Mechanics, 6, 523-541, 1959.

[21] CERCIGNANI, C., "The Method of Elementary Solutions for Kinetic Models with Velocity-dependent Collision Frequency", Annals of Physics, 40, 469-481, 1966.

[22] HOLWAY, L.H., Jr. "Approximation Procedures for Kinetic Theory", Ph. D. Thesis, Harvard, 1963.

[23] CERCIGNANI, C. and TIRONI, G., "Nonlinear Heat Transfer Between Two Parallel Plates According to a Model with Correct Prandtl Number", Rarefied Gas Dynamics, C.L. Brundin editor, vol. 1, 441-453, Academic Press, New York, 1967.

[24] GROSS, E.P. and JACKSON, E.A., "Kinetic Model and the Linearized Boltzmann Equation", Physics of Fluids, 2, 432-441, 1959.

[25] SIROVICH, L., "Kinetic Modeling of Gas Mixtures", Physics of Fluids, 5, 908, 1962.

[26] GRAD, H., "Asymptotic Theory of the Boltzmann Equation II. Rarefied Gas Dynamics, J.A. Laurmann, editor., vol. 1, 26-59, Academic Press, New York, 1963.

[27] CERCIGNANI, C., "On Boltzmann Equation with Cutoff Potentials", Physics of Fluids, 10, 2097, 1968.

[28] LOYALKA, S.K. and FERZIGER, J.A., "Model Dependence of the Slip Coefficient", Physics of Fluids, 10, 1833, 1967.

[29] GRAD, H., "Asymptotic Theory of the Boltzmann Equation I", Physics of Fluids, 6, 147, 1963.

[30] CERCIGNANI, C., "Elementary Solutions of the Linearized Gas-dynamics Boltzmann Equation and their Application to the Slip-flow Problem", Annals of Physics, 20, 219-233, 1962.

[31] MUSKHELISHVILI, N.I., "Singular Integral Equations", Noordhoff, Groningen, 1953.

[32] CERCIGNANI, C., "Elementary Solutions of Linearized Kinetic Models and Boundary Value Problems in the Kinetic Theory of Gases", Brown University Report, 1965.

[33] MAXWELL, J.C., "Scientific Papers", 704-705, Dover, New York, 1965.

[34] ALBERTONI, S., CERCIGNANI, C., GOTUSSO, L., "Numerical Evaluation of the Slip Coefficient", Physics of Fluids, 6, 993, 1963.

[35] CERCIGNANI, C., "On the General Solution of the Steady Linearized Boltzmann Equation", presented at the 9th Symposium on Rarefied Gas Dynamics, Göttingen, July 1974.

[36] CERCIGNANI, C., "Analytic Solution of the Temperature Jump Problem by Means of the BGK Model", Transport Theory and Stat. Physics (1977).

[37] CERCIGNANI, C., FORESTI, P., SERNAGIOTTO, F., "Dependence of the slip coefficient on the Form of the Collision Frequency", Nuovo Cimento, X, 57B, 297, 1968.

[38] CERCIGNANI, C. and TIRONI, G., "New Boundary Conditions in the Transition Regime", J. Plasma Physics, 2, 293, 1968.

[39] LANGMUIR, I., Jnl of Chemical Society, 37, 417, 1915.

[40] LOYALKA, S., Z. Naturforschung, 26a, 1708 (1971).

[41] CERCIGNANI, C. and SERNAGIOTTO, F., "The Method of Elementary Solutions for Time-dependent Problems in Linearized Kinetic Theory", Annals of Physics, 30, 154-167, 1964.

[42] CERCIGNANI, C., "Unsteady Solutions of Kinetic Model with Velocity Dependent Collision Frequency", Annals of Physics, 40, 454-468, 1966.

[43] GROSS, E.P., JACKSON, E.A., ZIERING, S., "Boundary Value Problems in Kinetic Theory of Gases", Annals of Physics, 1, 141-167, 1957.

[44] LEES, L., "A Kinetic Theory Description of Rarefied Gas Flows", Caltech, Memorandum, 1959.

[45] WILLIS, D.R., "A Study of Some Nearly Free Molecular Flow Problems", Ph. D Thesis, Princeton, 1958.

[46] LIEPMANN, H.W., NARASIMHA, R., CHAHINE, M.T., "Structure of a Plane Shock Layer", Physics of Fluids, 5, 1313-1324, 1962.

[47] WILLIS, D.R., "Heat Transfer in a Rarefied Gas Between Parallel Plates at Large Temperature Ratios", Rarefied Gas Dynamics, J.A. Laurmann, ed. vol. 1, 209-225, Academic Press, New York, 1963.

[48] ANDERSON, D., "On the Krook Kinetic Equation", Part 2, Jnl of Plasma Physics, 255, 1967.

[49] WILLIS, D.R., "Comparison of Kinetic Theory Analyses of Linearized Couette Flow", Physics of Fluids, 5, 127-135, 1962.

[50] CERCIGNANI, C. and TIRONI, G., "Some Applications of a Linearized Kinetic Model with Correct Prandtl Number", Nuovo Cimento, 43, 64-78, 1966.

[51] CERCIGNANI, C. and DANERI, A., "Flow of a Rarefied Gas Between Two Parallel Plates.', Jnl of Applied Physics, 34, 3509-3513, 1963.

[52] CERCIGNANI, C. and SERNAGIOTTO, F., "Cylindrical Poiseuille Flow of a Rarefied Gas", Physics of Fluids, 9, 40-44, 1966.

[53] CERCIGNANI, C. and TIRONI, G., "Alcune applicazioni di un nuovo modello linearizzato dell'equazione di Boltzmann", Atti del Congresso AIDA-AIR, AIDA-AIR, Roma, 1967.

[54] CERCIGNANI, C., "Reply to the Comments by A.S. Berman", Physics of Fluids, 10, 1859, 1967.

[55] BASSANINI, P., CERCIGNANI, C., SERNAGIOTTO, F., "Flow of a Rarefied Gas in a Tube of Annular Section", Physics of Fluids, 9, 1174-1178, 1966.

[56] BASSANINI, P., CERCIGNANI, C., SCHWENDIMANN, P., "The Problem of a Cylinder Rotating in a Rarefied Gas", Rarefied Gas Dynamics, C.L. Brundin, editor, vol. 1, Academic Press, New York, 1967, pp 505-516.

[57] CERCIGNANI, C. and SERNAGIOTTO, F., "Cylindrical Couette Flow of a Rarefied Gas", Physics of Fluids, 10, 1200-1204, 1967.

[58] BASSANINI, P., CERCIGNANI, C., PAGANI, C.D., "Comparison of Kinetic Theory Analyses Between Parallel Plates", Int. Jnl Heat and Mass Transfer, 10, 447-460, 1967.

[59] CERCIGNANI, C., J. of Statistical Physics, 1, 297 (1969).

[60] CERCIGNANI, C. and PAGANI, C.D., "Variational Approach to Rarefied Gas Dynamics", Physics of Fluids, 9, 1167-1173, 1966.

[61] CERCIGNANI, C., "Flows of Rarefied Gases Supported by Density and Temperature Gradients", University of California Report, Berkeley, 1964.

[62] CERCIGNANI, C., "Plane Poiseuille Flow According to the Method of Elementary Solutions", Jnl Mathematical Analysis and Applications, 12, 234, 1965.

[63] CERCIGNANI, C., "Plane Poiseuille Flow and Knudsen Minimum Effect", Rarefied Gas Dynamics, J.A. Laurmann ed., vol. II, 920101, 1963.

[64] KNUDSEN, M., "Die Gesetze der Molekularströmung under der inneren Reibungsströmung der Gase durch Rohren", Ann. Physik, 8, 75-130, 1909.

[65] CERCIGNANI, C. and PAGANI, C.D., "Variational Approach to Rarefied Flows in Cylindrical and Spherical Geometry. Rarefied Gas Dynamics", C.L. Brundin, ed. vol. 1, 555-573, Academic Press, New York, 1967.

[66] CERCIGNANI, C. and PAGANI, C.D., "Flow of a Rarefied Gas Past an Axisymmetric Body", I General remarks, Physics of Fluids, 11, 1395, 1968.

[67] CERCIGNANI, C., PAGANI, C.D., BASSANINI, P., "Flow of a Rarefied Gas Past an Axisymmetric Body", II Case of a Sphere. Physics of Fluids, 11, 1399, 1968.

[68] BASSANINI, P., CERCIGNANI, C. PAGANI, C.D., "Influence of the Accommodation Coefficient on the Heat Transfer in a Rarefied Gas", Int. Jnl Heat and Mass Transfer, 11, 1359, 1968.

TOPICS ON EXISTENCE THEORY OF THE BOLTZMANN EQUATION

J.P. Guiraud
Universitè P. & M. Curie

CHAPTER I

PRELIMINARIES

1.1. The Boltzmann's Equation

1.1.1. The Distribution Function

1.1.1.1. The basic object under study in kinetic theory is the so-called *distribution function* $F(t, x, \xi)$. At a given instant t and a given place x in ordinary affine eudidean space (in three dimensions) it gives the (relative) probability density of finding a molecule with velocity ξ. We shall say that x belongs to position space and that ξ belongs to velocity space. Latin characters will be used for vectors of position space and Greek ones for vectors of velocity space. From the distribution function F we can deduce the *number density* of molecules defined as the number of molecules found, at a given instant in the unit volume of (position) space, namely

$$n(t, x) = \int_{R^3} F(t, x, \xi)\, d\xi \tag{1}$$

1.1.1.2. Assume that we specify some unit of velocity C and some unit n_0 for n, then it is natural to measure F with unit $n_0 C^{-3}$ in order to preserve 1.1.1.(1) with nondimensional quantities for n, F, ξ.

1.1.1.3. It is usual to speak of n, as defined previously, as a *moment* of the distribution function. To each function of ξ, say $Q(\xi)$, we associate its moment $Q(t, x, \xi)$ defined by

$$nQ = \int_{R^3} Q(\xi)\, F(t, x, \xi)\, d\xi\,. \tag{1}$$

We may, if convenient, consider the moment Q of a function Q which depends also upon time and space. A special role is devoted to the so-called *hydrodynamic moments*. These are the number density which is the moment of $Q = 1$, the *momentum density* ρu which is the moment of $Q = m\xi$ when m is the mass of one molecule of the gas,

(2) $$\rho u = \overline{m\xi} = \int_{R^3} m\,\xi\, F(t, x, \xi)\, d\xi$$

and finally the *energy density* $\rho(e + 1/2\,|u|^2)$ which is the moment of $Q = 1/2\, m\, |\xi|^2$

(3) $$\rho(e + \frac{1}{2}|u|^2) = \frac{1}{2}\overline{m\,|\xi|^2} = \int_{R^3} \frac{1}{2} m\, |\xi|^2\, F(t, x, \xi)\, d\xi$$

of course ρ is the density

(4) $$\rho = n\,m$$

while e is the specific internal energy.

1.1.2. The Boltzman's Equation

1.1.2.1. The basic problem in kinetic theory is to solve an evolution problem: given, at some initial time, for which we set $t = 0$, an *initial distribution function* $F_0(x, \xi)$ the problem is to find how it evolves for $t > 0$, as a time dependent distribution function $F(t, x, \xi)$. The first problem, solved in 1872 by Boltzmann was to translate into a mathematical problem the elementary laws of physics which rule the evolution in time. The first physical law is that if it were not for any collision process within molecules or of molecules with the walls, the distribution function would remain constant along rays which are particles paths, namely

(1) $$x = x_0 + (t - t_0)\,\xi\ .$$

This law is translated mathematically by

(2) $$(\frac{\partial}{\partial t} + \xi \cdot \nabla) F = 0$$

which may be called the *collisionless Boltzmann's equation.* It was the discovery of Boltzmann to translate into mathematical terms the physical laws which govern the effects of molecular collisions on the evolution of F. We shall not go into the details of this derivation which is not free from difficulties, even a century after Boltzmann. We refer to Grad [1] for a thorough discussion of this topic and to Cercignani [2] for a discussion of the so called *molecular cahos hypothesis.* Here, and in what follows, we take for granted that the evolution of F is ruled *exactly* by the so called *Boltzmann's equation*

$$\left(\frac{\partial}{\partial t} + \xi \cdot \nabla\right)F = J(F, F) \; . \tag{3}$$

The collision operator $F \to J(F, F)$ is of local type and independent of time and position. It may be completely specified by defining it for functions of velocity only.

We observe that, from the physics, it is not required that $\partial/\partial t$ and ∇ be applicable separately to F. It is only required that the so-called *free-streaming operator* $\partial/\partial t + \xi \cdot \Delta$ be applicable.

We shall come back later to the effect of wall molecules collision and examine a bit more closely the form of the collision operator.

1.1.2.2. One of the basic physical assumptions underlying the derivation of 1.1.2.1.(3) is that only *binary* encounters or *collisions* enter into account for the evolution of F. As a consequence J(F, F) is a quadratic operator

$$J(a_1 F_1, a_2 F_2) = a_1 a_2 J(F_1, F_2) \; . \tag{1}$$

Let us describe briefly the collision operator. From elementary mechanics it is known that the global dynamics of the collision may be described in a frame of fixed orientation in which the molecule of velocity ξ is at rest, at 0. Suppose that this molecule is colliding with a second molecule of velocity ξ_1, which is approaching along a rectilinear trajectory P, directed parallel to $\xi_1 - \xi$. We set M the plane through 0 perpendicular to IP or OA. We call b, ϵ polar coordinates of the point of impact P in the plane M. The dynamics of the collision depends only on $\xi_1 - \xi$, b and ϵ, so cnat when ξ, ξ_1, b, ϵ are given, the velocities after

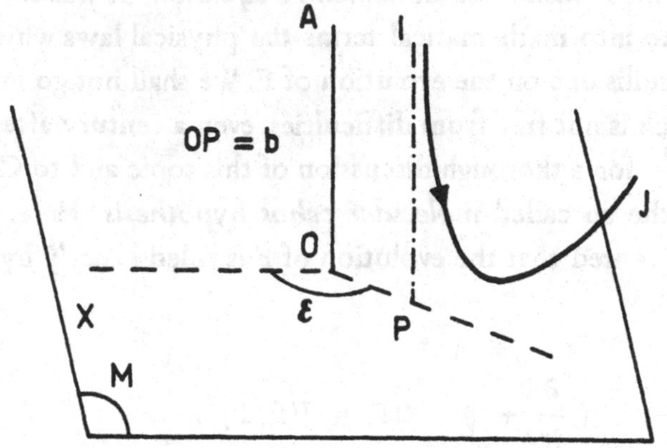

collision ξ' and ξ'_1 are known. The collision operator is written as

$$J(F,F) = \int_{R^3} |\xi_1 - \xi| \, d\xi_1 \int_0^\infty b \, db \int_0^{2\pi} d\epsilon \, (F' F'_1 - F F_1) \tag{2}$$

where $F_1 = F(\xi_1)$, $F' = F(\xi')$, $F'_1 = F(\xi'_1)$. Referring to figure 2 we may specify

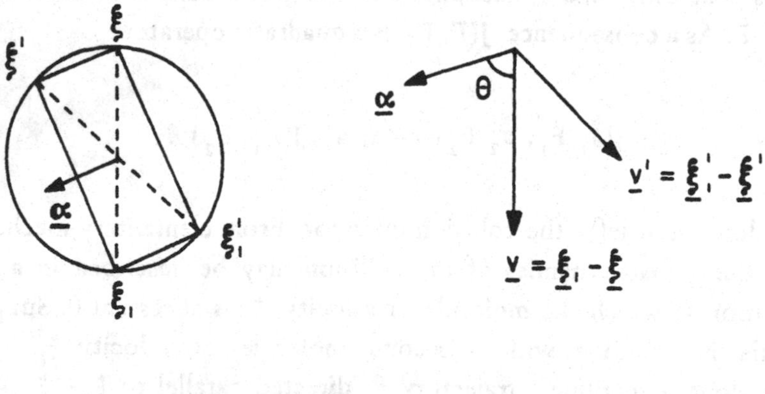

the global dynamics of the collision in a different way. For a fixed pair of entry velocities ξ and ξ_1, the possible velocities after collisions are at the extremities of any diameter of the sphere which has ξ and ξ_1 for one of its diameter. To specify ξ' and ξ'_1 it is only necessary to specify the unit vector α, such that

$$\begin{aligned} \xi' &= \xi + (\alpha \cdot V)\,\alpha\,, \\ \xi'_1 &= \xi_1 - (\alpha \cdot V)\,\alpha\,, \end{aligned} \tag{3}$$

for doing this we use ϵ and θ as polar spherical coordinates with $V = \xi_1 - \xi$ as polar axis. For molecules which are point center of repulsive forces, b is a monotone function of $\theta \in [0, \pi/2]$ for a fixed $|V| = V$, and, as a consequence we may set

$$|\xi_1 - \xi|\, b\, d\, b = B(\theta, V)\, d\theta\,, \tag{4}$$

and another form of (2) is

$$J(F, F) = \int_{R^3} d\xi_1 \int_0^{\pi/2} d\theta\ B(\theta, |\xi_1 - \xi|) \int_0^{2\pi} d\epsilon\ (F' F'_1 - F F_1)\,. \tag{5}$$

1.1.2.3. An obvious fact which must be emphasized is that conservation of momentum and energy imply

$$\begin{aligned} \xi + \xi_1 &= \xi' + \xi'_1 \\ |\xi|^2 + |\xi_1|^2 &= |\xi'|^2 + |\xi'_1|^2 \end{aligned} \tag{6}$$

1.1.2.4. We say a word about dimensional arguments. Obviously B has the dimension of area times velocity and a typical unit for it is Cb_o^2 when b_o is some typical unit for the parameter of impact b. A typical unit for J is then $n_o^2 b_o^2 C^{-2}$ (refer to 1.1.1.2. for C). Set L for unit of length and C^{-1} L for unit time. We get for the dimensionless Boltzmann's equation

$$\left(\frac{\partial}{\partial t} + \xi \cdot \Delta\right) F = m_o b_o L^2 J(F, F)$$

1.1.3. Elementary Properties of the Collision Operator

We refer to Grad[1] for a proof of the statements to be given under this heading.

1.1.3.1. Let $Q(\xi)$ be any function of ξ, we have

$$\int Q\, J(F, F)\, d\xi = \frac{1}{4} \int_{R^3} d\xi_1 \int_0^{\pi/2} d\vartheta\, B(\theta, V) \int_0^{2\pi} d\epsilon\, (F' F_1' - F F_1)(Q + Q_1 - Q' - Q_1')$$
(1)

and, as a consequence

(2) $$\int Q\, J(F, F)\, d\xi = 0 \quad \text{when Q is any of } 1,\ \xi \text{ or } \frac{1}{2} |\xi|^2 .$$

1.1.3.2. Set Q = Log F in 1.1.3.1.(1), we get

(1) $$\int_{R^3} J(F, F) \operatorname{Log} F\, d\xi = -\Delta(F) ,$$

with

(2) $$\Delta(F) = \frac{1}{4} \int_{R^3} d\xi_1 \int_0^{\pi/2} d\theta\, B(\theta, V) \int_0^{2\pi} d\epsilon\, (F' F_1' - F F_1) \operatorname{Log} \frac{F' F_1}{F F_1} \geqslant 0 .$$

This inequality is known as the *Boltzmann's H theorem.* We observe that it is a consequence of the obvious fact that $B(\theta, V) \geqslant 0$ and of the elementary inequality

$$x > 0 \Rightarrow (x-1) \log x \geqslant 0 \tag{3}$$

the equality holding if and only if $x = 1$.

1.1.3.3. Converting back 1.1.3.2.(2) to variables b, ϵ it is seen that $\Delta(F) > 0$ unless $F'F_1' = F F_1$ for any entry configuration. It may be shown (refer to Carieman[3] page 21-23) that this implies that F is Maxwellian

$$F_M = n(2\pi a^2)^{-3/2} \exp\{-\frac{1}{2}\frac{|\xi - u|^2}{a^2}\}, \qquad ma^2 = kT. \tag{1}$$

1.1.3.4. We observe that

$$\int_{R^3} F_M \, d\xi = n \tag{1}$$

$$\int_{R^3} F_M \, \xi \, d\xi = n\,u \tag{2}$$

$$\int_{R^3} F_M \frac{1}{2} |\xi|^2 \, d\xi = n\,(e + \frac{1}{2}|u|^2) \tag{3}$$

1.1.4. Further Considerations About the Collision Operator

1.1.4.1. An obvious formal point suggested by 1.1.2.2.(5) is to separate $J(F, F)$ as the difference between a creation and a destruction parts, namely

$$J(F, F) = C(F, F) - D(F)\, F \tag{1}$$

but, unless some cut-off is introduced the defining integral do not converge in general.

1.1.4.2. For rigid spheres of diameter σ, we have

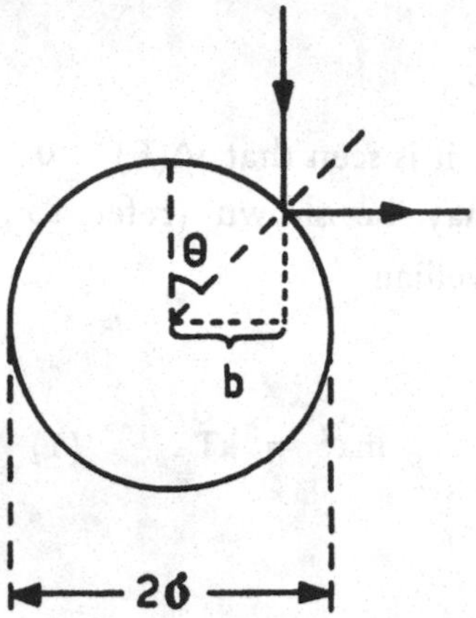

(1) $$B(\theta, V) = \sigma^2 \; V \sin\theta \; \cos\theta$$

and no cut-off is needed, or, more precisely, there is a natural cut-off at $b = \sigma$. We have obviously

(2) $$C(F, F) = \sigma^2 \int_{R^3} |\xi_1 - \xi| \, d\xi_1 \int_0^{\pi} d\theta \; \sin\theta \, \cos\theta \int_0^{\pi/2} d\epsilon \; F' \, F'_1 \; ,$$

(3) $$D(F) = \pi \sigma^2 \int_{R^3} |\xi_1 - \xi| \; F(\xi_1) \; d\xi_1 \; .$$

1.1.4.3. Let us assume that the molecules are point center of forces which repel each other according to the power law

(1) $$F(r) = \frac{K \, m}{r^s}$$

then, it can be shown (refer to Cercignani[4] page 37) that

(2) $$B(\theta, V) = (2K)^{2/(s-1)} \, V^{(s-5)/(s-1)} \; \beta \; \frac{d\beta}{d\theta}$$

when the function $\beta(\theta)$ is the inverse of $\theta(\beta)$ as defined by

$$\theta(\beta) = \int_0^{x_0} \frac{dx}{\sqrt{1 - x^2 - \frac{2}{s-1}\left(\frac{x}{\beta}\right)^{s-1}}} \quad , \quad 1 - x_0^2 - \frac{2}{s-1}\left(\frac{x_0}{\beta}\right)^{s-1} = 0 \tag{3}$$

It is readily checked that V is an increasing function of β which varies from 0 to $\pi/2$ when β grows from zero to infinity. It is also easily seen that $\theta(\beta)$ is continuously derivable and that there exists two positive constants C_1 and C_2 such that

$$\frac{c_1}{1 + \beta^s} \leqslant \frac{d\theta}{d\beta} \leqslant \frac{c_2}{1 + \beta^s} \quad , \qquad s > 3 \, , \tag{4}$$

from which it may be deduced that, for $s > 3$,

$$B(\theta, V) = K^{2/(s-1)} V^{(s-5)/(s-1)} \sin\theta(\cos\theta)^{-\frac{1}{s-1}} \left[1 + (\sin\theta(\cos\theta)^{-\frac{1}{s-1}})^s\right] \phi(\vartheta) \tag{5}$$

with

$$0 < m \leqslant \phi(\theta) \leqslant M < +\infty \, , \tag{6}$$

while $\phi(\theta)$ is continuous on $[0, \pi/2]$.

1.1.4.4. It is seen, from 1.1.4.3(5) that $B(\theta, V)$ has a non integrable singularity near $\theta = \pi/2$ that is for grazing collisions and that singularity precludes the decompositions 1.1.4.1.(1). As almost nothing is mathematically possible without the decomposition Grad[5] introduced a modified collision operator by introducing an angular cut-off near $\theta = \pi/2$. We define an angular cut-off power law potential by

$$B(\theta, V) = V^{(s-5)/(s-1)} \beta(\theta) \tag{1}$$

with

(2) $$\beta(\theta) \leqslant \beta_1 \sin\theta \cos\theta .$$

A more general cut-off is defined by the inequality ($\epsilon > 0$)

(3) $$B(\theta, V) \leqslant b_1 (V + V^{\epsilon-1}) \sin\theta \cos\theta .$$

For any such cut-off 1.1.4.(1) holds and

(4) $$D(F) = 2\pi \int_{R^3} d\xi_1 \; F(\xi_1) \int_0^{\pi/2} d\theta \; B(\theta, |\xi_1 - \xi|)$$

(5) $$C(F, F) = \int_{R^3} d\xi_1 \int_0^{\pi/2} B(\theta, |\xi_1 - \xi|) \; d\theta \int_0^{2\pi} d\epsilon \quad F' F'_2 \; .$$

1.1.4.5. Cercignani[4] (refer to p. 33-38 and [2]) has devised a radial cut-off which amounts to cut-down to zero the force $F(r)$ when $r > \sigma$. It is stated that, for $F(r)$ as in 1.1.4.3.(1), the following estimate holds

$$B(\theta, V) = \sigma_0^2 \Big(1 + \frac{\sigma^{s-1} V^2}{2K}\Big)^2 \Big\{1 + (s-1)\Big(1 + \frac{\sigma^{s-1} V^2}{2K}\Big)^2 \cos^2\theta\Big\}^{-s/s-1} .$$

$$\cdot \phi(\theta, V) \sin\theta \cos\theta$$

where ϕ is bounded as in 1.1.4.3.(6) and goes to 1 when $\theta \to \pi/2$, uniformly with respect to V.

1.1.4.6. Drange[7] has studied another cut-off which amounts to cut-down to zero $B(\theta, V)$ when $b > \sigma$, the precise definition being that B is cut to zero for $\theta > \theta_0(V)$ such that

$$2 \int_0^{\theta_0(V)} V^{-1} \; B(\theta, V) \, d\theta = \sigma^2 .$$

1.1.4.7. For any of the previously defined cut-offs, 1.1.4.1.(1) holds and

$$D(F) = 2\pi \int_{R^3} F(\zeta)\, d\zeta \int_0^{\pi/2} B(\theta, |\xi - \zeta|)\, d\theta \tag{1}$$

$$C(F, F) = \int_{R^3} d\xi_1 \int_0^{2\pi} d\epsilon \int_0^{\pi/2} d\theta\; B(\theta, |\xi_1 - \xi|)\; F' F_1' \,. \tag{2}$$

We observe that, for Drange's or Cercignani's cut-off $D(F)$ has the same value as for rigid spheres. For power law angular cut-off,

$$D(F) = \left(2\pi \int_0^{\pi/2} \beta(\theta)\, d\theta\right) \int_{R^3} |\xi - \zeta|^{\frac{s-5}{s-1}}\; F(\zeta)\, d\zeta \tag{3}$$

1.2. Surface Interaction

1.2.1. The Process of Reflection

Consider a point $\mathbf{x}_1$ on a wall which moves with material velocity $\mathbf{u}_w$, and let $\mathbf{n}$ be the unit normal to the wall directed towards the gas. Consider a ray 1.1.2.1.(1) which hits the wall at $\mathbf{x}_1$ at time t_1. According to the part of the ray which is within the gas, the molecule travelling along this ray carries information from the gas to the wall or from the wall to the gas. The first situation holds when $(\xi - u_w) \cdot n < 0$ while the second holds when $(\xi - u_w) \cdot n > 0$. We set

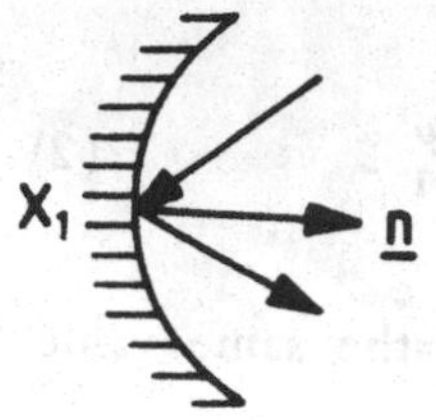

$$\xi_r = \xi - u_w \, , \tag{1}$$

and, on the wall

$$F(\xi) = \overset{+}{F}(\xi) + F^-(\xi) \, , \tag{2}$$

$$\overset{+}{F}(\xi) = \begin{cases} F(\xi) & \text{if} \quad \xi_r \cdot n \geqslant 0 \, , \\ 0 & \text{if} \quad \xi_r \cdot n < 0 \, , \end{cases} \tag{3}$$

$$F^-(\xi) = \begin{cases} 0 & \text{if} \quad \xi_r \cdot n \geqslant 0 \, , \\ F(\xi) & \text{if} \quad \xi_r \cdot n < 0 \, . \end{cases} \tag{4}$$

From what has been said we expect that the process of interaction gives no information concerning F^- while it gives full information concerning F^+ .

The most general situation would be that $F^+(t, x, \xi)$ depends on the history of the wall and of F^- up to time t and in a vicinity of the point x. We shall assume that the process of interaction is local in space, that is

$$\overset{+}{F}(t, x, \cdot) = K_{t,x} \{ F_t^-(\cdot \, , x, \cdot) \} \tag{5}$$

where $F_t^-(\cdot, x, \xi)$ means the history of F^- up to time t. The operator $K_{t,x}$ needs not be linear.

It is not our purpose to go into any discussion about the physics of gas surface interaction. The interested reader may find a few comments in Kogan [8] (p.88-89) and a review of current work in Barantsev [9]. We shall say the least which is necessary to understand present state mathematical theory of the Boltzmann equation.

1.2.2. The Simplest Models

1.2.2.1. The simplest and sometimes widely used model of gas surface interaction goes back to Maxwell. According to this model one has

$$F^+(t, x, \xi_r) = (1-\alpha)\ F^-(t, x, \xi_r - 2(\xi_r \cdot n)\, n)\ +$$

$$+\ \alpha\ n_r (2\pi a_r^2)^{-3/2} \left(\exp\left(-\frac{1}{2}\frac{|\xi_r|^2}{a_r^2}\right)\right)^+ \tag{1}$$

α is a number which specifies the fraction of molecules which are reflected diffusely, $1-\alpha$ being the fraction reflected specularly. The number density n_r is determined by the condition that there is no mass flux to the wall*

$$a_r n_r = \sqrt{2\pi} \int_{R^3_-} |\xi_r \cdot n|\, F^-(\xi)\, d\xi\,. \tag{2}$$

The temperature parameter a_r is at our disposal and we may think of it in terms of energy exchange with the wall. To this end we define the energy flux

$$E^\pm = \int_{R^3_\pm} \frac{1}{2} m\ |\xi_r|^2\ |\xi_r \cdot n|\ F^\pm(\xi)\ d\xi \tag{3}$$

*) When dealing with velocity integration at the wall we separate R^3 as $R^3_+ \cup R^3_-$, with R^3_+: $\{\xi \mid (\xi - u_w) \cdot n \geq 0\}$

and set

$$E_M(n, a) = \int |n \cdot \xi_r| \frac{m}{2} |\xi_r|^2 (2\pi a^3)^{-3/2} \exp(-\frac{1}{2} |\xi_r|^2) d\xi_r = \sqrt{\frac{2}{\pi}} n m a^3. \quad (4)$$

If a_w is the temperature parameter which corresponds to the temperature of the wall we define an energy accommodation coefficient as

$$(5) \qquad \alpha_e = \frac{E^- - E^+}{E^- - E_M(n_w, a_w)}, \qquad n_w a_w = \sqrt{2\pi} \int_{R^3_-} |\xi_r \cdot n| F(\xi) d\xi ,$$

and it is seen that

$$(6) \qquad \alpha_e = \frac{E^- + \sqrt{\frac{2}{\pi}} m n_r a_r^3}{E^- + \sqrt{\frac{2}{\pi}} m n_r a_r a_w^2} .$$

When α and α_e are given (1) (2) (6) completely specify F^+ in terms of F^-, that is the operator $K_{t,x}$ of 1.2.1.(5). The past history does not come into account, and the operator $K_{t,x}$ is local in time as well as in space. We observe that a_r depends in a non-linear way on F^- and that, accordingly, the operator $K_{t,x}$, in the Maxwell model is non-linear.

1.2.2.2. Other simple models of gas surface interactions have been proposed and used in the literature on rarefied gas dynamics. We refer to Barantsev [9] (p.41-57) and come back to the general frame work of 1.2.1.(5).

1.2.3. The Reflection Operator

1.2.3.1. From now on we assume that the reflection operator $K_{t,x}$ is linear and of integral type, namely

$$(1) \qquad F^+(t, x, \xi) = \int_{-\infty}^{t} d\tau \int_{R^3} K(t, \tau, x, \xi, \zeta) F^-(\tau, x, \zeta) d\zeta ,$$

according to Kuscer[10]. We note that for a wall whose properties are independent of time the kernel K depends only on the time interval $t-\tau$. If K is significantly different from zero for only a time interval which is small in comparison with the time interval over which F^- varies significantly it is possible to assume that $K_{t,x}$ is local in time as well as in space and that

$$\overset{+}{F}(t, x, \xi) = \int_{R^3} K(t, x, \xi, \zeta)\; F^-(t, x, \zeta)\, d\zeta \;, \tag{2}$$

with

$$K(t, x, \xi, \zeta) = \int_{-\infty}^{t} K(t, \tau, x, \xi, \zeta)\, d\partial \;. \tag{3}$$

The mathematical theory that we develop later will use a reflection operator of type (2) with a time independent kernel.

1.2.3.2. From its very definition, we expect that only non-negative distribution functions are physically admissible. The process of interaction must conserve this property. As a consequence we impose the first condition

$$F^- \geqslant 0 \Rightarrow K_{t,x}\; F^- \geqslant 0 \tag{1}$$

from which it is concluded that the kernel has to be non-negative

$$K(t, x, \xi, \zeta) \geqslant 0 \;. \tag{2}$$

Our second general and obvious condition is that there is no net mass flux to the wall, that is

$$\int_{R^3_+} |\xi_r \cdot n|\, \overset{+}{F}(\xi)\, d\xi = \int_{R^3_-} |\xi_r \cdot n|\, F^-(\xi)\, d\xi \;. \tag{3}$$

which implies that the kernel K has to satisfy the following conditions

$$(4) \qquad \int_{R_+^3} |\xi_r \cdot n| \, K(t, x, \xi, \zeta) \, d\xi = |\zeta_r \cdot n| \, 1^- ,$$

where 1^- is the function identically equal to one for $\zeta_r \cdot n < 0$ and to zero for $\zeta_r \cdot n \geqslant 0$. Our third condition is double. First the process of gas surface interaction perturbs only very slightly the wall which will be assumed to be in a local thermodynamic equilibrium, with a well defined temperature parameter a_w. Second we assume that the process of interaction cannot destroy equilibrium in the gas when is itself in equilibrium with the wall. Letting F_{M_w} for the Maxwellian 1.1.3.3.(1) with $u = u_w$, $a = a_w$ this hypothesis implies that

$$(5) \qquad F^+_{M_w} = K \, F^-_{M_w} ,$$

this identity holding at each point of the wall.

1.2.3.3. There is a fourth property of the operator K which results from basic principles, namely the so called reciprocity relation (see Kuscer [10] relation (12) and (13)). We state this property as

$$(1) \qquad F_{M_w}(\zeta) \, K(t, x, \xi, \zeta) = F_{M_w}(\xi) \, K(t, x, -\zeta, -\xi)$$

1.2.3.4. Cercignani [11] has devised a reflection operator which is simple enough, meets all the requirements of 1.2.3.2. and 1.2.3.3., and, moreover, may be derived from a physical model of the gas surface interaction (see Cercignani [12]). Letting

$$(1) \qquad \begin{cases} \xi_{rn} = (\xi - u_w) \cdot n & \zeta_{rn} = (\zeta - u_w) \cdot n \\ \xi_{rt} = \xi_r - \xi_{rn} \, n & \zeta_{rt} = \zeta_r - \zeta_{rn} \, n \end{cases}$$

the Cercignani kernel is

$$K(\xi, \zeta) = \frac{2\,|\zeta_{rn}|}{(2a^2)^2\ \pi\alpha_n\,\alpha_r(2-\alpha_t)} \exp\{-\frac{\xi_{rn}^2}{2a_w^2} - \frac{1-\alpha_n}{\alpha_n}\,\frac{\xi_{rn}^2+\zeta_{rn}^2}{2a_w^2}\}\cdot \tag{2}$$

$$\cdot\, I_o\,\{\frac{2\sqrt{1-\alpha_n}}{\alpha_n}\,\frac{\xi_{rn}\,\zeta_{rn}}{2a_w^2}\}\quad \exp\,\{-\frac{|\xi_{rt}-(1-\alpha)\,\zeta_{rt}|^2}{\alpha_t(2-\alpha_t)\,2a_w^2}\}$$

where I_o is the modified Bessel function.

1.2.4. Dissipativity of the Reflection Operator

1.2.4.1. The following result was stated by Darrozes, Guiraud [13] and rederived by Case[14], Cercignani[12]. Consider a particular point on the wall, F_{M_w} the (unique up to a constant factor)corresponding Mawwellian and K the corresponding reflection operator. Let $F = F_{M_w}\,G$ and assume that $F^+ = K\,F^-$, with of course $F^+_{M_w} = K\,F^-_{M_w}$. Then, if $J(G)$ is any strictly convex function, one has

$$\int_{R^3} \xi_r\cdot n\ F_{M_w}\ J(G)\,d\xi \leqslant 0\ , \tag{1}$$

with an equality sign which holds if and only if G is (almost everywhere) constant.*
To prove this result we normalize F_{M_w} according to

$$F^{(n)}_{M_w} = (2\pi a_w^2)^{-3/2}\ \exp\ \{-\frac{1}{2}\,\frac{|\xi-u_w|^2}{a_w^2}\}\ , \tag{2}$$

and transform the operator K to $\mathcal{K}$

$$\mathcal{K}(\cdot) = (F^{(n)^+}_{M_w})^{-1}\ K\ F^{(n)-}_{M_w}\{\cdot\}\ , \tag{3}$$

*) This statement requires in 1.2.3.2.(1) that $K_{t,x}\,F^-$ is not zero if F^- is not almost everywhere zero.

so that G satisfies

$$G^{+} = K\, G^{-} \; . \tag{4}$$

We set

$$d\mu = F_{M_w}^{(n)} \mid \xi_r \cdot n \mid d\xi \tag{5}$$

and the inequality to be proven is

$$\int_{R^3_+} J(G^{+})\, d\mu \leqslant \int_{R^3_-} J(G^{-})\, d\mu \, . \tag{6}$$

Now, from the second property of 1.2.3.2. we have that, for any h^-,

$$\int_{R^3_-} h^-\, d\mu = \int_{R^3_+} K\, h^-\, d\mu \tag{7}$$

and (6) is converted to

$$\int_{R^3_+} \{ K\, J(G^-) - J(K\, G^-) \}\, d\mu \geqslant 0 \; . \tag{8}$$

Now we prove (8) and the above argument implies that (1) will hold. It is clear that (8) will hold if the stronger statement that

$$K\, J(G^-) - J(K\, G^-) \geqslant 0 \tag{9}$$

holds for each ξ_r such that $\xi_r \cdot n \geqslant 0$. From convexity of J we have

$$J(x) - J(y) \geqslant \lambda(y)\; (x - y) \tag{10}$$

for any x and y in the domain of J. We apply (10) to the (variable) G^-, considered as x, and the (constant) KG^- considered as y and we get

$$J(G^-) - J(KG^-) \geqslant \lambda(KG^-)\,(G^- - KG^-)\ . \tag{11}$$

We repeat that $J(G^-)$ and G^- are considered as functions of ζ while $J(KG^-)$, $\lambda(KG^-)$, KG^- are constants; we make them functions of ζ of type minus by multiplying by 1^- getting

$$J(G^-) - 1^-\,J(KG^-) \geqslant \lambda(KG^-)\,(G^- - 1^-\,KG^-) \tag{12}$$

and we may, then, apply the operator K to (12) getting

$$K\,J(G^-) - J(KG^-) \geqslant \lambda(KG^-)\,(KG^- - KG^-) = 0 \tag{13}$$

taking into account 1.2.3.2.(1) which implies

$$h^- \geqslant 0 \Rightarrow K\,h^- \geqslant 0 \tag{14}$$

and 1.2.3.2.(5) which implies

$$1^+ = K\,1^-\ . \tag{15}$$

We observe that we have used linearity of K. From strict convexity of J, the equality in (13) does not hold unless $G^- = KG^-$ where G^- is considered as a function of ζ while KG^- is considered as a constant. Now if we make (14) a little bit more strict, namely that $K\,h^-$ cannot be zero for a non-negative h^- unless h^- is almost evcywhere zero, we conclude that the equality in (1) cannot hold unless G is a constant.

1.2.4.2. We apply the previous result with

$$J(g) = g\,\mathrm{Log}\,g - g + 1 \tag{1}$$

from which we deduce that for any $F > 0$, such that, at the specified point on the wall, $F^+ = K \cdot F^-$, we have

$$(2) \qquad - \int_{R^3} \xi_r \cdot n \; F(\mathrm{Log}\, F - 1) \; d\xi = \frac{\epsilon_w}{m\, a_w^2} + \delta(F) \; ,$$

with

$$(3) \qquad \epsilon_w = \int_{R^3} \frac{m}{2} \; |\xi_r|^2 \; \xi_r \cdot n \;\; F(\xi) \; d\xi \; ,$$

and

$$(4) \qquad \delta(F) \geqslant 0 \; ,$$

with equality holding if and only if $F = F_{M_w} = n_w \, F_{M_w}^{(n)}$. As a matter of fact

$$(5) \qquad \begin{aligned} \int_{R^3} \xi_r \cdot n \;\; F(\mathrm{Log}\, F - 1) \; d\xi &= \int_{R^3} F \, \{ - \frac{|\xi_r|^2}{2a_w^2} - \frac{3}{2} \mathrm{Log}\, (2\pi a_w^2) \} \; \xi_r \cdot n \, d\xi \\ &+ \int_{R^3} F_{M_w} \; (G \, \mathrm{Log}\, G - G) \; \xi_r \cdot n \; d\xi \end{aligned}$$

and from 1.2.3.2.(3), we have

$$(6) \qquad \int_{R^3} F_{M_w} \; G \; \xi_r \cdot n \; d\xi = \int_{R^3} F_{M_w} \;\; \xi_r \cdot n \; d\xi = 0$$

so that

$$(7) \quad \int_{R^3} \xi_r \cdot n \;\; F(\mathrm{Log}\, F - 1) \; d\xi = - \frac{\epsilon_w}{m a_w^2} + \int_{R^3} F_{M_w} \; (G \, \mathrm{Log}\, G - G + 1) \; \xi_r \cdot n \; d\xi$$

and (2), (3), (4) follow. We remark that ϵ_w is the net energy flux density from the wall to the gas.

1.2.4.3. With J as in 1.2.4.2.(1) we may take $\lambda(K G^-) = \mathrm{Log}(K G^-)$.

1.3. The H Theorem

1.3.1. Steady Flow in a Bounded Domain

We consider a positive solution of

$$(1) \qquad \xi \cdot \nabla F = J(F, F) \qquad \text{within } \Omega$$

$$(2) \qquad F^{+} = K F^{-} \qquad \text{on } \partial\Omega$$

where Ω is a fixed bounded domain in euclidean space, with a smooth boundary. From (1) we have

$$(3) \qquad \int_{R^3} d\xi \int_{\Omega} \xi \cdot \nabla [F(\operatorname{Log} F - 1)]\, dv = \int_{\Omega} dv \int_{R^3} J(F, F) \operatorname{Log} F \, d\xi$$

and integrating the inner integral of the left hand side along rays, we obtain thanks to 1.1.3.2.(1) and 1.2.4.2.(2)

$$(4) \qquad \int_{\partial\Omega} \frac{\epsilon_w}{m a_w^2}\, ds + \int_{\partial\Omega} \delta(F)\, ds + \int_{\Omega} \Delta(F)\, dv = 0 .$$

From this identity and from 1.1.3.3. and 1.2.4.2. we conclude that a steady state within a bounded isolating container cannot be maintained unless the wall is at constant temperature and the gas in in complete equilibrium with the wall. If the wall is not at constant temperature or is sliding over itself, so that the gas is set into motion, a positive amount of "entropy" is transfered from the wall to the gas, as a whole. This "entropy" flux* balances exactly internal and boundary dissipation through collisions of the molecules between themselves and of the molecules with the wall.

1.3.2. Unsteady Flow in a Bounded Domain

1.3.2.1. We consider a positive solution of

$$(1) \qquad \left(\frac{\partial}{\partial t} + \xi \cdot \nabla\right) F = J(F, F) \qquad \text{within } \Omega$$

*) We define it as the energy flux density, divided by temperature and integrated over the boundary.

$$F^{+} = K F^{-} \qquad \text{on } \partial\Omega \tag{2}$$

$$F|_{t=0} = F_o \tag{3}$$

where Ω is, again a bounded domain in euclidean space with a smooth boundary $\partial\Omega$. We do not assume Ω to be independent of time, which means that $u_w \cdot n$ needs not be zero. We start from

$$\int_{R^3} d\xi \int_0^{t_1} dt \int_{\Omega(t)} \left(\frac{\partial}{\partial t} + \xi \cdot \nabla\right) F(\operatorname{Log} F - 1)\, dv = \int_0^{t_1} dt \int_{\Omega(t)} dv \int_{R^3} J(F, F) \operatorname{Log} F \, d\xi \tag{4}$$

and we integrate, along rays, and inner integral according to the following scheme, with reference to the figure (we assume Ω convex for simplicity)

$$\int_{R^3} d\xi \int_0^{t_1} dt \int_{\Omega(t)} dv = \int_{R^3} d\xi \int_{A(\xi)} dS_m \int_0^{t_1} dt \int_{\sigma_1(t)}^{\sigma_2(t)} d\sigma \tag{5}$$

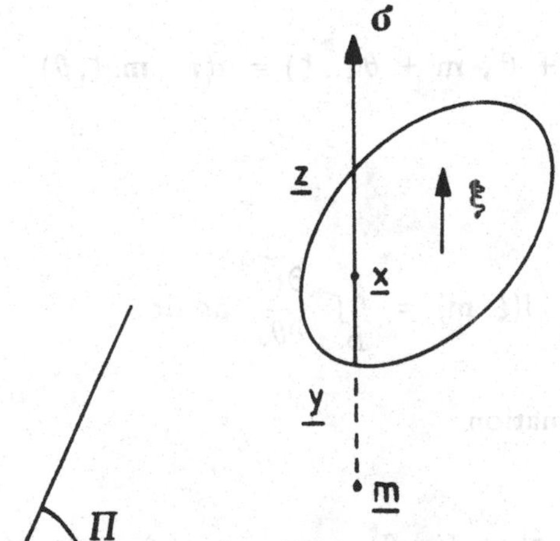

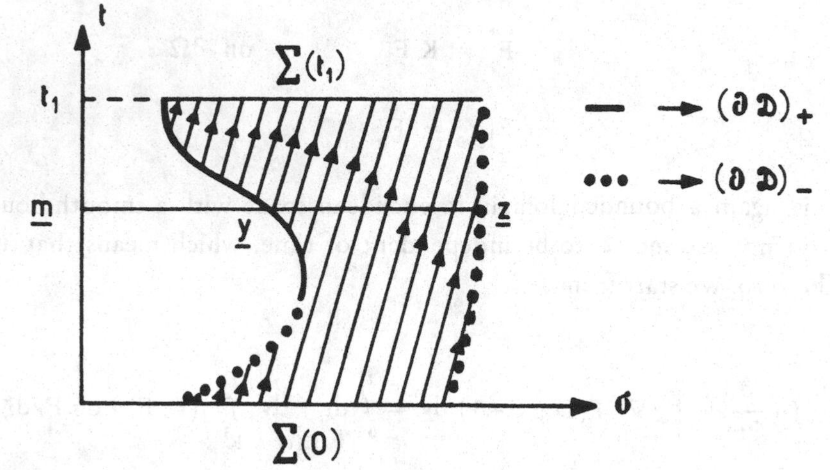

where $A(\xi)$ is the projection of $\bigcup_{t\in[0,t_1]} \Omega(t)$ on the plane Π perpendicular to the direction ξ. We want to evaluate

(6) $$I(\xi, m) = \int_0^{t_1} dt \int_{\sigma_1(t)}^{\sigma_2(t)} d\sigma \left(\frac{\partial}{\partial t} + \xi \cdot \nabla\right) f .$$

To this end, we set

(7) $$f(\tau + \theta, m + \theta\xi, \xi) = \tilde{f}(\tau, m, \xi, \theta)$$

and we have

(8) $$I(\xi, m) = \iint_D \frac{\partial \tilde{f}}{\partial \theta} \, d\sigma \, dt .$$

Now we use the transformation

(9) $$(\tau, \theta) \to (t, \sigma) = J(\tau, \theta) \; : \; t = \tau + \theta \; , \; \sigma = \theta \, |\xi|$$

and, noting that the Jacobian is $|\xi|$, we get

$$I = \int\int_{J^{-1}D} |\xi| \frac{\partial \tilde{f}}{\partial \theta} \, d\theta \, d\tau = \oint_{\partial(J^{-1} D)} |\xi| f \, d\tau \, . \tag{10}$$

In order to evaluate the right hand side of (10) we decompose ∂D as

$$\partial D = \Sigma(o) \cup \Sigma(t_1) \cup (\partial D)_+ \cup (\partial D)_- \tag{11}$$

where $\Sigma(t)$ is defined as the segment y, z at a given time, while $(\partial D)_+$ is the part of ∂D for which $0 < t < t_1$ and $(\xi - u_w) \cdot n \geqslant 0$, while $(\partial D)_-$ is defined in a similar way, with $(\xi - u_w) \cdot n < 0$. As a consequence we have

$$\begin{aligned} I = & \int_{\sigma_1(t_1)}^{\sigma_2(t_1)} f(t_1, m + \sigma u, \xi) \, d\sigma - \int_{\sigma_1(o)}^{\sigma_2(o)} f(o, m + \sigma u, \xi) \, d\sigma \\ & + \int_{J^{-1}\{(\partial D)_+ \cup (\partial D)_-\}} |\xi| f \, d\tau . \end{aligned} \tag{12}$$

All along $(\partial D)_+ \cup (\partial D)_-$, we have $\theta = \theta_*(\tau)$, multiply valued, and

$$\frac{d\theta}{d\tau} (\xi - u_w) \cdot n = u_w \cdot n \tag{13}$$

while, if we set

$$x = m + \xi \, \theta_* (\tau) \quad , \qquad \theta_* (\tau) + \tau = t \, , \tag{14}$$

we have

$$dS_m = |\xi \cdot n| \, dS_x \, , \tag{15}$$

$$\xi \cdot n \, d\tau = (\xi - u_w) \cdot n \, dt. \tag{16}$$

We note that, when remaining through $(\partial D)_+$, τ is increasing or decreasing according to $\xi \cdot n \geqslant 0$ or $\xi \cdot n < 0$, while, when remaining through $(\partial D)_-$ the reverse situation holds; accordingly we get

$$\int_{A(\xi)} dS_m \; I(\xi, m) = \int_0^{t_1} dt \int_{\partial\Omega} (\xi - u_w) \cdot n \, f \, dS_x \; . \tag{17}$$

From (5) (6) (10) (12) (17) we conclude that

$$\begin{aligned} &\int_{R^3} d\xi \int_\Omega dv \int_0 dt \left(\frac{\partial}{\partial t} + \xi \cdot \nabla \right) f = \left\{ \int_\Omega dv \int_{R^3} f \, d\xi \right\}_0^{t_1} + \\ &\qquad + \int_0^t dt \int_{\partial\Omega} dS \int_{R^3} (\xi - u_w) \cdot n \, f \, d\xi \; . \end{aligned} \tag{18}$$

We apply (18) to (4) and take into account (1.1.3.2.)(1) (1.2.4.2.)(2) with the result that

$$\begin{aligned} &\mathcal{H}(t) + \int_0^t dt_1 \int_{\partial\Omega} \frac{\epsilon_w}{m \, a_w^2} \, dS - \mathcal{H}(o) = \\ &= - \int_0^t dt_1 \int_{\partial\Omega} \delta(F) \, dS - \int_0^t dt_1 \int_\Omega \Delta(F) \, dv \leqslant 0 \; , \end{aligned} \tag{19}$$

where

$$\mathcal{H}(t) = \int_\Omega H(t, x) \, dv \quad , \tag{20}$$

$$H(t, x) = \int_{R^3} F(\operatorname{Log} F - 1)\, d\xi \; . \tag{21}$$

1.3.2.2. The result 1.3.2.1.(19) was proven by Darrozes, Guiraud [13]. If the wall is insulating, namely $\epsilon_w = 0$, we see that $\mathcal{H}(t)$ is a decreasing function of time. If the solution exists for all time either $H(t)$ tends to infinity or $H(t)$ tends to a definite limit when $t \to \infty$. In the second case it is necessary that $\delta(F)$ and $\Delta(F)$ tend to zero. As a consequence, if F tends to a definite limit F_∞, this limit must be a Maxwellian. From the equation, this Maxwellian has to be uniform in space and, from $\delta(F_\infty) = 0$ it has to be in equilibrium with the wall. For this to hold a_w must be a constant in a vicinity of infinity or, at least, it must tend to a constant value; in the same way u_w must be a constant in the vicinity of infinity or, at least, it must tend to a constant value. The gas must tend to a state or rest with respect to the container with the same temperature as its wall.

The mathematical problem with this argument is to prove the existence of solution F, non-negative, such that the formal operations of 1.3.2.1. are legitimate, such that it exists for all time, that it tends to a definite limit when $t \to \infty$, and that all the terms in 1.3.2.1.(19) tend to definite limit. A result of this nature, for a fixed domain Ω with a wall at uniform and constant temperature, has been proven very recently by Guiraud [15] under the hypothesis that the initial value of F is sufficiently close to a Maxwellian in equilibrium with the wall.

1.3.2.3. We assess the physical meaning of the H theorem by making the assumption that the solution F is everywhere close to equilibrium, namely

$$F = F_M\,(1 + \phi)\;, \tag{1}$$

with ϕ related to F_M through Chapman-Enskog functional relation. Let us set

$$\sigma = -k\,H = -\int_{R^3} k\; F(\operatorname{Log} F - 1)\; d\xi \tag{2}$$

where k is the Boltzmann's constant. We know that with F_M in place of F, σ is identified with the specific entropy

$$s = -k \int_{R^3} F_M (\text{Log}\, F_M - 1)\, d\xi \,. \tag{3}$$

Then we have, taking into account the fact that F and F_M have the same hydrodynamic moments,

$$\sigma = s - \frac{k}{2} \int_{R^3} F_M \, \phi^2 \, d\xi + O(|\phi|^3) \tag{4}$$

while

$$k\,\Delta(F) \cong \frac{k}{4} \int_{R^3} d\xi_1 \int_0^{2\pi} d\epsilon \int_0^{\pi/2} B(Q, |\xi_1 - \xi|) d\theta \; F_M F_{M_1} (\phi' + \phi'_1 - \phi - \phi_1)^2 + \tag{5}$$
$$+ O(|\phi|^3) \,.$$

From Chapman-Enskog theory we may evaluate the right hand side of (5) to $O(|\phi|^3)$ and conclude that

$$k\,\Delta(F) = \frac{\rho}{2p} \; P_{ij}(u_{i,j} + u_{j,i}) + q_i\left(\frac{1}{T}\right)_{,i} + O(|\phi|^3) \tag{6}$$

with standard notation : ρ is the density, p the pressure, P_{ij} the viscous part of the stress tensor, q_i the components of the heat flux density vector, T the temperature, u the mean velocity as they appear in

$$F_M = m^{-1} \rho (2\pi\, m^{-1}\, kT)^{3/2} \exp\left\{ - \left(\frac{m\,|\xi - u|^2}{2\,k\,T} \right) \right\} . \tag{7}$$

We define

$$c = \xi - u \tag{8}$$

and note that, F_w meaning the value of F at the wall,

$$- q_w = q \cdot n|_w = \int_{R^3} \frac{m\,|c|^2}{2}\, c \cdot n\, F_w\, d\xi \;, \tag{9}$$

so that, by defining

$$\tau_w = \int_{R^3} F_w\, mc\; c \cdot n\; d\xi \;, \tag{10}$$

we obtain

$$\epsilon_w = q_w + \tau_w \cdot (u - u_w) \;, \tag{11}$$

from which it is seen that the energy flux $-\epsilon_w$ is the sum of the heat flux q_w and the power of the shear stress against slip of the gas along the wall.

We rewrite the H theorem as

$$\frac{d}{dt}\int_\Omega \rho\sigma\; dv + \int_{\partial\Omega} \frac{-q_w + \tau_w \cdot (u_w - u)}{T_w}\; dS = \int_\Omega k\,\Delta(F)\; dv + \int_{\partial\Omega} k\,\delta(F)\; dS \tag{12}$$

and we compare it with the entropy equation of Navier-Stokes theory

$$\frac{d}{dt}\int_\Omega \rho\, s\; dv + \int_{\partial\Omega} \frac{-q_w}{T_w}\; dS = \int_\Omega \left\{ \frac{P_{ij}\,(u_{ij} + u_{j,i})}{2\,T} + q_i \frac{\partial}{\partial x_i}\left(\frac{1}{T}\right) \right\} d\vartheta \;. \tag{13}$$

The two time derivatives agree up to $O(\zeta)$ with

$$\zeta = \frac{P_{ij}\,P_{ij}}{p^2} + \frac{q_i\, q_i}{(p\,a)^2} \;, \tag{14}$$

again the first term on the right hand side of (12) and the right hand side of (13) agree to the leading order which is $O(\zeta)$, at least if one excepts the so called Knudsen layer. Finally we observe that two terms in (12) have not their counterparts in (13). The reason is that (12) incorporates the effect of the Knudsen layer while it is not so for (13). We observe that $[\tau_w \cdot (u_w - u)]/T_w$ is of order $k/m_1 |\tau_w|^2 / \rho_w a_w^3$, while $k\,\delta(F)$ is of order $(k/m)\rho_w a_w \zeta_w$ so that both terms are of the same order. On the other hand we estimate $P_\alpha \Delta(F)$ to be of order $k/m\,\rho\, a \zeta\, \ell^{-1}$ where ℓ stands for the mean free path. As a consequence the ratio of the second to the first in the two terms at the right hand side of (12) is estimated to be of the order of ℓ/D where D stands for the diameter Ω. As a consequence the two terms of (12) which have no counter part in (13) are of a smaller order than other terms in a ratio equal to the Knudsen number.

1.3.3. Steady Subsonic Flow Around a Body

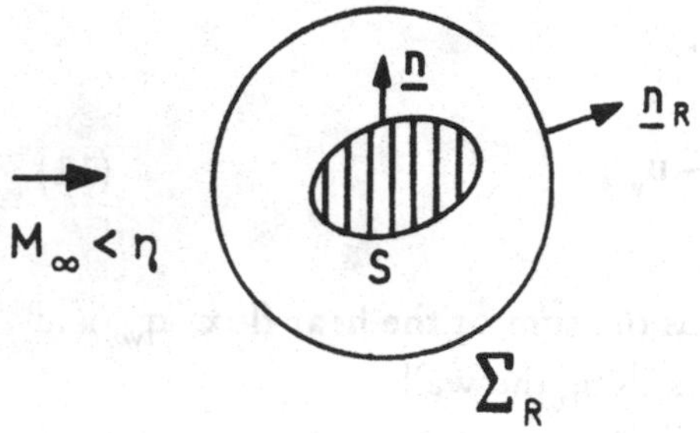

1.3.3.1. We consider now a gas flowing steadily, with subsonic speed, over a body at rest. We call S the boundary of the body, **n** the unit normal to S, pointing towards the gas. We must set

$$\lim_{|x|\to\infty} F = F_{M_\infty} = n_\infty (2\pi a_\infty^2)^{-3/2} \exp\left(-\frac{1}{2}\frac{|\xi - u_\infty|^2}{a_\infty^2}\right). \tag{1}$$

We let Σ_R be a sphere of (large) radius R centered on a fixed point inside the body. We call $\mathcal{E}_R$ the domain enclosed by S and Σ_R and we apply 1.3.1. by taking $\mathcal{E}_R$ for Ω then letting R go to infinity. Care must be taken of the fact that Σ_R is not a wall and we cannot apply 1.2.4.2. to it. For convenience we replace H(F) by

$$H(F, F_\infty) = \int_{R^3} \{ F(\mathrm{Log}\frac{F}{F_\infty} - 1) + F_\infty \}\, d\xi \tag{2}$$

and we repeat the argument of 1.3.1. starting with

$$\int_{R^3} d\xi \int_\Omega \xi \cdot \nabla \{ F(\mathrm{Log}\frac{F}{F_\infty} - 1) + F_\infty \}\, dv = \int_\Omega d\nu \int_{R^3} J(F,F)\,\mathrm{Log}\,F\, d\xi \tag{3}$$

where use has been made of 1.1.3.1.(2). From the argument of 1.2.4.2. we get

$$\int_S \{\epsilon_w (\frac{1}{T_w} - \frac{1}{T_\infty}) + \frac{\tau_w \cdot u_\infty}{T_\infty}\} \, dS + \int_S k \, \delta(F) \, dS +$$

$$+ \int_{\mathcal{E}_R} k \, \Delta(F) \, dv = \int_{\Sigma_R} k \, dS \int_{R^3} \xi \cdot n_R \, \{F(\mathrm{Log} \frac{F}{F_\infty} - 1) + F_\infty\} \, d\xi \,. \tag{4}$$

There exists no rigorous mathematical theory which would allow to perform safely the limit process $R \to \infty$ and we need to proceed heuristically.

1.3.3.2. In the vicinity of infinity we expect F to be almost Chapman Enskog like with hydrodynamic moments obeying the compressible Oseen equations. Namely, we set

$$F = F_{M_\infty} \{1 + \psi + \phi_{CE} + \chi\} \,, \tag{1}$$

$$\psi = \frac{\rho}{\rho_\infty} - 1 + \frac{m(u - u_\infty)}{k T_\infty} \cdot (\xi - u_\infty) + \frac{m}{k} (\frac{1}{T_\infty} - \frac{1}{T}) \frac{|\xi - u_\infty|^2}{2} \,, \tag{2}$$

$$\phi_{CE} = A_{ij} (u_{i,j} + u_{j,i}) + B_i T_{,i} \,, \tag{3}$$

with χ negligible in comparison with ψ and ϕ_{cE}. The fluid density is ρ , the velocity u and temperature T while A_{ij} and B_i are function of ξ which may be considered, in the leading approximation, to be independent of x. Remembering the fact that the speed of sound is $\sqrt{5/3}\, a$, setting $u_\infty = U_\infty \iota$, $|\iota| = 1$, and μ_∞, λ_∞ for the two coefficients of viscosity, k_∞ for the coefficient of heat conductivity we write

$$
(4)\quad
\begin{cases}
u = U_\infty(i + M_\infty^{-1}\,\bar{u})\,, \quad p = p_\infty + \gamma\,\rho_\infty\,a_\infty^2\,\bar{p}\,, \quad \rho = \rho_\infty(1+\bar{\rho}) \\[2mm]
T = T_\infty(1+\bar{T})\,, \quad x = L\,\bar{x}\,, \quad \bar{x}\cdot i = \bar{x} \\[2mm]
M_\infty = \dfrac{U_\infty}{\sqrt{\gamma}\,a_\infty}\,, \quad Re = \dfrac{\sqrt{\gamma}\,\rho_\infty\,a_\infty\,L}{\mu_\infty}\,, \quad P_r = \dfrac{\mu_\infty\,c_p}{k_\infty} \\[2mm]
\alpha = \dfrac{\lambda_\infty + \mu_\infty}{\mu_\infty} = \dfrac{1}{3} \qquad \gamma = \dfrac{c_p}{c_v} = \dfrac{5}{3}
\end{cases}
$$

and we get for the compressible Oseen equations, after suppressing the bars over the quantities,

$$
(5)\quad
\begin{cases}
M_\infty \dfrac{\partial u}{\partial x} + \nabla p = \dfrac{1}{Re}\,(\Delta u + \alpha\,\nabla\,\mathrm{div}\,(u)) \\[3mm]
M_\infty \dfrac{\partial \rho}{\partial x} + \mathrm{div}\,(u) = 0 \\[3mm]
M_\infty \dfrac{\partial T}{\partial x} + (\gamma-1)\,\mathrm{div}(u) = \dfrac{1}{Re\,Pr}\,\Delta T \\[3mm]
\gamma\,p = \rho + T
\end{cases}
$$

The asymptotic behaviour of the solutions of this system near infinity has been discussed for the supersonic case, by Chong, Sirovich[16]. Here we need the subsonic behaviour. We change u to

$$
(6)\qquad \omega = \nabla \wedge u\,, \qquad \theta = \mathrm{div}(u)
$$

and we obtain

$$\left\{\begin{array}{l}
(M_\infty \dfrac{\partial}{\partial x} - \dfrac{1}{Re}\Delta)\ \omega = 0\ , \\[2ex]
(M_\infty \dfrac{\partial}{\partial x} - \dfrac{\alpha + 1}{Re}\Delta)\ \theta + \Delta p = 0\ , \\[2ex]
(M_\infty \dfrac{\partial}{\partial x} - \dfrac{1}{Re\ Pr}\Delta)\ T + (\gamma - 1)\ \theta = 0\ , \\[2ex]
M_\infty \dfrac{\partial \rho}{\partial x} + \theta = 0\ , \\[2ex]
\gamma p = \rho + T\ .
\end{array}\right. \tag{7}$$

From standard arguments we get the following behaviour of ω near infinity

$$\omega \cong A\ \frac{\exp\{1/2\ M_\infty R_e(x-r)\}}{r} = A\ F(x) \tag{8}$$

where

$$r = |x| = (x^2 + y^2 + z^2)^{1/2} \tag{9}$$

and A is an arbitrary constant vector. We eliminate θ and choose p and s as basic variables, so that

$$\rho = p - s\ , \qquad T = (\gamma - 1)p + s\ , \qquad \theta = -\ M_\infty \frac{\partial}{\partial x}\ , \tag{10}$$

and we get the following system

$$(11)\quad \begin{cases} \left(M_\infty^2 \dfrac{\partial^2}{\partial x^2} - \Delta - \dfrac{M_\infty}{\mathrm{Re\,Pr}} \dfrac{\partial \Delta}{\partial x}\right) p = \dfrac{M_\infty}{(\gamma-1)\,\mathrm{Re\,Pr}} \dfrac{\partial \Delta s}{\partial x} , \\[2ex] \left(\gamma M_\infty \dfrac{\partial}{\partial x} - \dfrac{1}{\mathrm{Re\,Pr}} \Delta\right) s = \dfrac{\gamma-1}{R_e P_r} \Delta p . \end{cases}$$

We shall not investigate this system. We remark that the behaviour near infinity may be guessed by letting $\mathrm{Re} \to \infty$ from which we conclude that the leading term of p must be a solution of

$$(12)\quad M_\infty^2 \frac{\partial^2 p}{\partial x^2} - \Delta p = 0 ,$$

which is the equation of linearized steady compressible flow. The leading term of s is then solution of

$$(13)\quad M_\infty \gamma \frac{\partial s}{\partial x} - \frac{1}{\mathrm{Re\,Pr}} \Delta s = - \frac{\gamma-1}{\mathrm{Re\,Pr}} M_\infty^2 \frac{\partial^2 p}{\partial x^2} ,$$

from which we get

$$(14)\quad s \cong - \frac{(\gamma-1)\, M_\infty}{\gamma R_e P_r} \frac{\partial p}{\partial x} + \Sigma\, H(x) ,$$

$$(15)\quad H(x) = \frac{\exp\{(1/2)\, \gamma M_\infty \mathrm{Re\,Pr}\,(x-r)\}}{r} ,$$

where Σ is an arbitrary constant. From (6) we have

$$(16)\quad u = \nabla \phi + \nabla \wedge \psi$$

with

$$\Delta\phi = \theta\ , \quad \Delta\psi = -\omega, \ \nabla\cdot\psi = 0\ . \tag{17}$$

From

$$\Delta\frac{\partial\psi}{\partial x} = -\frac{\partial\omega}{\partial x} = -\frac{1}{M_\infty Re}\Delta\omega$$

we obtain

$$\psi = -\frac{1}{M_\infty Re}\chi\ , \quad \frac{\partial\chi}{\partial x} = \omega \tag{18}$$

which determines ψ with the additional condition that $\psi\to 0$ when $x\to-\infty$. On the other hand,

$$\Delta\phi = -M_\infty\frac{\partial p}{\partial x} + \frac{\gamma-1}{\gamma}\frac{M_\infty^2}{Re\ Pr}\frac{\partial^2 p}{\partial x^2} + M_\infty\,\Sigma\,\frac{\partial H}{\partial x}\ ,$$

so that, using the equations satisfied by p and H we obtain

$$\Delta\phi = -M_\infty\frac{\partial p}{\partial x} + \frac{\gamma-1}{\gamma}\frac{1}{Pr\ Re}\Delta p + \frac{\Sigma}{\gamma Re\,Pr}\Delta H\ ,$$

and

$$\nabla\phi = -\frac{1}{M_\infty}q + \frac{\gamma-1}{\gamma}\frac{1}{Pr\ Re}\nabla p + \frac{\Sigma}{\gamma Re\,Pr}\nabla H\ , \tag{19}$$

with

$$(20) \qquad \frac{\partial q}{\partial x} = \nabla p .$$

If (19) is to be used in (16) we may add to it $\Delta \varkappa$ where $\varkappa$ is an harmonic function. As a consequence we may write

$$u = v + \nabla \varkappa ,$$

$$v = -\frac{1}{M_\infty} q + \frac{\gamma - 1}{\gamma} \frac{1}{PrRe} \nabla p + \frac{\Sigma}{\gamma RePr} \nabla H - \frac{1}{M_\infty Re} \nabla \wedge \chi .$$

Referring to the first equation (5) we observe that

$$\Delta \mathbf{v} + \alpha \nabla \operatorname{div}(\mathbf{v}) = \Delta \mathbf{u} + \alpha \nabla \operatorname{div}(\mathbf{u})$$

and, as a consequence

$$\frac{\partial}{\partial x} (\Delta \varkappa) = 0 .$$

From the condition that $\mathbf{u}$ vanishes at infinity we conclude that

$$(21) \qquad u \cong -\frac{1}{M_\infty} q + \frac{\gamma - 1}{\gamma} \frac{1}{PrRe} \nabla p + \frac{\Sigma}{\gamma RePr} \nabla H - \frac{1}{M_\infty Re} \nabla \wedge \chi .$$

From (12) and the absence of source or sink at infinity we deduce the behaviour of p at infinity from which we deduce that of ρ, T using (10) and (14) (15) finally the behaviour of $\mathbf{u}$ results from (21) (20) (18)

$$\begin{cases} u = M_\infty^{-1}\, O(r^{-3}) + Re^{-1}\, Pr^{-1}\, O(r^{-4}) + \\ \quad + \Sigma\, Re^{-1}\, Pr^{-1}\, O\left[\dfrac{\exp\{k_1(x-r)\}}{r}\right] + M_\infty^{-1}\, Re^{-1}\, O\left[\dfrac{\exp\{k_2(x-r)\}}{r}\right] \\ p = O(r^{-3}) \\ \rho = O(r^{-3}) + M_\infty\, Re^{-1}\, Pr^{-1}\, O(r^{-4}) + \Sigma\, O\left[\dfrac{\exp\{k_1(x-r)\}}{r}\right] \\ T = O(r^{-3}) + M_\infty\, Re^{-1}\, Pr^{-1}\, O(r^{-4}) + \Sigma\, O\left[\dfrac{\exp\{k_1(x-r)\}}{r}\right] \\ k_1 = \dfrac{1}{2}\,\gamma\, M_\infty\, Re\, Pr\,, \qquad k_2 = \dfrac{1}{2}\, M_\infty\, Re \end{cases} \tag{22}$$

1.3.3.3. From 1.3.3.2.(1) (2) (3) (22) we see that

$$F = F_{M_\infty}\left\{ 1 + \left[O(r^{-3}) + O\left(\frac{\exp[k(x-r)]}{r}\right)\right] f(\xi - u_\infty) \right\} \tag{1}$$

and, through computation

$$F\left(\mathrm{Log}\,\frac{F}{F_\infty} - 1\right) + F_\infty = F_{M_\infty}\, O\left\{\left[\left(r^{-3} + \frac{\exp[k(x-r)]}{r}\right) f(\xi - u_\infty)\right]^2\right\} \tag{2}$$

so that

$$\int_{R^3} \xi \cdot \mathbf{n}_R \left\{ F\left(\mathrm{Log}\,\frac{F}{F_\infty} - 1\right) + F_\infty \right\} d\xi = O\left\{ r^{-6} + \frac{\exp[2k(x-r)]}{r} \right\} . \tag{3}$$

It is an easy matter to show that

$$\lim_{R\to\infty} \int_{\Sigma_R} \{r^{-6} + \frac{\exp[2k(x-r)]}{r^2}\} \, dS = 0 \tag{4}$$

on the other hand, from 1.2.3.2.(3)

$$\Delta(F) = O\{ \int_{R^3} F_{M_\infty} |\phi_{CE}|^2 \, d\xi \} = O\{ r^{-8} + \frac{\exp[2k(x-r)]}{r^2} \}$$

and, as

$$\int_{|x|\geq R} [r^{-8} + \frac{\exp[2k(x-r)]}{r^2}] \, dv < +\infty$$

we conclude that the integral over $\mathcal{E}_R$ in 1.3.3.1.(4) tends to a definite limit when $R\to\infty$. It is thus legitimate to proceed to the limit $R\to\infty$ in 1.3.3.1.(4).

1.3.3.4. Let $\mathcal{E}$ be the exterior of the body, τ_w the shear stress experienced by the gas from the wall, ϵ_w the energy flux density flowing from the wall to the gas, T_∞ the temperature of the gas at infinity, T_w the temperature of the wall, we have

$$\int_S \{\epsilon_w (\frac{1}{T_\infty} - \frac{1}{T_w}) - \frac{\tau_w \cdot u_\infty}{T_\infty}\} \, dS = \int_S k \, \delta(F) \, dS + \int_{\mathcal{E}} k \, \Delta(F) \, dv . \tag{1}$$

1.3.3.5. Assume that the temperature T_w is uniform over the body. Let D be the drag experienced by the body and $\dot{E}$ the rate of transfer of energy from the gas to the body, we have

$$\frac{U_\infty D}{T_\infty} + \dot{E}\left(\frac{1}{T_w} - \frac{1}{T_\infty}\right) = k \int_S \delta(F)\, dS + \int_{\mathcal{E}} k\, \Delta(F)\, dv \; . \tag{1}$$

The left hand side expresses the dissipation as the sum of two terms. There is first the dissipation by friction between the external world at temperature T_∞ and the body and second the dissipation due to the energy transfer $\dot{E}$ from the source at temperature T_∞ to the body at temperature T_w. This is the dissipation which would be computed from an elementary thermodynamic argument. We thus see that this dissipation at the thermodynamic level is exactly equal to the integrated dissipation evaluated through microscopic dissipative processes of collisions of molecules between themselves and of collisions of molecules with the wall.

1.3.3.6. The necessary material for an extension to steady supersonic flow is provided by Chong, Sirovich [16] but the details of the required analysis have not been worked out.

1.4 The Process of Linearisation

1.4.1. Linearisation of the Equation

1.4.1.1. We consider a definite Maxwellian. By a proper choice of the frame of reference and of the units we may specify this Maxwellian as

$$(1)\qquad \omega = (2\pi)^{-3/2} \exp\{-\frac{1}{2}|\xi|^2\}$$

Then, for any distribution function which is close to ω we set

$$(2)\qquad F = \omega(1+f),$$

and substitute into the Boltzmann equation with, as a result

$$(3)\qquad (\frac{\partial}{\partial t} + \xi\cdot\nabla) f + Lf = Q(f,f),$$

and

$$(4)\quad Lf = -\int_{R^3} \omega(|\xi_1|)\, d\xi_1 \int_0^{\pi/2} d\theta\, B(\theta, |\xi_1 - \xi|) \int_0^{2\pi} d\psi\, (f' + f_1' - f - f_1),$$

$$(5)\quad Q(f,f) = \int_{R^3} \omega(|\xi_1|)\, d\xi_1 \int_0^{\pi/2} d\theta\, B(\theta, |\xi_1 - \xi|) \int_0^{2\pi} d\psi\, (f'f_1' - ff_1).$$

We observe that $-\omega Lf$ is the Frechet derivative, at ω of $J(F,F)$.

1.4.1.2. One of the most powerful methods for studying the Boltzmann equation is to make first a thorough study of the so-called linear Boltzmann equation

$$(1)\qquad (\frac{\partial}{\partial t} + \xi\cdot\Delta) f + Lf = \varphi$$

where φ is supposed to be given.

1.4.1.3. Consider a cut-off collision operator for which 1.1.4.1.(1) holds with 1.1.4.7.(1) and (2). We have

$$D\{\omega(1+f)\} = \omega\,\nu(|\xi|) + D(\omega f)\ , \tag{1}$$

$$C\{\omega(1+f), \omega(1+f)\} = \omega\,\nu + \omega\{\nu f + D(\omega f) - Lf\} \tag{2}$$
$$+ \omega\{f\,D(\omega f) + Q(f,f)\}\ ,$$

with

$$\nu(|\xi|) = 2\pi \int_{R^3} \omega(|\zeta|)\, d\zeta \int_0^{\pi/2} B(\theta, |\xi-\zeta|)\, d\theta\ . \tag{3}$$

On the other hand

$$Lf = \nu(|\xi|)f - Af \tag{4}$$

with

$$Af = \int_{R^3} \omega(|\xi_1|)\, d\xi_1 \int_0^{\pi/2} d\theta\ B(\theta, |\xi-\xi_1|) \int_0^{2\pi} d\psi (f' + f'_1 - f_1) \tag{5}$$

1.4.2. Elementary Properties of the Linear Collision Operator

1.4.2.1. We shall speak of L as the linear collision operator. We state here without proof (refer to Grad[1]) the property analogous to 1.1.3.1. that, for any $g(\xi)$,

$$\int \omega\, g\, L\, f\, d\xi =$$

(1) $$\frac{1}{4} \int_{R^3} \omega(|\xi|)\, d\xi \int_{R^3} \omega(|\xi_1|)\, d\xi_1 \int_0^{\pi/2} d\theta\; B(\theta, |\xi_1 - \xi|)\, .$$

$$\cdot \int_0^{2\pi} d\epsilon\; (f' + f_1' - f - f_1)(g' + g_1' - g - g_1)\, .$$

An obvious consequence of this identity is that

(2) $$\int_{R^3} \omega\, g\, L f\, d\xi = \int_{R^3} \omega\, f\, Lg\, d\xi\, ,$$

and another equally obvious consequence is that

(3) $$\int_{R^3} \omega\, f\, L f\, d\xi = \frac{1}{4} \int_{R^3} \omega(|\xi|)\, d\xi \int_{R^3} \omega(|\xi_1|)\, d\xi_1\, .$$

$$\int_0^{\pi/2} d\theta\; B(\theta, |\xi_1 - \xi|) \int_0^{2\pi} d\epsilon\; (f' + f_1' - f - f_1)^2$$

1.4.2.2. The linearity of L allows the consideration of complex valued f. If we specify that f is real we have

(4) $$\int \omega f\; Lf\; d\xi \geqslant 0$$

and, obviously, the equality sign holds if and only if $f' + f_1' = f + f_1$. From the argument of 1.1.3.3. this implies that f is a linear combination of the five so-called collisional invariants ψ_α, $\alpha = 0, 1, \ldots 4$

(5) $$\psi_o = 1\, ;\; \psi_i = \xi_i,\; i = 1,2,3\, ;\; \psi_4 = 6^{-1/2}\,(|\xi|^2 - 3)$$

which have been so chosen for later convenience.

1.4.2.3. If $Lf = 0$, by considering separately the real and imaginary parts, we may assume beforehand that f is real, then, from 1.4.2.2. it is seen that f must be a linear combination of the collisional invariants.

1.4.3. The Hilbert Space H

1.4.3.1. From 1.4.2. it is natural to consider the real or complex Hilbert space of measurable function f which are square integrable with respect to the measure $\omega \, d\, \xi$. As a norm we use

$$\| f \| = \{ \int_{R^3} \omega \, | f |^2 \, d\xi \}^{1/2} , \tag{1}$$

and as a scalar product

$$(f, g) = \int_{R^3} \omega \; f \, \bar{g} \; d\xi , \tag{2}$$

where an overbar stands for complex conjugation.

1.4.3.2. From 1.4.2.1.(2) we deduce trivially that L is a symmetric operator, namely

$$(Lf, g) = (f, Lg) , \tag{1}$$

and from 1.4.2.2., that it is non-negative

$$(Lf, f) \geqslant 0 \tag{2}$$

the equality sign holding if and only if f is a linear combination of the collisional invariants 1.4.2.2.(5).

1.4.3.3. We may reword 1.4.2.3. by saying that zero is an eigenvalue of the operator L and that the corresponding eigenspace, the so-called kernel of L, is five dimensional. It is readily shown, through computation, that the five eigenfunctions

1.4.2.2.(1) form an orthonormal set in the eigenspace, namely

(1) $$(\psi_\alpha, \psi_\beta) = \delta_{\alpha\beta}$$

1.4.3.4. We shall need at some stage later the quantities

(1) $$A_{k,\alpha\beta} = (\xi_k \psi_\alpha, \psi_\beta)$$

which are readily computed. We state the result

(2) $$\begin{cases} A_{k,o\beta} = A_{k,\beta o} = \delta_{k\beta}\,, \\ A_{k,4\beta} = A_{k,\beta 4} = \sqrt{2/3}\ \delta_{k\beta}\,, \\ A_{k,i\beta} = A_{k,\beta i} = (\delta_{ok} + \sqrt{2/3}\ \delta_{4\beta})\ \delta_{ik}\,, \qquad i = 1,2,3\,. \end{cases}$$

1.4.3.5. We observe that the norm $\|\,.\,\|$ provides a measure of smallness of the deviation from ω. This measure of smalness is not entirely satisfactory from a physical point of view. As a matter of fact we may have an f such that $\|f\|$ is as small as we please, yet F can assume negative values. This occurs in the Chapman Enskog theory. This remark is related to a non-uniformity in the velocity space and has far reaching consequences in the asymptotic theory of the Boltzmann equation. An example of the way in which one can deal with such a non-uniformity and its consequences is provided by Narasimha[17].

1.4.4. The Reflection Operator

1.4.4.1. We set ω for the normalized Maxwellian $F_{M_w}^{(n)}$ of 1.2.4.1.(2), namely

(1) $$\omega_w = (2\pi\, a_w^2)^{-3/2} \exp\!\left(-\frac{1}{2}\,\frac{|\xi - u_w|^2}{a_w^2}\right)$$

and we represent it as

$$\omega_w = \omega(1 + f_{M_w}) . \tag{2}$$

We observe that f_{M_w} depends on a_w and $\mathbf{u}_w$ and that, for small $|a_w - 1|$ and $|\mathbf{u}_w|$, f_{M_w} is small in H. Moreover f_{M_w} is differentiable, near zero, with respect to $a_w - 1$ and $\mathbf{u}_w$, the result is

$$(1 + f_{M_w})^{-1} \, df_{M_w} = \frac{(\xi - \mathbf{u}_w)}{a_w^2} \cdot d\mathbf{u}_w + \Big(\frac{|\xi - \mathbf{u}_w|^2}{a_w^2} - 3\Big) \, da_w . \tag{3}$$

On the other hand, for small $|a_w - 1|$ and $|\mathbf{u}_w|$ we have

$$f_{M_w} = \xi \cdot \mathbf{u}_w + (|\xi|^2 - 3)(a_w - 1) + O(|\mathbf{u}_w|^2 + |a_w - 1|^2) \tag{4}$$

the error estimate standing in H.

1.4.4.2. The reflection operator K of 1.2.4.1.(3) depends on a_w and $\mathbf{u}_w$. We restate that K is related to K by

$$K = (\overset{+}{\omega}_w)^{-1} \, \mathrm{K}(\omega_w) . \tag{5}$$

For a given material of the wall, we may assume that, indeed, K depends only on a_w and $\mathbf{u}_w$, and of course the direction n. We shall set G for the operator K which corresponds to $a_w = 1$ and $\mathbf{u}_w = 0$ and

$$(\omega^+)^{-1} \, \mathrm{K}(\omega^-) = \mathrm{G} + \mathrm{G}_1 \tag{6}$$

and we expect G_1 to be small in a function space, to be defined later, when $|a_w - 1| + |\mathbf{u}_w|$ is small.

From 1.2.3.2.(5) we have

$$\overset{+}{1} = K \, 1^- \tag{7}$$

and, from this, we conclude that

(8) $$1^{+} = G\,1^{-}$$

and

(9) $$f_{M_w}^{+} - G\,f_{M_w}^{-} = G_1(1 + f_{M_w})^{-}$$

1.4.4.3. From 1.4.4.2., the condition $F^{+} = K F^{-}$ is converted to

(1) $$f^{+} = G f^{-} + G_1(1 + f)^{-}$$

1.4.4.4. To the operator G corresponds a kernel $\Gamma(x, \xi, \zeta)$, independent of time, which is related to the kernel $K_o(x, \xi, \zeta)$ of the operator K for $a_w = 1$, $u_w = 0$ by

(1) $$\Gamma(x, \xi, \zeta) = \{\omega(|\xi|)\}^{-1} K_o(x, \xi, \zeta)\ \omega(|\zeta|)\ ,$$

and we have

(2) $$Gf^{-}(\zeta, x, \xi) = \int_{R^3_-} \Gamma(x, \xi, \zeta)\ f^{-}(\zeta)\ d\zeta\ ,$$

where R^3_- stands for the half space $\zeta \cdot n < 0$.

1.4.4.5. We restate the three basic properties of 1.2.3.2. in terms of the operator G as

(1) $$f^{-} \geqslant 0 \Rightarrow G f^{-} \geqslant 0\ ,$$

(2) $$f^{+} = G f^{-} \Rightarrow \int_{R^3} \xi \cdot n\ \omega(|\xi|)\ f(\xi)\ d\xi = 0\ ,$$

$$\overset{+}{1} = G\ 1^- . \tag{3}$$

In terms of the kernel I, the corresponding properties read as follows

$$\Gamma(x, \xi, \zeta) \geq 0 , \tag{4}$$

$$\int_{R^3_+} \omega(|\xi|)\ \xi \cdot n\ \Gamma(x, \xi, \zeta)\ d\xi = \omega(|\zeta|)\ |\zeta \cdot n|\ 1^- , \tag{5}$$

$$\overset{+}{1} = \int_{R^3} \Gamma(x, \xi, \zeta)\ d\zeta . \tag{6}$$

1.4.4.6. We restate the reciprocity relation of 1.2.3.3. In terms of the kernel K_o it reads as

$$\omega(|\zeta|)\ K_o(x, \xi, \zeta) = \omega(|\xi|)\ K_o(x, -\zeta, -\xi) \tag{1}$$

while, in terms of the kernel Γ, it reads

$$\omega(|\xi|)\ \Gamma(x, \xi, \zeta) = \omega(|\zeta|)\ \Gamma(x, -\zeta, -\xi) \tag{2}$$

1.4.5. Dissipativity of the Reflection Operator

1.4.5.1. According to Darrozes, Guiraud[13], Cercignani[11][12] Case[14] the dissipativity of 1.2.4. induces a corresponding property for the operator G. From 1.2.4.1. we may state that, for any strictly convex function J(f) one has, for a real f,

(1) $$\overset{+}{f} = G f^- \Rightarrow \int_{R^3} \xi \cdot n \quad \omega(|\xi|) \ J\{ f(\xi)\} \, d\xi \leqslant 0 \ ,$$

equality holding if and only if f is almost everywhere a constant provided that 1.4.4.4.(4) is slightly strengthened in that Γ is zero only on a set of zero measure.

1.4.5.2. Referring to the argument of 1.2.4.1. and setting

(1) $$d\mu = \omega(|\xi|) \ |\xi \cdot n| \ d\xi$$

We have, for $f^+ = Gf^-$, f real

(2) $$-\int_{R^3} \xi \cdot n \ \omega(|\xi|) \ J\{f(\xi)\} \, d\xi = \int_{R^3_+} \{G\, J(f^-) - J(G\, f^-)\} \, d\mu \geqslant 0$$

1.4.5.3. For $J(g) = g^2$, f real, from the argument of 1.2.4.3. we have,

(1) $$\int_{R^3_+} \{ G(f^-)^2 - (G\, f^-)^2 \} \, d\mu = \int_{R^3_+} G(f^- - G\, f^-)^2 \quad d\mu \geqslant 0$$

which must be read as follows

(2) $$\int_{R^3_+} \{ G\,(f^-)^2 - (G\, f^-)^2 \} \, d\mu = \int_{R^3_+} d\mu(\xi) \int_{R^3_-} \Gamma(\xi, \zeta) \ d\zeta \ . \\ \{ f^-(\zeta) - \int_{R^3} \Gamma(\xi, \zeta_1) f^-(\zeta_1) d\zeta_1 \}^2 \ .$$

We note that, for $J(x) = x^2$, the inequality 1.2.4.3.(1) has to be replaced by the corresponding equality. We transform this identity to a more convenient form for later work. We set

$$I(\xi) = \int_{R^3_-} \Gamma(\xi,\zeta)\,\{ f^-(\zeta) - \int_{R^3_-} \Gamma(\xi,\zeta_1) f^-(\zeta_1)\, d\zeta_1 \}^2\, d\zeta \tag{3}$$

and we use 1.4.4.5.(6) which allows to write

$$I(\xi) = \int_{R^3_-} \Gamma(\xi,\zeta)\, d\zeta \int_{R^3_-} \Gamma(\xi,\zeta_1) d\zeta_1 \int_{R^3_-} \Gamma(\xi,\zeta)\, [f(\zeta) - f(\zeta_1)[f(\zeta) - f(\zeta_2)]\, d\zeta_2 \tag{4}$$

then, setting

$$\phi(s) = f^-(\zeta) - f^-(s) \tag{5}$$

We transform the right hand side of (4) to

$$I = \int_{R^3_-} \Gamma(\xi,\zeta)\, d\zeta \int_{R^3_-} ds \int_{R^3_-} ds_1\, \Gamma(\xi,s)\,\Gamma(\xi,s_1)\, \phi(s)\, \phi(s_1)$$

$$= \int_{R^3_-} \Gamma(\xi,\zeta)\, d\zeta \int_{R^3_-} ds \int_{R^3_-} ds_1\;\; \Gamma(\xi,s)\,\Gamma(\xi,s_1)\, (\phi(s))^2 \; +$$

$$+ \int_{R^3_-} \Gamma(\xi,\zeta)\, d\zeta \int_{R^3_-} ds \int_{R^3_-} ds_1\, \Gamma(\xi,s)\,\Gamma(\xi,s_1)\, \{\phi(s)\,\phi(s_1)\, (\phi(s))^2\}$$

Again we use 1.4.4.5.(6) and we obtain

$$I = \int_{R^3_-} \Gamma(\xi,\zeta)\, d\zeta \int_{R^3_-} \Gamma(\xi,\zeta)\; |\,\phi(\zeta)\,|^2\, ds \; -$$

$$- \frac{1}{2} \int_{R^3_-} \Gamma(\xi,s)\, ds \int_{R^3_-} d\zeta \int_{R^3_-} d\zeta_1\;\; \Gamma(\xi,\zeta)\,\Gamma(\xi,\zeta_1)\, [\phi(\zeta) - \phi(\zeta_1)]^2$$

and, from (5) and 1.4.4.5.(6) again

$$I = \frac{1}{2} \int_{R^3_-} d\zeta \int_{R^3_-} d\zeta_1 \, \Gamma(\xi, \zeta) \; P(\xi, \zeta_1) \, | f^-(\zeta) - f^-(\zeta_1) |^2$$

so that

$$\int_{R^3_+} \{ G(f^-)^2 - (Gf^-)^2 \} \, d\mu = \frac{1}{2} \int_{R^3_+} d\mu(\xi) \int_{R^3_-} d\zeta \int_{R^3_-} d\zeta_1 \, \cdot$$

(6)

$$\cdot \; \Gamma(\xi, \zeta) \, \Gamma(\xi, \zeta_1) \, | f^-(\zeta) - f^-(\zeta_1) |^2 \; .$$

We note that a direct proof of (6), which does not rely on the argument of 1.2.4.3., is given by Guiraud [18] (page 464).

1.4.5.4. Following Guiraud[18] we state, as a basic hypothesis concerning the kernel Γ that there exists some positive constant γ such that

(1) $$\int_{R^3_+} d\mu(\xi) \; \Gamma(\xi, \zeta) \, \Gamma(\xi, \zeta_1) \geqslant \gamma \, \omega(|\zeta|) \; \omega(|\zeta_1|) \; |\zeta \cdot n| \; |\zeta_1 \cdot n| \, ,$$

and, from 1.4.5.2. and 1.4.5.3.(6) we conclude that

(2) $$f^+ = Gf^- \Rightarrow - \int_{R^3} \omega \; \xi \cdot h \; |f|^2 \, d\xi \geqslant \frac{\gamma}{2} \int_{R^3_-} d\mu(\zeta) \int_{R^3_-} d\mu(\zeta_1) | f^-(\zeta) - f^-(\zeta_1) |^2 \, ,$$

for any real f .

1.4.5.5. We are lead to consider the hilbert space H of measurable functions which are square integrable with respect to the measure $d\mu$. As a norm we take

(1) $$\| f \|_{\bar{H}} = [f] = \{ \int_{R^3} |f|^2 \; d\mu \}^{1/2}$$

and the corresponding scalar product is

$$(f, g)_{\bar{H}} = \{ f, g\} = \int_{R^3} f\bar{g}\ d\mu \tag{2}$$

1.4.5.6. It will prove useful to introduce other Hilbert spaces. Let $\rho(\xi)$ be a g non negative function, we let

$$[f]_\rho = [\rho^{1/2} f] , \tag{1}$$

$$\{f, g\}_\rho = \{\rho^{1/2} f, \rho^{1/2} g\} \tag{2}$$

and we call H_ρ the corresponding Hilbert space.

1.4.5.7. We observe that the right hand side of 1.4.5.4.(2) is unchanged if we substitute f^- + const to f^-. We take advantage of this to change f^- to Πf^- defined as

$$\Pi f^- = f^- - \sqrt{2\pi}\,(f^-, 1^-)_{\bar{H}} , \tag{1}$$

so that

$$(\Pi f^-, 1^-)_{\bar{H}} = 0 , \tag{2}$$

and, as a consequence

$$\int_{R^3_-} d\mu(\zeta) \int_{R^3_-} d\mu(\zeta_1)\, |\Pi f^-(\zeta) - \Pi f^-(\zeta_1)|^2 = 2 \int_{R^3_-} d\mu(\zeta) \int_{R^3_-} d\mu(\zeta_1) |\Pi f^-(\zeta)|^2 . \tag{3}$$

We set

$$\text{(4)} \qquad \gamma \int d\mu(\xi) = \frac{\gamma}{\sqrt{2\pi}} = 2k > 0 \ ,$$

and we may state the final form of the dissipation inequality, which holds under the hypothesis of 1.4.5.4., as

$$\text{(5)} \qquad f^{+} = Gf^{-} \Rightarrow -\int_{R^3} \omega\ \xi \cdot n\, |f|^2\, d\xi \geqslant 2k \, \| \Pi f^{-} \|^2_{\tilde{H}} = 2k[\Pi f^{-}]^2$$

for any real f^-. This statement has been proven for a continuous f^-. Now let us assume that the kernel Γ is such that G may be extended to a continuous operator in H. Let $f \in H$ we shall say that $f^+ = g f^-$ holds if, for any sequence f_n of continuous functions such that $\| f^- - f_n^- \|_{\tilde{H}} \to 0$ we have $\| f^+ - G f_n^- \|_{\tilde{H}} \to 0$. Let us define $f_n = f_n^- + G f_n^-$, then the inequality (5) holds for f_n and letting $n \to \infty$ we conclude that (5) is extended to the whole of H.

1.4.5.7. Let $f = f_1 + i f_2$ with f_1 and f_2 real, wherever $f^+ = G f^-$, we have $f_i^+ = G f_i^-$, $i = 1,2$ and 1.4.5.6.(5) apply to f_1 and f_2 separately. We conclude that, if the kernel Γ is such that G may be extended to a continuous operator in $\tilde{H}$, whatever be the, complex, or real, f such that $f^+ = g f^-$, 1.4.5.6(5) holds

1.4.5.8. Now let

$$\text{(1)} \qquad f^{+} = Gf^{-} + \phi^{+}$$

and consider $\tilde{f} = f^- + Gf^-$ so that $\tilde{f} = f + \phi^+$, we have

$$\text{(2)} \qquad \begin{aligned} -\int_{R^3} \omega\, \xi \cdot n\, |f|^2\, d\xi &= -\int_{R^3_-} \omega\, \xi \cdot n\, |\tilde{f}|^2\, d\xi - \int_{R^3_+} \omega\, \xi \cdot n\, |\phi^+|^2\, d\xi \\ &\quad - 2\,\mathrm{Re} \int_{R^3_+} \omega\, \xi \cdot n\ \phi^+\, Gf^-\, d\xi \end{aligned}$$

and, observing that

$$Gf^- = \Pi G f^- + \sqrt{2\pi}\ (Gf^-, 1^+)_{\tilde{H}}\ 1^+ \tag{3}$$

while, from 1.4.4.5.(2)

$$(Gf^-, 1)_{\tilde{H}} = (f^-, 1^-)_{\tilde{H}} \tag{4}$$

We conclude that, provided G is continuous with the norm of $\tilde{H}$, we have

$$f^+ = Gf^- + \phi^+ \Rightarrow - \int_{R^3_-} \omega\, \xi \cdot n\, |f|^2\, d\xi + \|\phi^+\|^2_{\tilde{H}} +$$

$$+ 2\,\mathrm{Re}(\phi^+, \Pi G f^-)_{\tilde{H}} + 2\sqrt{2\pi}\ \mathrm{Re}(\phi^+, 1^+)_{\tilde{H}}\, (1^-, f^-)_{\tilde{H}} \geqslant \tag{5}$$

$$2k\|\Pi f^-\|^2_{\tilde{H}}$$

1.4.5.9. We state our basic set of hypothesis which will be preserved throughout the work. First, to G is associated a kernel $\Gamma(x, \xi, \zeta)$ which is continuous within its domain of definition, which meets the requirements 1.4.4.5.(4)(5)(6) and which is such that G is continuous with the norm of $\tilde{H}$.

1.4.6. Some Basic Identities and Inequalities

1.4.6.1. We set

$$\mathcal{L} = \xi \cdot \nabla + L = \mathcal{M} - A \quad , \qquad \mathcal{M} = \xi \cdot \nabla + \nu \tag{1}$$

$$\mathcal{L}^* = - \xi \cdot \nabla + L = \mathcal{M}^* - A \quad , \qquad \mathcal{M}^* = \xi \cdot \nabla + \nu \tag{2}$$

where ν and A are as in 1.4.1.3. Let Ω be a bounded three dimensional domain with smooth boundary $\partial\Omega$ and let n be the unit normal to $\partial\Omega$ pointing towards the interior of Ω. We set

$$(3) \qquad ((f, g)) = \int_{\Omega} (f, g)\, dv, \quad |||f|||^2 = ((f, f))$$

$$(4) \qquad \{f, g\} = \int_{\partial\Omega} (|\xi \cdot n| f, g)_{\tilde{H}}\, ds, \quad \int_{\partial\Omega} (|\xi \cdot n| f, f)_{\tilde{H}}\, ds = [f]^2,$$

then for any f which is continuous, continuously derivable along rays and vanish in some vicinity of infinity in velocity space, we have

$$(5) \qquad ((\xi \cdot \nabla f, g)) + ((f, \xi \cdot \nabla g)) + \int_{\partial\Omega} (\xi \cdot n\, Bf, Bg)_H\, dS = 0$$

where Bf stands for the boundary value of f on $\partial\Omega$. From the symmetry of L, 1.4.3.2.(1) we deduce that

$$(6) \qquad ((\mathcal{L}f, g)) - ((f, \mathcal{L}^* g)) + \int_{\partial\Omega} (\xi \cdot n\, Bf, Bg)_H\, dS = 0$$

and, directly from (5), that

$$(7) \qquad ((\mathcal{M}f, g)) - ((f, \mathcal{M}^* g)) + \int_{\partial\Omega} (\xi \cdot n\, Bf, Bg)_H\, dS = 0$$

1.4.6.2. Consider a point on a wall, we ask the following question: how to choose h such that

$$f^+ = Gf^- \Rightarrow (f, h)_{\tilde{H}} = 0,$$

or stated otherwise, what must be the relation between h^-, and h^+ in order that

$$\int_{R^3_+} h^+ \, Gf^- \, d\mu = \int_{R^3_-} h^- \, f^- \, d\mu \tag{1}$$

holds for any f^-? Written in full, (1) states that

$$\int_{R^3_-} \{ \omega(|\zeta|) \, |\zeta \cdot n| \, h^-(\zeta) - \int_{R^3_+} \omega(|\xi|) \, |\xi \cdot n| \, \Gamma(\xi, \zeta) \, h^+(\xi) \, d\xi \} f^-(\zeta) d\zeta = 0 \, , \tag{2}$$

holds for any f^- and from this we conclude that

$$h^-(\xi) = \int_{R^3_+} P^*(\xi, \zeta) \, h^+(\zeta) \, d\zeta \, , \tag{3}$$

with

$$|\xi \cdot n| \, \omega(|\xi|) \, \Gamma^*(\xi, \zeta) = |\zeta \cdot n| \, \omega(|\zeta|) \, \Gamma(\zeta, \xi) \, . \tag{4}$$

Let us set G^* the operator defined by (3) such that $h^- = G^* h^+$ we have proven that

$$f^+ = Gf^- \, , h^- = G^* h^+ \Rightarrow (f, h)_{\tilde{H}} = 0 \, , \tag{5}$$

and G^* may be considered as the formal adjoint of G in H. We observe that under the assumption that the reciprocity relation of 1.2.3.3. holds we have, from (4) and 1.4.4.6.(2),

$$|\xi \cdot n| \, \Gamma^*(\xi, \zeta) = |\zeta \cdot n| \, \Gamma(-\xi, -\zeta) \tag{6}$$

1.4.6.3. From 1.4.6.2.(4) and 1.4.4.5.(4) (5) (6) we deduce that the kernel $\Gamma^*(\xi, \zeta)$ meets the following requirements

(1) $$\Gamma^*(\xi,\zeta) \geqslant 0 \ ,$$

(2) $$\int_{R^3_-} \omega(|\xi|)\ |\xi\cdot n|\ \Gamma^*(\xi,\zeta)\ d\xi = \omega(|\zeta|)\ \zeta\cdot n\ 1^+ \ ,$$

(3) $$1^- = \int_{R^3_+} \Gamma^*(\xi,\zeta)\ d\zeta \ ,$$

which are necessary and sufficient in order that G^* satisfies the conditions analogous to the ones for G, namely

(4) $$\overset{+}{h} \geqslant 0 \Rightarrow G^*\ \overset{+}{h} \geqslant 0 \ ,$$

(5) $$\bar{h} = G^*\overset{+}{h} \Rightarrow \int_{R^3} \xi\cdot n\ \ \omega(|\xi|)\ \ h(\xi)\ d\xi = 0 \ ,$$

(6) $$1^- = G^*\ \overset{+}{1} \ .$$

This allows to repeat with G^* the argument of 1.4.5.2. and 1.4.5.3., and to state that if G^* is a bounded operator in $\widetilde{H}$ we have

(7) $$h^- = G^*\overset{+}{h} \Rightarrow \int_{R^3} \omega\ \ \xi\cdot n\ |h|^2\ d\xi \geqslant 0$$

the equality holding if and only if h is almost everywhere a constant. Under the additional hypothesis that

(8) $$\int_{R^3} d\mu(\xi)\ \Gamma^*(\xi,\zeta)\ \Gamma^*(\xi,\zeta) \geqslant \gamma^*\ \omega(|\zeta|)\ \omega(|\zeta|)\ |\zeta\cdot n|\ |\zeta\cdot n|$$

with a positive constant γ^* we have

$$h^- = G^* h^+ + \psi^- \Rightarrow \int_{R^3} \omega \xi \cdot n \, |h|^2 \, d\xi + \| \psi^- \|^2_{\tilde{H}} +$$

$$+ 2\mathrm{Re}(\psi^-, \Pi G^* h^-)_{\tilde{H}} + 2\sqrt{2\pi}\, \mathrm{Re}(\psi^-, 1^-)_{\tilde{H}} (1^+, h^+)_{\tilde{H}} \geqslant \quad (9)$$

$$2k^* \| \Pi h^- \|^2_{\tilde{H}}$$

with $2k^* = \gamma^* / \sqrt{2\pi}$.

1.4.6.4. Let us set J^+ (resp J^-) the projection operator as defined in $\tilde{H}$, on the subspace of functions which vanish almost everywhere in $\xi \cdot n < 0$ (resp. $\xi \cdot n > 0$). We set $\tilde{H}^+ = J^+ \tilde{H}$, $\tilde{H}^- = J^- \tilde{H}$. The relation defining G^* reads as

$$\forall f^- \in \tilde{H}^-, \; (h^+, G f^-)_{\tilde{H}} = (h^-, f^-)_{\tilde{H}} \Rightarrow h^- = G^* h^+, \quad (1)$$

and we see that G^* is the adjoint of G, considered as an operator from $\tilde{H}^-$ into $\tilde{H}^+$. From

$$|(h^-, f^-)_{\tilde{H}^-}| \leqslant |G|_{L(\tilde{H}^-, \tilde{H}^+)} \| f^- \|_{\tilde{H}^*} \| h^+ \|_{\tilde{H}^+}, \quad (2)$$

we conclude that

$$|G^*|_{L(\tilde{H}^+, \tilde{H}^-)} \leqslant |G|_{L(\tilde{H}^-, \tilde{H}^+)} \quad (3)$$

and no additional hypothesis is necessary to make of G^* a bounded operator if we assume before hand that G is.

Let us define G^{**} by

$$(4) \qquad \forall h^+ \in H^+, \quad (f^-, G^* h^+)_{\tilde{H}^-} = (f^+, h^+)_{\tilde{H}^+} \Rightarrow f^+ = G^{**} f^- ,$$

we have, of course

$$(5) \qquad |G^{**}|_{L(\tilde{H}^-, \tilde{H}^+)} \leqslant |G^*|_{L(\tilde{H}^+, \tilde{H}^-)} \leqslant |G|_{L(\tilde{H}^-, \tilde{H}^+)} .$$

on the other hand, from (4) and (1) we have, for any $f^- \in \tilde{H}^-$ and any $h^+ \in \tilde{H}^+$,

$$(6) \qquad \begin{aligned} (G^{**} f^-, h^+)_{\tilde{H}^+} &= (f^-, G^* h^+)_{\tilde{H}^-} = \overline{(G^* h^+, f^-)}_{\tilde{H}^-} = \overline{(h^+, G f^-)}_{\tilde{H}^+} \\ &= (G f^-, h^+)_{\tilde{H}} , \end{aligned}$$

and, as a consequence,

$$(7) \qquad G^{**} = G ,$$

then, from (5),

$$(8) \qquad |G^*|_{L(\tilde{H}^+, \tilde{H}^-)} = |G|_{L(\tilde{H}^-, \tilde{H}^+)} .$$

We may state that, under 1.4.5.9., and under the two further conditions 1.4.5.4.(1) and 1.4.6.3.(8), the two basic inequalities 1.4.5.8.(5) and 1.4.6.3.(9) both hold in H.

1.4.6.5. Let f and h belong to $\tilde{H}$ and set

$$\bar{f} = f^- + Gf^- , \qquad \bar{h} = h^+ + G^*h^+ , \tag{1}$$

so that

$$f = \bar{f} + \phi^+ , \qquad h = \bar{h} + \psi^- . \tag{2}$$

$$\begin{aligned}(\xi \cdot n\, f, h)_H &= (\xi \cdot n[\bar{f} + \phi^+] , \bar{h} + \psi^-)_{\tilde{H}} = (\xi \cdot n\, \bar{f}, \bar{h})_{\tilde{H}} + \\ &+ (\xi \cdot n\, \bar{f}, \psi^-)_{\tilde{H}} + (\xi \cdot n\, \phi^+, \bar{h})_{\tilde{H}} + (\xi \cdot n\, \phi^+, \psi^-)_{\tilde{H}} ,\end{aligned} \tag{3}$$

but, from either 1.4.6.4.(1) or (4) we have $(\xi \cdot n\, \tilde{f}, \tilde{h})_{\tilde{H}} = 0$ while $(\xi \cdot n\, \phi^+, \psi^-)_{\tilde{H}} = 0$ trivially, and

$$\begin{aligned}(\xi \cdot n\, f, h)_{\tilde{H}} &= (\xi \cdot n\, \bar{f}, \psi^-)_{\tilde{H}} + (\xi \cdot n\, \phi^+, \bar{h})_{\tilde{H}} = \\ &= (\xi \cdot n\, f^{--}, \psi^-)_{\tilde{H}} + (\xi \cdot n\, \phi^+ , h^+)_{\tilde{H}}\end{aligned} \tag{4}$$

1.4.6.6. We apply 1.4.6.5.(4) to 1.4.6.1.(6) and (7) this time under smooth f and g, according to the requirements of 1.4.6.1. and we obtain

$$\begin{aligned}&((\mathcal{L}f, g)) - ((f, \mathcal{L}^*g) + \{ J^+Bf - GJ^-Bf, J^+Bg\} - \\ &- \{J^-Bf, J^-Bg - G^* J^+Bg\} = 0\end{aligned} \tag{1}$$

$$((Mf, g)) - ((f, M^*g)) + \{J^+ Bf - GJ^- Bf, J^+ Bg\}$$

(2)

$$- \{J^- Bf, J^- Bg - G^* J^+ Bg\} = 0$$

1.5 Further Properties of the Linear Collision Operator

1.5.1. Referring to 1.4.1.3. we state here a number of properties of the operator A which will be essential for the existence theory of the next chapters. All the proofs of our statements are contained in Grad[5] Cercignani[8] (also Cercignani[4] Chapter 3) Drange[7] .

1.5.2. Properties of ν

1.5.2.1. Referring to 1.4.1.3.(3) the simplest situation corresponds to rigid spheres, for which, from 1.1.4.2.(1) we have

(1) $$\nu(|\xi|) = \pi\sigma^2 \int_{R^3} |\xi - \zeta|\ \omega(|\zeta|)\ d\zeta$$

or

$$\nu(|\xi|) = -\pi\sigma^2 \int_0^\infty d\omega(|\zeta|) \int_{|u|=1} |\zeta|\ |\ \xi - |\zeta|\ u|\ du$$

$$= \pi\sigma^2 \int_0^\infty \omega(|\zeta|)\ d|\zeta|\ \frac{\partial}{\partial|\zeta|} \int_{|u|=1} |\zeta|\ |\xi - |\zeta|u|\ du$$

(2) $$= \frac{2}{3}\pi^2\sigma^2 \int_0^\infty d|\zeta|\ \omega(|\zeta|)\ |\xi|^{-1} \frac{\partial}{\partial|\zeta|} \{(|\xi| + |\zeta|)^3 - |\,|\xi| - |\zeta|\,|^3\}$$

$$= 2\pi^2\sigma^2 \int_0^\infty \omega(|\zeta|)\ |\xi|^{-1}\ \{(|\xi| + |\zeta|)^2 + |\,|\xi| - |\zeta|\,|\ (|\xi| - |\zeta|)\}\ d|\zeta|$$

and

$$\nu(|\xi|) = 8\pi^2\sigma^2 \int_0^\infty |\zeta|\ \omega(|\zeta|)\ d(\zeta) + 4\pi^2\sigma^2\ |\xi|^{-1} \int_0^\infty (|\xi|^2 + |\zeta|^2)\ \omega(|\zeta|)\ d|\zeta|$$

$$= 8\pi^2\sigma^2\ \omega(|\xi|) + 4\pi^2\sigma^2\ |\xi| \int_0^\infty \omega(|\zeta|)\ d(\zeta) + 4\pi^2\sigma^2\ |\xi|^{-1}\ (-|\xi|\omega(|\xi|)\ (3)$$

$$+ \int_0^\infty |\xi|\ \omega(|\zeta|)\ d|\zeta|)$$

and finally

$$\nu(|\xi|) = 4\pi^2\sigma^2 \{\omega(|\xi|) + (|\xi| + |\xi|^{-1}) \int_0^{|\xi|} \omega(|\zeta|)\, d|\zeta|\} \tag{4}$$

From this formula it is seen, first that

$$\nu_o = \operatorname*{Inf}_{|\xi|} \nu(|\xi|) > 0 \tag{5}$$

and second that there exists two constants a and b both positives such that

$$a(1 + |\xi|^2)^{1/2} \leqslant \nu(|\xi|) \leqslant b(1 + |\xi|^2)^{1/2} \tag{6}$$

We observe that $\nu(|\xi|)$ is obviously continuous and even indefinitely differentiable and that it is a monotonely increasing function of $|\xi|$, from which we compute

$$\nu_o = 2\sqrt{2\pi}\,\sigma^2 \tag{7}$$

1.5.2.2. For the radial cut-off of Cercignani (refer to 1.1.4.5) $\nu(|\xi|)$ has the same expression as with rigid sphere with diameter equal to the cut-off radius σ, at least for repulsive interactions for which the impact parameter b is a monotone function of θ as is supposed to be the case in the whole of our treatment. The proof is very simple

$$B(\theta, V) = V b \frac{\partial b}{\partial \theta}\, d\theta \tag{1}$$

$$\nu(|\xi|) = 2\pi \int_{R^3} \omega(|\zeta|)\, |\xi - \zeta|\, d\zeta \int_{b=0}^{b=\sigma} b \frac{\partial b}{\partial \theta}\, d\theta \tag{2}$$

The same property holds for Drange's[7] cut-off (refer to 1.1.4.6).

1.5.2.3. For angular cut-off power law potentials one has, referring to 1.1.4.7.(3)

$$\nu(|\xi|) = (2\pi)^{3/2}\, \beta_o \int_{R^3} |\xi - \zeta|^\gamma\; \omega(|\zeta|)\; d|\zeta| \;, \tag{3}$$

with

$$\beta_o = (2\pi)^{-1/2} \int_0^{\pi/2} \beta(\theta)\; d\theta \;, \tag{4}$$

$$\gamma = \frac{s-5}{s-1} \;. \tag{5}$$

It has been shown by Grad[5] (refer to page 39 of that reference) that $\nu(|\xi|)$ has the same property of monotonity as

$$B_0(V) = \int_0^{\pi/2} B(\theta, V)\; d\theta \;. \tag{6}$$

As a consequence of this, for hard power-law potentials, defined as the ones for which $s > 5$, $\nu(|\xi|)$ is monotonely increasing from $\nu_o = \nu(o)$ to infinity, while for power-law soft potentials, defined as the ones for which $s < 5$, $\nu(|\xi|)$ is monotonely decreasing from $\nu(o) = \nu_o$ to 0. For power-law hard potentials it is a simple matter to prove the existence of two positive constants a and b such that

$$a(1 + |\xi|^2)^{\gamma/2} \leqslant \nu(|\xi|) \leqslant b(1 + |\xi|^2)^{\gamma/2} \;. \tag{7}$$

We observe that for $s = 5$ (Maxwell angular cut-off), ν is a constant.

1.5.3. Properties of the Operator A

1.5.3.1. The first statement, the proof of which may be found in Grad [5] pages 35-36 is that

$$Af = A_2 f - A_1 f \tag{1}$$

with the property that

(2) $$f \geqslant 0 \Rightarrow A_2 f \geqslant 0 \text{ and } A_1 f \geqslant 0,$$

whenever $B(\theta, V) \geqslant 0$ which holds for repulsive interactions. A second statement which is proven in the same place is that A_2 and A_1 have explicit representations in terms of kernels, namely

(3) $$A_{1,2} f(\xi) = \int_{R^3} A_{1,2}(\xi, \zeta) f(\zeta) \, d\mu$$

with

(4) $$A_1(\xi, \zeta) = 2\pi \int_{R^3}^{\pi/2} B(\theta, |\xi - \zeta|) \, d\theta$$

as it is readily got from 1.4.1.3.(5). It is less simple to compute the kernel A_2 from 1.4.1.3.(5) and we merely state the result because the proof is easily accessible in the reference quoted previously. We define $Q(v, w)$ as

(5) $$Q(V \cos\theta, V \sin\theta) = |\sin\theta|^{-1} B^*(\theta, V)$$

where

(6) $$B^*(\theta, V) = \begin{cases} \frac{1}{2}\{B(\theta, V) + B(\frac{\pi}{2} - \theta, V)\} & \text{if } 0 \leqslant \theta \leqslant \pi/2 \\ B^*(\pi - \theta, V) & \text{if } \pi/2 < \theta < \pi \end{cases}$$

then referring to 1.1.2.2. we set

(7) $$\xi' = \xi + V, \quad \xi_1 = \xi + V + W, \quad \zeta = V + \xi, \quad \eta = \frac{1}{2}(\xi + \zeta)$$

and we have

$$A_2(\xi, \zeta) = 2(2\pi)^{-3/2} \, |v|^{-2} \, \omega^{-1/2}(|\xi|) \, \omega^{1/2}(|\zeta|) \exp(-\frac{1}{8}|v|^2) \, .$$

$$\int_{\Pi(v)} \exp\{-\frac{1}{2}|w + \zeta|^2\} \, Q(|v|, |w|) \, dw \tag{8}$$

where $\Pi(v)$ stands for the plane perpendicular to v.

1.5.3.2. We state the expression of the kernels for rigid spheres

$$\omega^{1/2}(|\xi|) \, A_1(\xi, \zeta) \, \omega^{-1/2}(|\zeta|) = \frac{\sigma^2}{2\sqrt{2\pi}} \, |\xi - \zeta| \exp\{-\frac{1}{4}(|\xi|^2 + |\zeta|^2)\} \, , \tag{9}$$

$$\omega^{1/2}(|\xi| \, A_2(\xi, \zeta) \, \omega^{-1/2}(|\zeta|) = \frac{\sigma^2}{2\sqrt{2\pi}} |\xi - \zeta|^{-1} \exp\{-\frac{1}{8}|\xi - \zeta|^2 - \frac{1}{8}\frac{(|\zeta|^2 - |\xi|^2)^2}{|\zeta - \xi|^2}\} \, . \tag{10}$$

1.5.3.3. For any power-law angular cut-off, and rigid sphere, the operator A is bounded in H, namely

$$\| Af \|_H \leqslant k'_o \, \| f \|_H \tag{1}$$

with some positive constant k'_o. Drange[7] has proven that this result no longer holds for the two cut-offs of 1.1.4.4. and 1.1.4.5.

1.5.3.4. Grad[5] (pages 46-49) has proven that A is compact in H for power law angular cut-off and rigid sphere. The proof has been simplified by Drange[7] . From 1.5.3.3., a fortioré, the result does not extend to the other cut-offs.

1.5.3.5. In order to solve the equation

$$\nu f - Af = g \tag{1}$$

by the Fredholm alternative one writes

(2) $$\hat{f} = \nu^{1/2}\, f \qquad \hat{g} = \nu^{-1/2}\, g$$

(3) $$\hat{A} = \nu^{-1/2}\, A \nu^{-1/2}$$

and one sets

(4) $$\hat{f} = \hat{A}\hat{f} = \hat{g}$$

Compactness of A in H allows to apply the Fredholm alternative and to conclude that (3) has a unique solution in the sub-space orthogonal to the collisional invariants if g belongs to this sub-space. As a matter of fact $f = A\,f$ implies that $Lf = 0$ which does not hold in this sub-space. For hard power law angular cut-off and rigid sphere compactness of $\hat{A}$ in H is a consequence of that of A. Compactness of $\hat{A}$ in H for soft power law potentials has been proven by Drange [7]. For the two other cut-offs, $\hat{A}$ is shown to be bounded by Drange [7] but compactness is an open question. Cercignani [6] has considered

(5) $$A^{(\sigma)} = \nu^{-\sigma}\, A\, \nu^{-\sigma}$$

and stated that $A^{(\sigma)}$ is compact for $\sigma \geqslant 2$. Drange [7] has shown that compactness holds for $\sigma \geqslant 1/2$ boundedness for $\sigma \geqslant 1/2$, and unboundedness for $\sigma < 1/2$.

1.5.3.6. For a study of the nonlinear Boltzmann equation a maximum norm is useful. We shall consider a space of continuous functions (of ξ) that we call B_r with the norm

(1) $$<f>_r = \sup_{\xi \in R^3} \omega^{1/2}(|\xi|)(1 + |\xi|^2)^{r/2} |f(\xi)|\,.$$

It may prove convenient to deal with a space of measurable functions, then. in (1) the supremum has to be replaced by the essential supremum. It has been shown by Grad[5] (page 42-44) that for any power-law angular cut-off and rigid spheres, A_1 is bounded from H to B_o and from B_r to B_{r+1}, namely

$$< f >_o \leq k_o \| f \|_H , \tag{2}$$

$$< Af >_r \leq k_r < f >_{r-1} , \qquad r \geq 1 . \tag{3}$$

This holds in the space of measurable as well as the one of continuous functions.

1.5.4. Properties of the Operator L

1.5.4.1. We examine L as an operator from H into itself. We assume an angular or rigid sphere cut-off, then A is bounded, thus the domain of L is the same as the one of ν, namely

$$\text{Domain of L}: \ \{ f \,|\, f \in H, \ \nu^{1/2} f \in H \} . \tag{1}$$

Through a standard argument Schechter[19] (page 58 and 249), the symmetry 1.4.3.2.(1) of L implies from (1) that it is a self-adjoint operator.

1.5.4.2. Again for angular cut-off or rigid sphere the spectrum of L, from self adjointness, is real and from 1.4.3.2.(2) it is restricted to the positive real axis plus the origin. The origin belongs to the point spectrum, being an eigenvalue of multiplicity five with the space of collisional invariants as a corresponding eigenspace. To investigate further the spectrum we consider the operator ν. The spectral properties of ν are evident from the consideration of the equation

$$[\sigma - \nu(|\xi|)] \ f(\xi) = g(\xi) . \tag{1}$$

It is clear that (1) is uniquely selvable in H with a bounded $(\sigma - \nu)^{-1}$ in H provided that σ does not belong to the range of values of ν, that is $[\nu_o, +\infty[$ for a hard potential, $[0, \nu_o]$ for a soft one. Any σ within the range of ν belongs to the spectrum but it is not an eigenvalue because $(\sigma - \nu)f = 0$ implies that $f = 0$

almost everywhere. The spectrum of ν is a pure continuous spectrum. As a matter of fact, the set of g for which (1) is solvable (uniquely) in H does not coincide with all of H but is in it. As a matter of fact, let φ be (arbitrary) in H and σ belong to the range of values of ν. We define $\varphi_{\sigma,\epsilon}$ to be equal to φ when $|\nu-\sigma| \geqslant \epsilon$ and zero everywhere. Evidently it belongs to the subset of H for which (1) is solvable in H and this holds for every $\epsilon > 0$. But the measure (with respect to $\omega d\xi$) of the set where $|\nu-\sigma| < \epsilon$ tends to zero when $\epsilon \to 0$ and, as a consequence, for fixed φ, $\|\varphi - \varphi_{\sigma,\epsilon}\|_H \to 0$.

1.5.4.3. Consider again an angular cut-off or rigid sphere, we know by 1.5.3.4. that A is compact, and this has a far reaching consequence concerning the spectrum of L. The point spectrum of L is the set of σ for which $(\sigma 1 - L) f = 0$ has a non zero solution H. Let σ belong to the spectrum but not to the point spectrum and assume that the range of $\sigma 1 - L$ is not dense in H then there exist some $g \in L$ such that, for any $f \in H$ we have $0 = ((\sigma 1 - L)f, g)_H = (f, (\sigma 1 - L)g)_H$ but $\sigma \epsilon$ spectrum so that $\sigma = \bar{\sigma}$ and as σ does not belong to the point spectrum, $(\sigma 1 - L)g$ is not zero if g is not, leading to a contradiction. As a consequence, the range of $\sigma 1 - L$ is dense in H. On the other hand $(\sigma 1 - L)^{-1}$ which is (well) defined in a dense subset of H, cannot be continuous otherwise it could be extended by continuity to the whole of H leading to a contradiction with the assumption that σ belongs to the spectrum of L and that the range of $\sigma 1 - L$ not the whole of H. As a consequence the spectrum of L consists in a point spectrum plus a continuous spectrum, consisting in a set of values of σ for which the kernel of $\sigma 1 - L$ is void while its range is dense in H but distinct from the whole of H. From Riesz and Nagy [20](page 364 and 361) the continuous spectrum of L plus the set of limit points of the point spectrum is identical with the range of values of $\nu(|\xi|)$. The same same result may be obtained from Kato[21] (theorem IV 5.35 page 244) with the remark that, for self adjoint operators the essential spectrum consists in the part of the spectrum which is not reduced to the point spectrum by restricting the point spectrum to the set of isolated eigenvalues with a finite multiplicity.

1.5.4.4. Consider further an angular cut-off or rigid sphere, and restrict ourselves to hard potentials for which $\nu(|\xi|)$ is an increasing function of $|f|$ ranging from $\nu_0 > 0$ to $+\infty$. From 0 to ν_0 the spectrum of L consists only of isolated eigenvalues with finite multiplicty. Zero is an eigenvalue of multiplicity five with, as eigenspace, the space of collisional invariants. Let μ be the next closest to zero eigenvalue or ν_0 if the rest of the point spectrum is embedded within the continuous spectrum, we have

$$(Lf, f)_H \geqslant \mu \, \| f \|^2 \quad \text{if} \quad (f, \psi_\alpha)_H = 0, \qquad \alpha = 0,1,2,3,4. \tag{1}$$

where the ψ_α are the five collisional invariants of 1.4.2.2. The proof rests on the existence of a spectral measure (refer to Riesz, Nagy[20] pages 312-318) $d\,\epsilon\,(\lambda)$ over the positive real axis, such that, for any measurable set a of the real axis

$$E(a)f = \int dE(\lambda) \quad f = 0\,, \qquad \text{if } a \;\eta\; \text{spectrum of } L = \phi\,. \tag{2}$$

For every a having a non void intersection with the spectrum $E(a)$ is a continuous projection depending only on the intersection of a with the spectrum. For $a \cap B = \phi$ we have $E(a) \cdot E(a) = E(B) \cdot E(a) = 0$. When a consists of an eigenvalue $E(a)$ is the orthogonal projection on the corresponding eigenspace. As an example

$$E(o)\, f = \sum_{\alpha=0}^{4} (f, \psi_\alpha)\; \psi_\alpha\,. \tag{3}$$

on the other hand, if Σ stands for the whole of the spectrum,

$$E(\Sigma)\, f = f\,, \tag{4}$$

$$\int_\Sigma \lambda \, dE(\lambda)\, f = Lf\,. \tag{5}$$

From (5) we conclude that

$$(Lf, f)_H = \int_{\mu-0}^{+\infty} \lambda (dE(\lambda)\, f, f)_H \geqslant \mu \int_{\mu-0}^{\infty} (d\,E(\lambda)\, f, f)_H\,. \tag{6}$$

whenever (f, ψ_α) 0, $\alpha = 0, 1, 2, 3, 4$, we have, from

(7) $$f = \int_{\Sigma} dE(\lambda)\, f = \sum_{\alpha=0}^{4} (f, \psi_\alpha)\, \psi_\alpha + \int_{\mu-0}^{+\infty} dE(\lambda)\, f \,,$$

that (1) holds.

We set

$$Pf = \int_{\mu=0}^{+\infty} dE(\lambda)\, f \,, \qquad QF = \sum_{\alpha=0}^{4} (f, \psi_\alpha)\, \psi_\alpha \,,$$

with the properties

(9) $$P^2 f = Pf \,, \quad Q^2 f = Qf \,, \quad PQf = QPf = 0 \,, \quad f = Pf + Qf \,,$$

holding for any $f \in H$. We may reset (1) as

(10) $$(Lf, f) \geqslant \mu \, \| Pf \|^2_H \,.$$

1.5.4.5. Cercignani[6] has proven that the basic property 1.5.4.4.(10) holds for radial cut-off although A is then not bounded and a fortori not compact, so that the previous method of proof does not work. This result proves that for radial cut-off, zero is an isolated eigenvalue separated from the rest of the spectrum. There is the question about what is the domain of L for radial cut-off.

CHAPTER 2

THE LINEAR BOUNDARY VALUE PROBLEM

2.1 Statement of the Interior B-V Problem*. The Method of Approach

2.1.1. Statement of the Problem. A Few Remarks

2.1.1.1. We consider a bounded domain Ω in three dimensional euclidean space, with a smooth boundary $\partial\Omega$. We make the assumption that $\partial\Omega$ is a wall and that to each point $x \in \partial\Omega$ and each time t, there corresponds a reflection operator $K_{t,x}$. We further assume that $|a_w - 1| + |u_w|$ is small and, consequently, we linearize the reflection operator according to 1.4.4.3. and neglect g, f setting $\phi^+ = G_1 1$. Defining B as the operator which, for continuous functions in $\Omega \times R^3$ associates their restrictions to $\partial\Omega \times R^3$ and the operators J^+, J^- defined as

$$\begin{cases} J^+ f(x, \xi) = f(x, \xi)\,, & \text{if } \xi \cdot n(x) \geq 0\,, \quad x \in \partial\Omega\,, \\ J^+ f(x, \xi) = 0 & \text{if } \xi \cdot n(x) < 0\,, \quad x \in \partial\Omega\,, \end{cases} \tag{1}$$

$$\begin{cases} J^- f(x, \xi) = 0\,, & \text{if } \xi \cdot n(x) > 0\,, \quad x \in \partial\Omega\,. \\ J^- f(x, \xi) = f(x, \xi)\,, & \text{if } \xi \cdot n(x) \leq 0\,, \quad x \in \partial\Omega \end{cases} \tag{2}$$

We state the boundary condition imposed to f as

$$J^+ Bf = G J^- Bf + \phi^+ \tag{3}$$

2.1.1.2. Linearizing the equation as in 1.4.1.1. and restricting attention to the steady case we are lead to the linear equation

*) Our treatment of the LIBVP is the same as in Guiraud [18].

$$\xi \cdot \nabla f + Lf = \varphi \, . \tag{1}$$

For later application to the non-linear problem we consider φ as an arbitrary known function. Later we shall replace φ by Q(f, f) and try to solve the non-linear problem as the research of a fixed point in a Banach space.

2.1.1.3. With the notations of 1.4.6.1. our basic problem may be stated as: given φ and ϕ^+ in suitable function spaces, to find f in another suitable function space such that

$$\text{(1)} \quad \begin{cases} \mathcal{L}f = \varphi \, , \\ J^+ Bf = G J^- Bf + \phi^+ \, . \end{cases}$$

2.1.1.4. From the discussion of 1.4.6. and especially the relation 1.4.6.6.(1) we find some interest in investigating the boundary value problem adjoint to 2.1.1.3.(1), namely to find f such that

$$\text{(1)} \quad \begin{cases} \mathcal{L}^* f = \varphi \, , \\ J^- Bf = G^* J^+ Bf + \phi^- \, . \end{cases}$$

2.1.1.5. A word of caution is in order about the discussion of 2.1.1.1. We cannot expect having a steady solution if Ω is not independent of time and, then, we must have $\mathbf{u}_w \cdot \mathbf{n} = 0$, that is the wall may move, but provided it is only gliding along itself.

2.1.1.6. We observe that any constant is a solution of 2.1.1.3.(1) with $\varphi = 0$ and $\phi^+ = 0$ and the same holds for 2.1.1.4.(1) with $\varphi = 0$ again and $\phi^- = 0$. Uniqueness, then, is not expected. Consequently, existence is not expected either, for quite arbitrary φ and ϕ^+ or φ and ϕ^-. This is easily visualized through use of 1.4.6.6.(1) where we set $g = 1$ and a solution of 2.1.1.3.(1) for f, getting

$$((\varphi, 1)) + \{\phi^+, 1^+\} = 0 \tag{1}$$

a relation which amounts to be a compatibility condition on the data of 2.1.1.3.(1) for existence to hold. The companion condition for 2.1.1.4.(1) is derived from 1.4.6.6.(1) as

$$((\varphi, 1)) - \{1^-, \phi^-\} = 0 \tag{2}$$

2.1.1.7. From 1.4.6.1.(5) we have, for a piecewise continuous f, continuously differentiable, along rays, with respect to x,

$$2\mathrm{Re}\,((\xi \cdot \nabla f, f)) + \int_{\partial\Omega} dS \int_{R^3} \omega\, \xi \cdot n\, |Bf|^2\, d\xi = 0\,, \tag{1}$$

and, as a consequence

$$((\mathcal{L}f, f)) = ((Lf, f)) - \frac{1}{2} \int_{\partial\Omega} dS \int_{R^3} \omega\, \xi \cdot n\, |Bf|^2\, d\xi + i\,\mathrm{Im}\,((\xi \cdot \nabla f, f)). \tag{2}$$

Now we use 1.4.5.8.(5) and 1.5.4.4.(10) and we obtain the basic inequality

$$\begin{aligned} \mu|||\, Pf\, |||^2 + k [\![\Pi J^- Bf]\!]^2 \leqslant{}& |||\, \mathcal{L}f\, |||\ |||\, f\, ||| + \frac{1}{2} [\![J^+Bf - GJ^- Bf]\!]^2 + \\ & \mathrm{Re}\{ J^+ Bf - G J^- Bf, G\Pi J^- Bf \} + \\ & + \sqrt{2\pi} \int_{\partial\Omega} \mathrm{Re}(J^+Bf - G J^- Bf, 1^+)_{\hat{H}}\ \overline{(J^- Bf, 1^-)_{\hat{H}}}\ dS. \end{aligned} \tag{3}$$

Using 1.4.6.3.(9) we obtain

$$\mu |||\, Pf \,|||^2 + k [\![\Pi J^+ Bf |||^2 \leqslant |||\, \mathcal{L}^* f \,|||\ |||\, f \,||| + \frac{1}{2} [\![J^- Bf - G^* J^+ Bf]\!]^2 +$$

$$(4) \qquad + \operatorname{Re} \{\{ J^- Bf - G^* J^+ Bf, G^* \Pi J^+ Bf \}\} +$$

$$+ \sqrt{2\pi} \int_{\partial\Omega} \operatorname{Re} (J^- Bf - G^* J^+ Bf, 1^-)_{\tilde{H}} \overline{(J^+ Bf, 1^+)_{\tilde{H}}} \, dS \, .$$

Without any new computation we have

$$\nu_0 |||\, f \,|||^2 + k [\![\Pi J^- Bf]\!]^2 \leqslant |||\mathcal{M} f|||\ |||\, f \,||| + \frac{1}{2} [\![J^+ Bf - G J^- Bf]\!]^2$$

$$(5) \qquad + \operatorname{Re} \{ J^+ Bf - G J^- Bf, G \Pi J^- Bf \} +$$

$$+ \sqrt{2\pi} \int_{\partial\Omega} \operatorname{Re} (J^+ Bf - G J^- Bf, 1^+)_{\tilde{H}} \overline{(J^- Bf, 1^-)_{\tilde{H}}} \, dS \, ,$$

$$\nu_0 |||\, f \,|||^2 + k [\![\Pi J^+ Bf]\!]^2 \leqslant |||\mathcal{M}^* f|||\ |||\, f \,||| + \frac{1}{2} [\![J^- Bf - G^* J^+ Bf]\!]^2 +$$

$$(6) \qquad + \operatorname{Re} \{ J^- Bf - G^* J^+ Bf, G^* \Pi J^+ Bf \} +$$

$$+ \sqrt{2\pi} \int_{\partial\Omega} \operatorname{Re} (J^- Bf - G^* J^+ Bf, 1^-)_{\tilde{H}} \overline{(J^+ Bf, 1^+)_{\tilde{H}}} \, dS.$$

2.1.1.8. From 2.1.1.7. we may examine the problem 2.1.1.3.(1) and 2.1.1.4.(1) with respect to uniqueness. Consider 2.1.1.3.(1) with $\varphi = 0$, $\phi^+ = 0$ and assume that this homogeneous problem has a solution f, which is piecewise continuous and continuously differentiable, along rays, with respect to x, we obtain, from 2.1.1.7.(3)

$$|||Pf||| = [\![\Pi J^- Bf]\!] = 0 . \tag{1}$$

From $\mathcal{L}f = 0$ we have, thanks to $\mathcal{L}f = LPf$, that $\xi \cdot \nabla f = 0$. This means that f is a combination of collisional invariants which is independent of x. Then $[\![\Pi J^- Bf]\!] = 0$ implies that f is a constant. We state

$$\mathcal{L}f = 0 , \quad J^+Bf = G J^- Bf \Rightarrow f = \text{const} , \tag{2}$$

$$\mathcal{L}^*f = 0 , \quad J^- Bf = G^* J^+ Bf \Rightarrow f = \text{const.} \tag{3}$$

2.1.2. The Method of Approach

2.1.2.1. We consider the problem of finding f such that

$$\mathcal{M}f = \varphi , \qquad J^+Bf = \phi^+ , \tag{1}$$

with φ and ϕ^+ given as continuous functions. We set

$$x = y + su , \qquad \xi = |\xi| u , \tag{2}$$

with $y \in \partial\Omega$ and $u \cdot n_y > 0$. We write

$$f(y + su, \xi) = \tilde{f}(s, y, \xi) \tag{3}$$

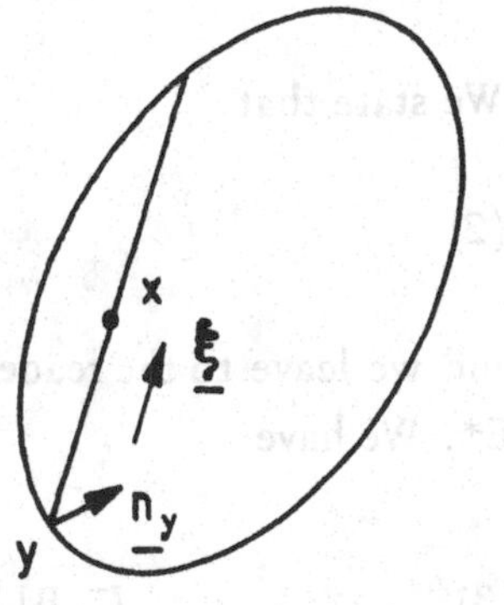

and (1) may be rewritten as

$$\xi \frac{\partial \tilde{f}}{\partial s} + \nu \tilde{f} = \varphi , \tag{4}$$

$$f(0, y, \xi) = \phi^+(y, \xi) . \tag{5}$$

For a fixed ξ , we integrate (4) as an ordinary differential equation with (5) as an initial condition and we find

$$(6)\qquad f(s, y, \xi) = \frac{1}{|\xi|} \int_0^s e^{-\frac{\nu(|\xi|)(s-\sigma)}{|\xi|}} \varphi(\sigma, y, \xi)\, d\sigma + \phi^+(y, \xi)\, e^{-\frac{\nu(|\xi|)s}{|\xi|}} .$$

We rewrite (6) as

$$(7)\qquad f = U\varphi + E\phi^+,$$

with well defined operators U and E. We shall extend later the definition of U and E to some suitably chosen function spaces. We have

$$(8)\qquad J^+ BU\varphi = 0 \qquad J^+ B E \phi^+ = \phi^+$$

2.1.2.2. For

$$(1)\qquad \begin{cases} \mathcal{M}^* f = \varphi , \\ J^- Bf = \psi^- , \end{cases}$$

We state that

$$(2)\qquad f = U^* \varphi + E^* \psi^- ,$$

and we leave to the reader the task of writing an explicit representation for U* and E*. We have

$$(3)\qquad J^- BU^* \varphi = 0 , \qquad J^- B E^* \psi^- = \psi^- .$$

2.1.2.3. Consider the problem

$$(1)\qquad \mathcal{M} f = \varphi , \qquad J^+ Bf = J^- Bf + \phi^+$$

and try to solve it by considering $f^+ = J^+ B f$ as an intermediate unknown, we get

$$f = U\varphi + E f^+ . \tag{2}$$

From the second of (1) and defining an operator $\mathcal{V}$ as

$$\mathcal{V} = G \mathcal{J}^- BE , \tag{3}$$

we have, for f^+, the equation

$$(1 - \mathcal{V})\, f^+ = G \mathcal{J}^-\, BU\varphi + \phi^+ . \tag{4}$$

2.1.2.4. Assume that the equation

$$(1 - \mathcal{V})\, f^+ = \psi^+ \tag{1}$$

is uniquely solvable as

$$f^+ = T\, \psi^+ \ , \tag{2}$$

we obtain, from 2.1.2.3

$$f = U\varphi + E\ T(G \mathcal{J}^-\ BU\varphi + \phi^+) \tag{3}$$

so that the solution of 2.1.2.1.(1) may be written as

$$f = \mathcal{W}\varphi + \mathcal{E}\, \phi^+ , \tag{4}$$

with two operators $\mathcal{W}$ and $\mathcal{E}$ defined by

(5) $$\mathcal{W} = \mathrm{U} + \mathrm{E\,T\,G}\,J^{-}\,\mathrm{B\,U} \ ,$$

(6) $$\mathcal{E} = \mathrm{E\,T} \ .$$

2.1.2.5. We state the definitions

(1) $$\mathcal{V}^{*} = \mathrm{G}^{*}\,J^{+}\,\mathrm{B\,E}^{*} \ ,$$

(2) $$(1-\mathcal{V}^{*})^{-1} = \mathrm{T}^{*} \ ,$$

(3) $$\mathcal{W}^{*} = \mathrm{U}^{*} + \mathrm{E}^{*}\,\mathrm{T}^{\phi}\,\mathrm{G}^{*}\,J^{+}\,\mathrm{B\,U}^{*} \ ,$$

(4) $$\mathcal{E}^{*} = \mathrm{E}^{*}\,\mathrm{T}^{*} \ ,$$

and the result that

(5) $$\mathrm{f} = \mathcal{W}^{*}\varphi + \mathcal{E}^{*}\,\psi^{-} \ ,$$

solves, formally,

(6) $$\mathcal{M}^{*}\mathrm{f} = \varphi \ , \quad J^{-}\,\mathrm{Bf} = \mathrm{G}^{*}\,J^{+}\,\mathrm{Bf} + \psi^{-} \ .$$

2.1.2.6. Consider

(1) $$\mathrm{f} = \mathcal{W}\varphi + \mathcal{E}\,\phi^{+} \ ,$$

(2) $$\mathrm{g} = \mathcal{W}^{*}\chi + \mathcal{E}^{*}\,\psi^{-} \ ,$$

we have

$$\mathcal{M}f = \varphi \ , \qquad J^+Bf - G\,J^-Bf = \phi^+ \ , \tag{3}$$

$$\mathcal{M}^*g = \chi \ , \qquad J^-Bf - G^*\,J^+Bf = \psi^- \ . \tag{4}$$

For continuous $\varphi, \phi^+, \chi, \psi^+$, we have piecewise continuous, and continuously differentiable (along rays) with respect to x, f and g, to which we may apply 1.4.6.6.(2) which gives

$$\begin{aligned}&((\varphi, W^*\chi)) - ((W\varphi, \chi)) + ((\varphi, \mathcal{E}^*\psi^-)) - ((\mathcal{E}\,\phi^+, \chi)) + \\ &+ \{\phi^+, J^+BW^*\chi\} - \{J^-BW\varphi, \psi^-\} \\ &+ \{\phi^+, J^+B\mathcal{E}^*\psi^-\} - \{J^-B\mathcal{E}\phi^+, \psi^-\} = 0\end{aligned} \tag{5}$$

2.1.2.7. Our strategy to solve 2.1.1.3.(1) will be to set

$$f = \tilde{f} + \mathcal{E}\phi^+ \ , \tag{1}$$

from which

$$\mathcal{L}f = \mathcal{L}\tilde{f} + (\mathcal{M} - A)\,\mathcal{E}\phi^+ \ , \tag{2}$$

and

$$\mathcal{L}\tilde{f} = \varphi + A\mathcal{E}\phi^+ \ , \qquad J^+B\tilde{f} = G\,J^-Bf \ , \tag{3}$$

so that it will be sufficient to solve 2.1.1.3.(1) with $\phi^+ = 0$. As a consequence if we can define an operator $\mathcal{T}$ such that

(4) $((\varphi, 1)) = 0 \Rightarrow f = \mathcal{T}\varphi$ solves $\mathcal{L}f = \varphi$, $J^+Bf = GJ^-Bf$, $((f, 1)) = 0$,

the solution of 2.1.1.3.(1), provided 2.1.1.6.(1) holds will be

(5) $$f = \mathcal{T}(\varphi + A\,\&\,\phi^+) + \&\phi^+ + \text{const.}$$

To be consistent we must prove that

(6) $$((\varphi, 1)) + \{\phi^+, 1^+\} = 0 \Rightarrow ((\varphi + A\&\phi^+, 1)) = 0\,.$$

Taking into account the symmetry of A and the, obvious, relation $A1 = \nu$, we see that it is sufficient to prove

(7) $$((\&\phi^+, \nu)) = \{\phi^+, 1^+\}\,.$$

The proof relies on 2.1.2.6.(5) with $\varphi = \psi^- = 0$, $\chi = \nu$, which gives

(8) $$((\&\phi^+, \nu)) = \{\phi^+, J^+BW^*\nu\}\,,$$

and the remark that

(9) $$W^*\nu = 1\,.$$

As a matter of fact

(10) $$\mathcal{M}^*1 = \nu\,, \qquad J^+B1 - GJ^-B1 = 0\,,$$

and the solution of 2.1.2.5.(6) is unique within the class of continuous, continuously differentiable, along rays, with respect to x, functions. This last statement is a consequence of 2.1.1.7.(6).

2.1.3. Function Spaces. Properties of U, E, U* *and* E*

2.1.3.1. Our basic function space will be $L^2(\Omega, H)$. It consists of (complex) functions $f(x, \xi)$ defined almost everywhere in $\Omega \times R^3$ which are measurable and square integrable with respect to the measure $\omega\, d\xi\, dx$. As a norm we use

$$|||f||| = \{ \int_\Omega dx \int_{R^3} d\xi\, \omega(|\xi|)\, |f(x,\xi)|^2 \}^{1/2} . \tag{1}$$

For almost every x, $f(x, \cdot)$ is a well defined element of the hilbert space H and $x \to f(x, \cdot)$ is measurable and square integrable with

$$|||f|||^2 = \int \| f(x, \cdot) \|_H^2 \, dx . \tag{2}$$

This explains the notation $L^2(\Omega, H)$. We convert this space to a hilbert space by using 1.4.6.1.(3) as a scalar product.

2.1.3.2. For functions on $\partial\Omega \times R^3$ we use essentially two basic function spaces, namely $L^2(\partial\Omega, H)$ and $L^2(\partial\Omega, \tilde{H})$. The definition goes like in 2.1.3.1. with $\partial\Omega$ standing for Ω and the surface measure on $\partial\Omega$ standing for the volume one. As a scalar product we use

$$(f, g)_{L^2(\partial\Omega, X)} = \int_{\partial\Omega} (f, y)_X \, ds , \quad X = H \text{ or } \tilde{H} , \tag{1}$$

and for a norm

$$\| f \|_{L^2(\partial\Omega, H)} = \{ \int_{\partial\Omega} \| f \|_H^2 \, ds \}^{1/2} . \tag{2}$$

$$\| f \|_{L^2(\partial\Omega, \tilde{H})} = [\![f]\!] = \{ \int_{\partial\Omega} \| f \|^2_{\tilde{H}} \, dS \}^{1/2} . \tag{3}$$

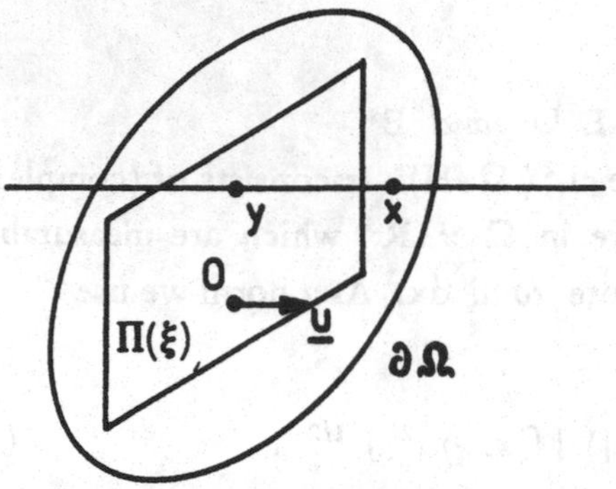

2.1.3.3. Let us consider a continuous f, and let us extend it to a piecewise continous $\tilde{f}$ on $E^3 \times R^3$ by setting $\tilde{f} = f$ on Ω and $f = 0$ outside of Ω. We define $\tilde{U}$ the extension of U such that

(1) $$1_\Omega \tilde{U} \tilde{f} = Uf \,,$$

with the convention that $1_\Omega(x)$ is the characteristic function of Ω. To any $g(x, \xi)$ defined on $E^3 \times R^3$ and piecewise continuous we associate

(2) $$\omega_{\xi,y}\, g(s) = g(y + su, \xi) \,, \quad \xi = |\xi| u \,, \quad y \in \Pi(\xi) \,,$$

where $\Pi(\xi)$ is the plane through the origin which is orthogonal to ξ. Define

(3) $$e_\xi(s) = \begin{cases} |\xi|^{-1} \exp[-|\xi|^{-1} \nu(|\xi|)\, s] & \text{for } s > 0 \\ 0 & \text{for } s < 0 \end{cases}$$

We observe, from the very definition of $\tilde{U}$ that

(4) $$\tilde{U} g(x, \xi) = (e_\xi * \omega_{\xi,y}\, g)(s) \,,$$

where $*$ means a convolution on the real line. Then

(5) $$Uf = 1_\Omega (e_\xi * \omega_{\xi,y}\, f) \,.$$

obviously we have

$$||| \nu U f |||^2 \leqslant \int_{R^3} \nu^2 \omega \, d\xi \int_{\Pi(\xi)} dS_y \, \| e_\xi * \omega_{\xi,y} f \|^2_{L^2(R)} \; . \tag{6}$$

We use a basic theorem of integration theory that if $f \in L^1(R)$, $g \in L^2(R)$, the convolution $f * g \in L^2(R)$ and

$$\| f * g \|_{L^2(R)} \leqslant \| f \|_{L^1(R)} \, \| g \|_{L^2(R)} \; . \tag{7}$$

The reader may find the statement and a proof in the Dunford-Schwartz[24] vol. 2 page 951-953. We set, in (7), $f = e_\xi$, $g = \omega_{\xi,y} f$ and we observe that

$$| e_\xi |_{L^1(R)} = \nu^{-1}(|\xi|) \tag{8}$$

then

$$\begin{aligned} ||| \nu U f |||^2 &\leqslant \int_{R^3} \omega \, d\xi \int_{\Pi(\xi)} dS_y \, \| \omega_{\xi,y} \tilde{f} \|^2_{L^2(R)} \\ &= \int_{R^3} \omega \, d\xi \int_{R^3} | \tilde{f}(x,\xi) |^2 \, dx = ||| f |||^2 \; . \end{aligned} \tag{9}$$

Now let $f \in L^2(\Omega, H)$, we can find a sequence $\{ f_n \}$ such that each f_n is continuous and belongs to $L^2(R, H)$, then $\{ \nu U f_n \}$ is a cauchy sequence in $L^2(R, H)$ and converges to a unique element of $L^2(R, H)$ which we can write as $\nu U f$. We conclude that U may be extended to a bounded operator* in $L^2(R, H)$ and that νU is even contradicting

$$||| \nu U f ||| \leqslant ||| f ||| \; . \tag{10}$$

*) We assume ν bounded away from zero.

The same type of reasoning applies to U^* which may be extended to a bounded operator in $L^2(\Omega, H)$ such that even νU^* is contracting

(11) $$|||\nu U^* f||| \leqslant |||f|||\ .$$

2.1.3.4. For a continuous f, Uf is continuously differentiable along rays and we have

(1) $$\xi \cdot \nabla Uf + Uf = f\ ,$$

so that

(2) $$|||\xi \cdot \nabla Uf||| \leqslant 2\,|||f|||\ .$$

From the reasoning of 2.1.3.3. we see that Uf admits $\xi \cdot \nabla$ in a strong sense for any $f \in L^2(\Omega, H)$ and that (2) holds for any such f. Again U^*f admits $\xi \cdot \nabla$ in a strong sense and

(3) $$|||\xi \cdot \nabla U^* f||| \leqslant 2\,|||f|||\ .$$

2.1.3.5. Let f be continuous and refer to the figure, we have

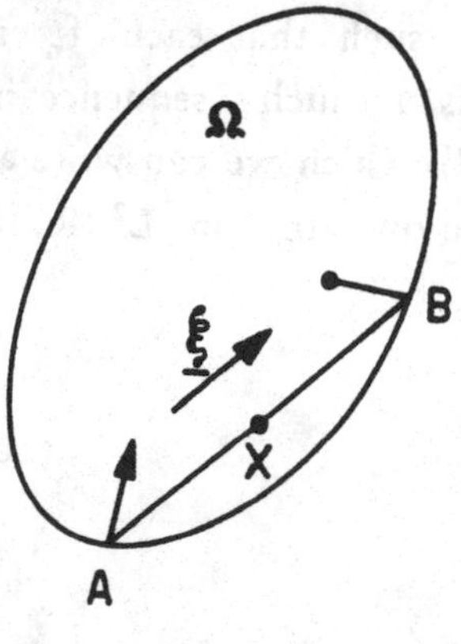

(1) $$|Uf|_B^2 = \left|\int_A^B |\xi|^{-1} e^{-\nu|x_B - x|/|\xi|} f(x, \xi)\, d\sigma\right|^2$$

with $x = A + \sigma u$, $\xi = |\xi|\, u$. From Schwarz inequality

$$|Uf(x_B,\xi)|^2 \leqslant \left(\int_A^B |\xi|^{-2}\, e^{-2\nu|x_B-x|/|\xi|}\, d\sigma\right)\left(\int_A^B |f(x,\xi)|^2\, d\sigma\right) \tag{2}$$

and as a consequence

$$|J^- BUf(x,\xi)|^2 \leqslant \frac{1}{2\nu|\xi|} \int_y^x |f(x,\xi)|^2\, d\sigma_z\, . \tag{3}$$

Now from the definition

$$[\![\nu^{1/2} J^- BUf]\!]^2 = \int_{\partial\Omega} dS_x \int_{R^3_-} \omega\nu|\xi\cdot n_x|\, |J^- BUf(x,\xi)|^2\, d\xi \tag{4}$$

and using (3)

$$[\![\nu^{1/2} J^- BUf]\!]^2 \leqslant \frac{1}{2} \int_{\partial\Omega} \cos\vartheta_x\, dS_x \int_{R^3_-} \omega\, d\xi \int_y^x |f(z,\xi)|^2\, d\sigma_z\, . \tag{5}$$

We interchange the integrations and, to do this, we define $(\partial\Omega)^+_{(\xi)}$ as that part of $\partial\Omega$ for which ξ is pointing towards the interior, that is the set of points $x \in \partial\Omega$ for which $\xi\cdot n_x > 0$. In the same way $(\partial\Omega^-)(\xi)$ is that part of $\partial\Omega$ for which $-\xi$ is pointing towards the interior. As a consequence

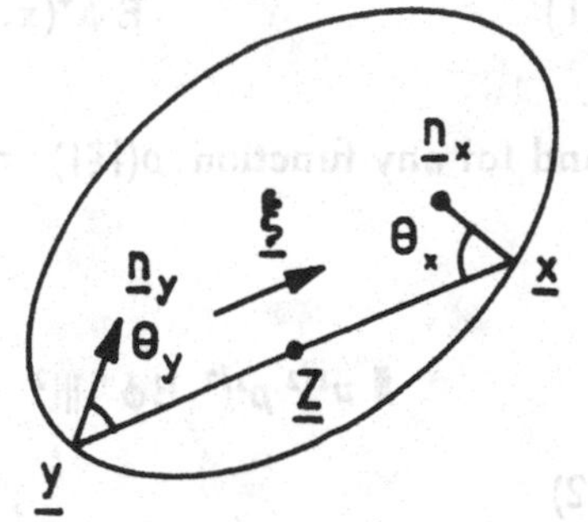

$$[\![\nu^{1/2} J^- BUf]\!]^2 \leqslant \frac{1}{2} \int_{R^3} \omega\, d\xi \int_{(\partial\Omega)^-(\xi)} \cos\theta_x\, dS_x \int_y^x |f(z,\xi)|^2\, d\sigma_z\, , \tag{6}$$

and finally

$$(7) \qquad [\![\nu^{1/2} J^- BUf]\!] < 2^{-1/2} \; |||f||| \; .$$

This relation holds for a continuous f and we remark that, then, $J^+ Bf = 0$. Through the same continuity argument as in 2.1.3.3. we see that, for any $f \in L^2(\Omega, H)$, BUf is well defined* in $L^2(\partial\Omega, H)$ and

$$(8) \qquad J^+ BUf = 0 \; , \quad [\![\nu^{1/2} J^- BUf]\!] < 2^{-1/2} \; |||f||| \; .$$

Again, for any $f \in L^2(\Omega,H)$, BU^*f is well defined in $L^2(\partial\Omega, H)$ and

$$(9) \qquad J^- BU^*f = 0 \; , \quad [\![\nu^{1/2} J^+ BU^* f]\!] < 2^{-1/2} \; |||f||| \; .$$

2.1.3.6. For a continuous ϕ^+ we have (refer to the figure, previous page)

$$(1) \qquad E\phi^+(x, \xi) = e^{-\nu|x-y||\xi|^{-1}} \; \phi^+(y, \xi)$$

and for any function $\rho(|\xi|)$ making reference to the figure, we have

$$(2) \qquad |||\, \nu^{1/2} \rho^{1/2} E\phi^+ \,|||^2 = \int_{R^3} \rho \nu \omega \, d\xi \; \cdot$$

$$\cdot \int_{\Pi(\xi)} |\phi^+(z, \xi)|^2 \, dS_y \int_z^t e^{-2\nu|x-z||\xi|^{-1}} \, d\sigma_x \; .$$

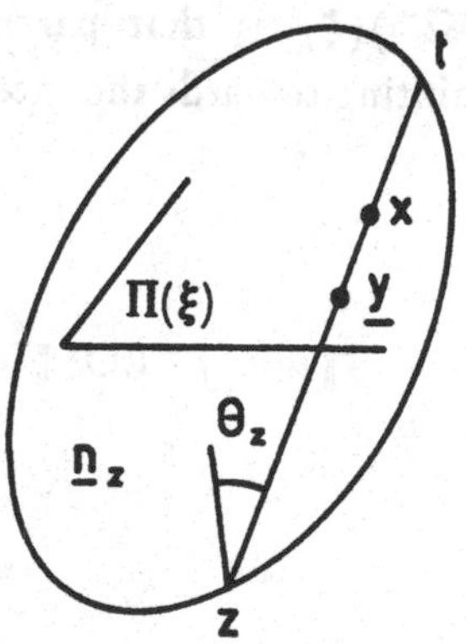

*) We assume ν bounded away from zero.

Referring to the technique of integration used in 2.1.3.5. we obtain

$$|||\nu^{1/2}\rho^{1/2}E\phi^+|||^2 = \int_{R^3}\rho\nu\omega\,d\xi\int_{(\partial\Omega)^+(\xi)}\cos\theta_z\,|\phi^+(z,\xi)|^2\,dS_z\,.$$

$$\cdot\int_z^t e^{-2\nu|x-z||\xi|^{-1}}\,d\sigma_x \tag{3}$$

$$\leqslant \frac{1}{2}\int_{R^3}\rho\,\omega\,d\xi\int_{(\partial\Omega)^+(\xi)}|\,\xi\cdot n_z\,|\,|\phi^+(z,\xi)|^2\,dS_z$$

and, as a consequence

$$|||\nu^{1/2}\rho^{1/2}\,E\phi^+|||\leqslant 2^{-1/2}\,[\![\rho^{1/2}\phi^+]\!]\,. \tag{4}$$

Again through a continuity argument we see that, for any $\phi^+\in L^2(\partial\Omega,\tilde{H}), E\phi^+$ is well defined* as an element of $L^2(\Omega,H)$ and

$$|||\nu^{1/2}\,E\phi^+|||\leqslant 2^{-1/2}\,[\![\phi^+]\!]\,. \tag{5}$$

In the same way $E^*\psi^-$ is well defined in $L^2(\Omega,H)$ when $\psi^-\in L^2(\partial\Omega,\tilde{H})$ and

$$|||\nu^{1/2}E^*\psi^-|||\leqslant 2^{-1/2}\,[\![\psi^-]\!]\,. \tag{6}$$

When $\rho^{1/2}\phi^+$ and $\rho^{1/2}\psi^-$ are in $L^2(\partial\Omega,H)$ as well, (4) and the corresponding inequality for E^* hold.

---

*) We assume ν bounded away from zero.

2.1.3.7. We come back to 2.1.3.6.(3) and majorize

(1) $$\cos\theta_z \int_z^t e^{-2\nu|x-z||\xi|^{-1}}\, d\sigma_x \leqslant \left(\frac{2\nu}{|\xi|}\right)^{-1} \left(1 - e^{-\frac{2\nu}{|\xi|}|z-t|}\right) \leqslant |z-t|$$

so that, for a bounded Ω ,

(2) $$\|\rho\, E\phi^+\| \leqslant (\operatorname{diam}\Omega)^{1/2}\ \|\rho\phi^+\|_{L^2(\partial\Omega,\, H)}$$

(3) $$\|\rho\, E^*\phi^+\| \leqslant (\operatorname{diam}\Omega)^{1/2}\ \|\rho\,\psi^-\|_{L^2(\partial\Omega,\, H)}$$

2.1.3.8. Referring to the figure of 2.1.3.5. we have

(1) $$(J^-\, B\, E\phi^+)\,(x,\xi) = \phi^+(y,\xi)\, \exp(-\nu\,|x-y|\,|\xi|^{-1})$$

then

(2)
$$[\![\rho^{1/2} J^-\, BE\phi^+]\!]^2 = \int_{\partial\Omega} dS_x \int_{R^3_+} \rho\omega|\xi\cdot n_x|\,|\phi^+(y,\xi)|^2\, e^{-2\nu|x-y|\,|\xi|^{-1}}\, d\xi$$
$$= \int_{\partial\Omega} dS_x \int_{R^3_+} (\cos\theta_x)\,\rho\,\omega\,|\xi|\,|\phi^+|^2\, e^{-2\nu|x-y|\,|\xi|^{-1}}\, d\xi\ ,$$

Setting $\xi = |\xi|\, u$ we have

(3) $$[\![\rho^{1/2} J^-\, BE\phi^+]\!]^2 = \int_{\partial\Omega} dS_x \int_{\substack{|u|=1 \\ u\cdot n_x<0}} \cos\theta_x\, du \int_0^\infty \rho\,\omega\,|\xi|^3|\phi^+|^2\, e^{-2\nu|x-y|\,|\xi|^{-1}}\, d|\xi|$$

Let us define $(\partial\Omega)^\circ_{(x)}$ as the set of points of $\partial\Omega$ which are seen, from x, for the first time, namely

(4) $$(\partial\Omega)^\circ(x) = \{y \mid y\in\partial\Omega\,;\ \rho x + (1-\rho)\, y\in\Omega \text{ for } \rho\in\,]0,1[\,\}$$

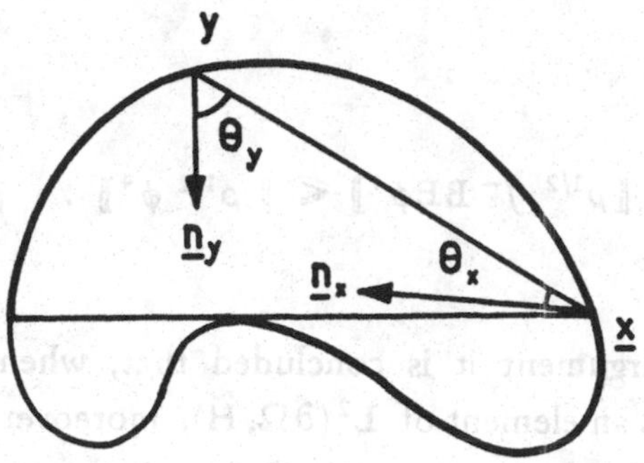

then we have

$$[\![\rho^{1/2} J^{-} \mathrm{B E} \phi^{+}]\!]^2 = \int_{\partial\Omega} dS_x \int_{(\partial\Omega)^{o}(x)} \cos\theta_x \cos\theta_y \, |x-y|^{-2} \, dS_y \, .$$

$$\cdot \int_0^{\infty} \rho\,\omega\, |\xi|^3 \, |\phi^{+}(y,\xi)|^2 \exp(-2\nu\, |x-y|\, |\xi|^{-1}) \, d|\xi| \, \cdot$$

$$= \int_{\partial\Omega} dS_y \int_{(\partial\Omega)^{o}(y)} \cos\theta_x \cos\theta_y \, |x-y|^{-2} \, dS_x \int_0^{\infty} \rho\,\omega\, |\xi|^3 \, |\phi^{+}(y,\xi)|^2 \, e^{-2\nu|x-y||\xi|^{-1}} \, d|\xi| \tag{5}$$

$$= \int_{\partial\Omega} dS_y \int_{\substack{|u|=1 \\ u\cdot n_y > 0}} \cos\theta_y \, du \int_0^{\infty} \rho\,\omega\, |\xi|^3 \, |\phi^{+}(y,\xi)|^2 \, e^{-2\nu|x-y||\xi|^{-1}} \, d|\xi|$$

$$= \int_{\partial\Omega} dS_y \int_{R^3_+} \rho\,\omega\, |\xi\cdot n_y| \, |\phi^{+}(y,\xi)|^2 \, e^{-2\nu|x-y||\xi|^{-1}} \, d\xi$$

and

(6) $$[\![\rho^{1/2} J^- BE\phi^+]\!] \leqslant [\![\rho^{1/2} \phi^+]\!] .$$

Through a continuity argument it is concluded that, whenever $\phi^+ \in L^2(\partial\Omega, \widetilde{H})$ $B\,E\,\phi^+$ is well defined as an element of $L^2(\partial\Omega, \widetilde{H})$, moreover

(7) $$J^+ B\,E\,\phi^+ = \phi^+$$

and (6) holds. Analogously, whenever $\psi^- \in L^2(\partial\Omega, \widetilde{H})$, $B\,E^*\,\psi^- \in L^2(\partial\Omega, \widetilde{H})$, moreover

(8) $$J^- B\,E^*\,\psi^- = \psi^-$$

and

(9) $$[\![\rho^{1/2} J^+ BE^* \psi^-]\!] \leqslant [\![\rho^{1/2} \psi^-]\!] .$$

2.1.3.9. From what has been shown, for any $\varphi \in L^2(\Omega, H)$, $\phi^+ \in L^2(\partial\Omega, \widetilde{H})$, $\psi^- \in L^2(\partial\Omega, \widetilde{H})$ it is completely legitimate to write

(1) $$\mathcal{M}(U\varphi + E\phi^+) = \varphi \;,\; \mathcal{M}^*(U^*\varphi + E^*\phi^+) = \varphi \;,$$

(2) $$J^+ B(U\varphi + E\phi^+) = \phi^+ \;, \qquad J^- B(U^*\varphi + E^*\psi^-) = \psi^- \;,$$

while $J^* \mathrm{B}(\mathrm{U}\varphi + \mathrm{E}\phi^+)$ and $J^+ \mathrm{B}(\mathrm{U}^*\varphi + \mathrm{E}^*\psi^-)$ are well defined in $\mathrm{L}^2(\partial\Omega, \tilde{\mathrm{H}})$. In (1) we apply M and M^ϕ each time as a whole; on the other hand, it is legitimate to write

$$\xi \cdot \nabla\, \mathrm{U}\varphi + \nu\, \mathrm{U}\varphi = \varphi\ , \quad \varphi \in \mathrm{L}^2(\Omega, \mathrm{H})\ , \tag{3}$$

$$-\xi \cdot \nabla\, \mathrm{U}^*\varphi + \nu \mathrm{U}^*\varphi = \varphi\ , \quad \varphi \in \mathrm{L}^2(\Omega, \mathrm{H})\ , \tag{4}$$

while

$$\xi \cdot \nabla\, \mathrm{E}\phi^+ + \nu\, \mathrm{E}\phi^+ = 0\ , \quad \nu^{1/2}\, \phi^+ \in \mathrm{L}^2(\partial\Omega, \bar{\mathrm{H}})\ , \tag{5}$$

$$-\xi \cdot \nabla\, \mathrm{E}^*\psi^- + \nu\, \mathrm{E}^\phi\, \psi^- = 0\ , \quad \nu^{1/2}\, \psi^- \in \mathrm{L}^2(\partial\Omega, \bar{\mathrm{H}})\ . \tag{6}$$

2.1.4. Comments

Now that we have defined appropriate function spaces, our strategy will be to prove the invertibility of $1 - \mathcal{V}$ in $\mathrm{L}^2(\partial\Omega, \mathrm{H})$ in order that the operators W, & W^* and &* be well defined. Then we legitimate 2.1.2.6.(5) for $\varphi, \chi \in \mathrm{L}^2(\Omega, \mathrm{H})$ and $\phi^+, \psi^- \in \mathrm{L}^2(\partial\Omega, \mathrm{H})$. At the same time it will be shown that the appropriate function space for solutions of

$$\mathcal{L}\mathrm{f} = \varphi\ , \qquad J^+\mathrm{Bf} = \mathrm{G}\, J^-\, \mathrm{Bf}\ , \tag{1}$$

will be

$$\mathrm{V} = \{\, \mathrm{f} = W\varphi\, ,\ \varphi \in \mathrm{L}^2(\Omega, \mathrm{H})\, \}\ , \tag{2}$$

and, analogously, the appropriate function space for solutions of

(3) $$\mathcal{L}^* f = \varphi \ , \qquad J^- B f = G^* J^+ B f \ ,$$

will be

(4) $$V^* = \{ f = W^* \varphi \ , \quad \varphi \in L^2(\Omega, H) \} .$$

Then, having legitimated 1.4.6.6.(1) we shall see that a necessary and sufficient condition for f to solve (1) is that

(5) $$((f, \mathcal{L}^* h)) = ((\varphi, h)) \ ,$$

for any $h \in V^*$. We remark that this statement is the mathematical analogue of the well known method of moments in rarefied gas dynamics. The last step to a proof of existence for a solution of (1) will be an a priori estimate for solutions of (1) and the existence will then be obtained by a standard argument using projection in Hilbert space.

2.2. Existence Theory for the Linear Interior Boundary Value Problem. First Step

2.2.1. An a Priori Estimate

2.2.1.1. For any real $\lambda > 0$ we set

$$\mathcal{M}_\lambda \equiv \xi \cdot \nabla + \lambda^{-1} \nu(|\xi|) \tag{1}$$

and define U_λ and E_λ as were defined U and E but with $\lambda^{-1}\nu$ in place of ν. Again we set

$$\mathfrak{V}_\lambda = G J^- B E_\lambda . \tag{2}$$

In an analogous way we define $\mathcal{M}^*_\lambda$, U^*_λ E^*_λ, $\mathfrak{V}^*_\lambda$. We state the hypothesis

$(\mathcal{H}_{X,Y})$: G is bounded from $L^2(\partial\Omega, X)$ into $L^2(\partial\Omega, Y)$,

and set

$$|G|_{X,Y} = |G|_{\mathcal{L}(L^2(\partial\Omega,X),\, L^2(\partial\Omega,Y))} \tag{3}$$

where X and Y stand for some of H, $\tilde{H}$, $\tilde{H}_\nu$ where

$$H_\nu = \{f \mid f = f(\xi),\ \nu^{1/2} f \in \tilde{H}\} . \tag{4}$$

In the same way we state the hypothesis

$(\mathcal{H}^*_{X,Y})$: G^* is bounded from $L^2(\partial\Omega, X)$ into $L^2(\partial\Omega, Y)$,

and set

(5) $$|G^*|_{x,y} = |G^*|_{\mathcal{L}(L^2(\partial\Omega, H),\, L^2(\partial\Omega, Y))} .$$

Consider any $f^+ \in L^2(\partial\Omega, H)$ and define

(6) $$f = E_\lambda f^+ \qquad \phi^+ = (1 - \mathcal{V}_\lambda) f^+ ,$$

we have, if $(\mathcal{H}_{\tilde{H},\tilde{H}})$ holds,

(7) $$\mathcal{M}_\lambda E_\lambda f^+ = 0 , \quad J^+ Bf - G J^- Bf = \phi^+$$

as relations holding (for the first one in a strong sense) in $L^2(\Omega, H)$ and $L^2(\partial\Omega, \tilde{H})$ respectively. Now let f^+ be continuous and belong to $L^2(\partial\Omega, \tilde{H})$, f is continuous and continuously differentiable along rays and ϕ^+ is continuous. We may apply the analogue of 2.1.1.7.(5) to get*

(8) $$\lambda^{-1}\nu_0 |||f|||^2 + k[\![\Pi J^- Bf]\!]^2 \leq \frac{1}{2}[\![\phi^+]\!]^2 + R_e\{\phi^+, G\Pi J^- Bf\} + \sqrt{2\pi} \int_{\partial\Omega} \mathrm{Re}(\phi^+, 1^+)_{\tilde{H}} \overline{F(x)}\, dS_x$$

with

(9) $$F(x) = (J^- Bf, 1^-)_{\tilde{H}} , \qquad x \in \partial\Omega .$$

*) We assume a hard potential angular cut-off, or rigid sphere or radial cut-off.

Let $\{f_n^+\}$ be a sequence of continuous functions such that $[\![f_n^+ - f^+]\!] \to 0$ as $n \to \infty$. Define $f_n = E_\lambda f_n^+$, $\phi_n^+ = (1-\mathcal{V}_\lambda) f_n^+$ and apply (8) to f_n, ϕ_n^+, we have, from 2.1.3.8.(6)

$$[\![J^- B f_n - J^- Bf]\!] \leqslant [\![f_n - f]\!] , \tag{10}$$

and

$$[\![\phi_n^+ - \phi^+]\!] \leqslant (1 + |G|_{\tilde{H},\tilde{H}}) [\![f_n - f]\!] , \tag{11}$$

So that (8) holds equally well for any $f^+ \in L^2(\partial\Omega, H)$ provided $(\mathcal{H}_{\tilde{H},\tilde{H}})$ holds and under the assumption of a hard potential angular cut-off, rigid spheres or radial cut-off.

2.2.1.2. Our goal will be to solve 2.2.1.1.(6) for f^+ given $\phi^+ \in L^2(\partial\Omega, \tilde{H})$, and our basic tool for this will be an a priori estimate of $[\![f^+]\!]$ in terms of $[\![\phi^+]\!]$. In order to obtain this estimate we use the obvious inequality

$$[\![f^+]\!] \leqslant [\![\phi^+]\!] + |G|_{\tilde{H},\tilde{H}} ([\![\Pi J^- Bf]\!] + \sqrt{2\pi}\, [\![F(\cdot)\, 1^-]\!]) , \tag{1}$$

with

$$[\![F(\cdot)\, 1^-]\!] = \| 1^- \|_{\tilde{H}} \, \| F \|_{L^2(\partial\Omega)} = (2\pi)^{-1/2} \| F \|_{L^2(\partial\Omega)} . \tag{2}$$

and 2.2.1.1.(8) that we restate as

$$\lambda^{-1} \nu_0 \, |\!|\!| f |\!|\!|^2 + k [\![\Pi J^- Bf]\!]^2 \leqslant \frac{1}{2} [\![\phi^+]\!]^2 + |G|_{\tilde{H},\tilde{H}} [\![\phi^+]\!] \, [\![\Pi J^- Bf]\!] + [\![\phi^+]\!] \, \| F \|_{L^2(\partial\Omega)} \tag{3}$$

and it is clear that we need a separate estimate for $\| F \|_{L^2(\partial\Omega)}$.

2.2.1.3. From 2.2.1.1.(6) and (9) we obtain

$$f^+ = \phi^+ + G\Pi J^- BE_\lambda f^+ + \sqrt{2\pi}\, F\, 1^+ \quad , \tag{1}$$

and we use that in 2.2.1.1.(9) with 2.2.1.1.(6) to get

$$F = \mathcal{C}_\lambda F + G + H \tag{2}$$

as an equation holding in $L^2(\partial\Omega)$ with

$$\mathcal{C}_\lambda F = \sqrt{2\pi}\, (J^- BE_\lambda F\, 1^+, 1^-)_{\tilde{H}} \; , \tag{3}$$

and

$$G = (J^- BE_\lambda \phi^+, 1^-)_{\tilde{H}} \; , \tag{4}$$

$$H = (J^- BE_\lambda G\Pi J^- BE_\lambda f^+, 1^-)_{\tilde{H}} \; . \tag{5}$$

Assume, for a moment, that $1-\mathcal{C}_\lambda$ is invertible by a bounded operator $(1-\mathcal{C}_\lambda)^{-1}$, in $L^2(\partial\Omega)$, we have then from 2.1.3.8.(6)

$$\|F\|_{L^2(\partial\Omega)} \leqslant (2\pi)^{-1/2} \, |(1-\mathcal{C}_\lambda)^{-1}|_{L^2(\partial\Omega)} ([\![\phi^+]\!] + |G|_{\tilde{H},\tilde{H}} [\![\Pi J^- Bf]\!]) \tag{6}$$

and it is clear that this inequality, with 2.2.1.2.(1) (3) will allow to estimate $[\![f^+]\!]$ in terms of $[\![\phi^+]\!]$.

2.2.1.4. Referring to the figure we have

$$\mathcal{T}_\lambda F(x) = \sqrt{2\pi} \int_{R^3} \omega\, |\xi \cdot n_x|\, e^{-\lambda^{-1}|\xi|^{-1}\nu|x-y|} F(y)\, d\xi$$

$$= \sqrt{2\pi} \int_0^\infty \omega |\xi|^3\, d|\xi| \int_{\substack{|u|=1 \\ u\cdot n_x<0}} |u \cdot n_x|\, e^{-\lambda^{-1}|\xi|^{-1}\nu|x-y|} F(y)\, du \qquad (1)$$

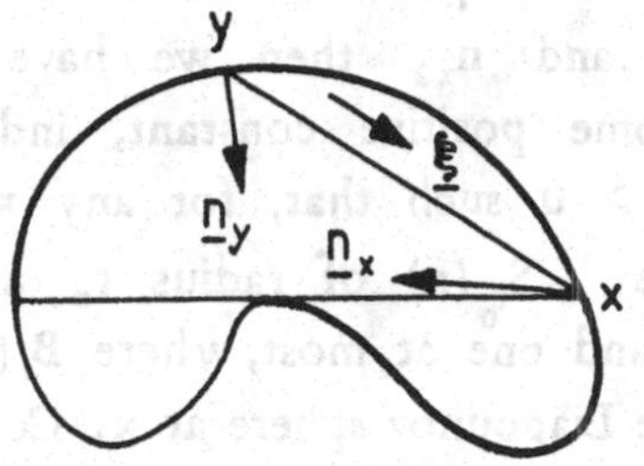

where we have set $\xi = |\xi|\, u$. Now we refer to 2.1.3.8.(4) and we obtain

$$\mathcal{T}_\lambda F(x) = \sqrt{2\pi} \int_0^\infty \omega |\xi|^3\, d|\xi| \int_{(\partial\Omega)^0(x)} \frac{|u\cdot n_x|\,|u \cdot n_y|}{|x-y|^2}\, e^{-\lambda^{-1}\nu|\xi|^{-1}|x-y|} F(y)\, ds_y \qquad (2)$$

then

$$\mathcal{T}_\lambda F(x) = \int_{(\partial\Omega)^0(x)} \frac{|(x-y)\cdot n_x|\,|(x-y)\cdot n_y|}{|x-y|^4}\, H_\lambda\, (|x-y|)\, F(y)\, ds_y \qquad (3)$$

with

(4) $$H_{\lambda}(s) = \sqrt{2\pi} \int_0^{\infty} |\xi|^3 \, \omega(|\xi|) e^{-\lambda^{-1}\nu(|\xi|)|\xi|^{-1} s} \, d|\xi| .$$

2.2.1.5. We State the Hypothesis
(L_α): $\partial\Omega$ satisfies Liapunov conditions with exponent $\tau \in]0,1[$, and comment the content of this hypothesis. At each point $x \in \partial\Omega$ there exists a well defined tangent plane and unit normal n_x pointing towards the interior of Ω. Let θ_{12} be the angle between n_{x1} and n_{x2} then we have $|\sin\theta_{12}| < E\, r_{12}^{\alpha}$, with $r_{12} = |x_1 - x_2|$ and E some positive constant, independent on x_1 and x_2 in $\partial\Omega$. There exists a $\gamma_0 > 0$ such that, for any $x \in \partial\Omega$ the parallels to n_x which interest the sphere $S_{r_0}(x)$ of radius r_0 and center x, intersect $\partial\Omega \cap B_{r_0}(x)$ at one point and one at most, where $B_r(x)$ is the ball interior to $S_{r_0}(x)$. We call $S_{r_0}(x)$ the Liapounov sphere at x. Consider a point P on $\partial\Omega$ and choose rectangular axes M x, y, z such that z = 0 is the plane tangent at $\partial\Omega$ through P. For any $M(x, y, z) \in \partial\Omega \cap B_{r_0}(M)$ we have

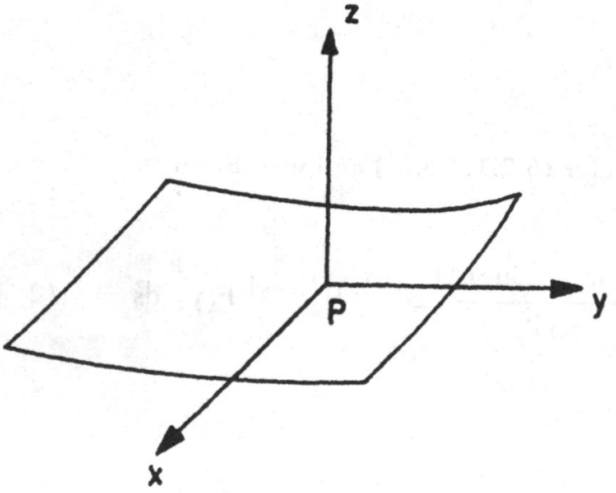

(1) $$z = f(x, y) .$$

From the existence of the tangent plane at M we conclude that $\partial f/\partial x$ and $\partial f/\partial y$ exist and are continuous (we use (L_α)) within $B_{r_0}(M)$. As a consequence

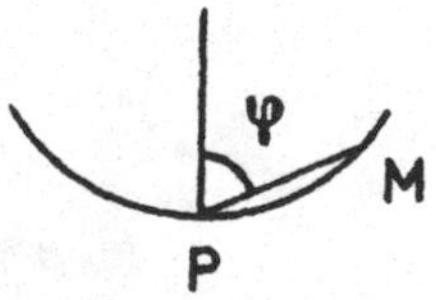

(2) $$z = \bar{p}\, x + \bar{q}\, y ,$$

and

$$\cos\varphi = \frac{z}{\sqrt{x^2 + y^2 + z^2}} \leqslant \frac{(\tilde{p}^2 + \tilde{q}^2)^{1/2}}{\{1 + \tilde{p}^2 + \tilde{q}^2\}^{1/2}} \leqslant E|PM|^{\alpha},\ 0 < \alpha < 1. \tag{3}$$

2.2.1.6. We come back to 2.2.1.4. and note that, from 2.2.1.5.(3) we have

$$|(\mathbf{x}-\mathbf{y}) \cdot \mathbf{n}_x||(\mathbf{x}-\mathbf{y}) \cdot \mathbf{n}_y| \leqslant E^2 |x-y|^{2(1+\alpha)} \tag{1}$$

and, consequently

$$\mathcal{T}_\lambda F(\mathbf{x}) = \int_{(\partial\Omega)^\circ(x)} \frac{M_\lambda(\mathbf{x}, \mathbf{y})}{|\mathbf{x}-\mathbf{y}|^{2-2\alpha}} F(\mathbf{y})\, dS_y . \tag{2}$$

We note that M_λ is continuous on $\partial\Omega \times \partial\Omega$ and bounded uniformly with respect to λ. We may write (2) as

$$\mathcal{T}_\lambda F(x) = \int_{\partial\Omega} \frac{N_\lambda(\mathbf{x}, \mathbf{y})}{|\mathbf{x}-\mathbf{y}|^{2-2\alpha}} F(y)\, dS_y \tag{3}$$

with

$$N_\lambda = M_\lambda \quad \text{if} \quad \rho x + (1-\rho)\, \mathbf{y} \in \Omega \ \text{ for } \rho \in]0,1[\tag{4}$$
$$N_\lambda = 0 \quad \text{otherwise.}$$

We conclude that N_λ is piecewise continuous on $\partial\Omega \times \partial\Omega$ and bounded uniformly with respect to λ. Iterating we get

$$(5) \qquad \mathcal{T}_\lambda^{(n)} F(x) = \int_{\partial\Omega} \frac{N_\lambda^{(n)}(x,y)}{|x-y|^{2-\beta_n}} F(y)\, dS_y \,, \qquad \beta_n = (n+1)\alpha \,,$$

with

$$(6) \qquad N_\lambda^{(n+1)}(x,y) = |x-y|^{2-(n+2)\alpha} \int_{\partial\Omega} \frac{N_\lambda^{(n)}(x,z)\, N_\lambda(z,y)}{|x-z|^{2-(n+1)\alpha}\, |z-y|^{2-\alpha}}\, dS_z \,,$$

and we may show that $N_\lambda^{(n)}$ is piecewise continuous and bounded uniformly with respect to λ. The proof is through recurrence on n. Let C be the uniform bound of N_λ and $C^{(n)}$ be that of $N_\lambda^{(n)}$ we need prove that

$$(7) \qquad \sup_{\substack{x\in\partial\Omega \\ y\in\partial\Omega}} \int_{\partial\Omega} \frac{dS_z}{|x-z|^p\, |z-y|^q} < K_{p,q}^{(\partial\Omega)} (|x-y|^{2-p-q} + 1), 0 \leqslant p < 2,\ 0 \leqslant q < 2$$

and we shall have

$$(8) \qquad C^{(n+1)} = K_{2-(n+1)\alpha,\, 2-\alpha}(\partial\Omega)\; C^{(n)}\, C \,,$$

which will prove the statement about boundedness. To prove the statement about piecewise continuity we proceed as follows. Choose $\alpha_1 \in]0,\alpha[$ as close to α as we want and set

$$(9) \qquad N_\lambda(x,y)\, |x-y|^{2(\alpha-\alpha_1)} = P_\lambda(x,y)$$

so that

$$\mathcal{T}_\lambda F(x) = \int_{\partial\Omega} \frac{P_\lambda(x,y)}{|x-y|^{2-2\alpha_1}} F(y)\, dS_y \tag{10}$$

and

$$\mathcal{T}_\lambda^{(n)} F(x) = \int_{\partial\Omega} \frac{P_\lambda^{(n)}(x,\zeta)}{|x-y|^{2-\gamma_n \cdot 0}} F(y)\, dS_y\,, \quad \gamma_n = (n+1)\alpha_1 \tag{11}$$

with

$$P_\lambda^{(n)} = |x-y|^{(n+1)(\alpha-\alpha_1)} N_\lambda^{(n)}(x,y)\,. \tag{12}$$

Now we observe that, on the set where we have the transition from $N_\lambda = M_\lambda$ to $N_\lambda = 0$, we have $P_\lambda = 0$, and, consequently, P_λ is continuous on $\partial\Omega \times \partial\Omega$ and, thanks to boundedness of Ω, it is bounded, uniformly with respect to λ. Now it is evident that $P_\lambda^{(2)}(x,y)$ is continuous for $x \neq y$ and from boundedness of $N_\lambda^{(2)}$ we see that $P_\lambda^{(2)}(x,y)$ is continuous on $\partial\Omega \times \partial\Omega$. The argument may be iterated to prove continuity of $P_\lambda^{(n)}$, on $\partial\Omega \times \partial\Omega$. Then, from (12), $N_\lambda^{(n)}$ is again continuous on $\partial\Omega \times \partial\Omega - J$ where J is the set of points $(\{x,x\}|x \in \partial\Omega)$ on $\partial\Omega \times \partial\Omega$.

2.2.1.7. Now we have to prove (7). From 2.2.1.5.(12) it is sufficient to prove, on account of boundedness of Ω,

$$\sup_{x\in\partial\Omega}\ \sup_{y\in B_{\frac{r_0}{k}}(x)} \int_{B_{r_0/2}} \frac{dS_z}{|x-z|^p\,|z-y|^q} \leqslant C_{p,q}\left(|x-y|^{-(p+q)+2} + 1\right) \tag{1}$$

with some constant $k \geqslant 3$ for example. We refer to 2.2.1.5.(1) and set

(2) $$\mathbf{x} = (0, 0, 0)\ ,\quad \mathbf{y} = (x, y, f(x, y)),\ \mathbf{z} = (\xi, \zeta, f(\xi, \zeta))\ .$$

We define

(3) $$p = f_x(x, y)\ ,\quad q = f_y(x, y)\ ,\ \omega = p(\xi, \zeta)\ ,\ \rho = q(\xi, \zeta)$$

and we have

(4) $$dS_z = (1 + \omega^2 + \rho^2)^{1/2}\ d\xi d\zeta\ .$$

From (L_α) we have

(5) $$\frac{\sqrt{p^2 + q^2}}{\sqrt{1 + p^2 + q^2}} \leqslant E\{x^2 + y^2 + |f(x, y)|^2\}^{\alpha/2}$$

for

(6) $$r^2 = x^2 + y^2 + |f(x, y)|^2 \leqslant r_0^2\ .$$

We choose r_o such that

(7) $$E r_0^\alpha \leqslant 2^{-1/2}\ ,$$

and we have

(8) $$x^2 + y^2 + |f(x, y)|^2 \leqslant r_0^2 \Rightarrow \sqrt{p^2 + q^2} \leqslant \sqrt{2} E\ r^\alpha\ ,$$

and from 2.2.1.5.(2)

$$x^2 + y^2 + |f(x, y)|^2 \leqslant r_0^2 \Rightarrow |f(x, y)| \leqslant \sqrt{2}\, E\, r^{\alpha}(x^2 + y^2)^{1/2} , \tag{9}$$

then, provided (7) holds we have

$$x^2 + y^2 + |f(x, y)|^2 \leqslant r_0^2 \Rightarrow (x^2 + y^2)^{1/2} \leqslant r \leqslant \sqrt{2}(x^2 + y^2)^{1/2} . \tag{10}$$

Again with (7)

$$B_{r_0/k}(x) \sqsubset \{(x, y, f(x, y)) \mid (x^2 + y^2)^{1/2} \leqslant \frac{\gamma_0}{\sqrt{2}k}\} , \quad k \geqslant 2 , \tag{11}$$

and the proof of (1) is reduced to the one of

$$\sup_{(x^2+y^2)^{1/2} \leqslant \frac{r_0}{k}} \iint_{(\xi^2+\zeta^2)^{1/2} \leqslant \frac{r_0}{2}} \frac{d\xi\, d\zeta}{(\xi^2 + \zeta^2)^{p/2} \{(x-\xi)^2 + (y-\zeta)^2\}^{q/2}} \leqslant \tag{12}$$

$$\leqslant C_{p,q}\left((x^2 + y^2)^{1-(p+q)/2} + 1\right)$$

or

$$\sup_{r \leqslant \frac{r_0}{K}} \int_0^{r_0/2} \frac{d\rho}{\rho^{p-1}} \int_0^{2\pi} \frac{d\psi}{\{r^2 + \rho^2 - 2r\rho\cos\psi\}^{q/2}} \leqslant C_{p,q}\left(r^{2-p-q} + 1\right) \tag{13}$$

or restricting to $p + q > 2$

$$\sup_{\theta \in [\frac{k}{2}, \infty]} \int_0^{\theta} t^{1-p}\, dt \int_0^{2\pi} \frac{d\psi}{(1 + t^2 - 2t\cos\psi)^{q/2}} \leqslant C_{p,q} \tag{14}$$

which is readily checked.

2.2.1.8. From 2.2.1.6. we conclude that some iterate $\mathcal{T}_\lambda^{(n)}$ of $\mathcal{T}_\lambda$ has a bounded kernel; from boundedness of the measure of $\partial\Omega$ (which results from 2.2.1.5 ($L\lambda$)) we see that this iterate is compact and, as such, $\mathcal{T}_\lambda$ itself is compact for any λ (refer to Zaanen [23] p. 317 for a proof that A is compact if A^n is). This implies that, in $L^2(\partial\Omega)$, $1-\mathcal{T}_\lambda$ is invalible by a bounded operator $1-\mathcal{T}_\lambda$ provided we prove that

$$(1) \qquad (1-\mathcal{T}_\lambda)\, F = 0 \quad \Rightarrow \quad F = 0 \,.$$

Suppose that (1) does not hold and let F be a solution of $F = \mathcal{T}_\lambda F$, define

$$(2) \qquad f^+ = \sqrt{2\pi}\; F\, 1^+, \qquad f = E_\lambda\, f^+, \qquad \phi^+ = (1-\mathcal{V}_\lambda)\, f^+$$

and observe that, from (2) and $F = \mathcal{T}_\lambda F$ (refer to 2.2.1.3.(3))

$$(3) \qquad \phi^+ + G\Pi J^- B E_\lambda\, f^+ = 0 \,.$$

From (3) we derive, thanks to $G^*\, 1^+ = 1^-$,

$$(4) \qquad (\phi^+, 1^+)_{\tilde{H}} = -(G\Pi\, J^- BE_\lambda\, f^+, 1^+)_{\tilde{H}} = -(\Pi\, J^- BE_\lambda\, f^+, G\, 1^+)_{\tilde{H}} = 0 \,,$$

$$(5) \qquad \frac{1}{2}[\![\phi^+]\!]^2 + \mathrm{Re}\,\{\phi^+, G\Pi J^- BE_\lambda\, f^+\} = -\frac{1}{2}[\![\phi^+]\!]^2 \,.$$

We substitute (4) and (5) into 2.2.1.1.(8) with f, ϕ^+, F as in (2) and we get

$$(6) \qquad 0 \leqslant \lambda^{-1}\nu_0\, \|f\|^2 + k\,[\![\Pi J^- Bf]\!]^2 + \frac{1}{2}[\![\phi^+]\!]^2 \leqslant 0$$

from which we conclude that $f = 0$ and, consequently, $F = 0$ so that (1) is proven.

As a consequence of the compacity of $\mathcal{T}_\lambda$ and of (1) we have

$$(7) \qquad (1-\mathcal{T}_\lambda)\, F = \phi \quad \Rightarrow \quad \|F\|_{L^2(\partial\Omega)} \leqslant C(\lambda)\, \|\phi\|_{\partial\Omega} \,.$$

Now we prove that $\mathcal{T}_\lambda$ depends continuously on λ as an operator in $L^2(\partial\Omega)$. Referring to 2.2.1.4.(4) we have

$$|H_{\lambda_2}(s)-H_{\lambda_1}(s)| \leqslant \sqrt{2\pi}\, s \left| \frac{1}{\lambda_2} - \frac{1}{\lambda_1} \right| \int_0^\infty |\xi|^2 \mathfrak{V}^{-1} \omega\, e^{-\nu|\xi|^{-1} s\lambda_0^{-1}}\, d|\xi| \quad (8)$$

where $\lambda_o = \sup(\lambda_1, \lambda_2)$. As a consequence of (8) and 2.2.1.4.(3) we have

$$\|\mathcal{T}_{\lambda_2} F - \mathcal{T}_{\lambda_1} F_1\|_{L^2(\partial\Omega)} \leqslant (\operatorname{diam}\Omega)\, D_1 \left| \frac{1}{\lambda_2} - \frac{1}{\lambda_1} \right| \|F\|_{L^2(\partial\Omega)}\,, \quad (9)$$

where D_1 is some constant independent of λ, and we observe that

$$\|\mathcal{T}_\lambda F\|_{L^2(\partial\Omega)} \leqslant D_2(\lambda)\, \|F\|_{L^2(\partial\Omega)} \quad (10)$$

where $D_2(\lambda)$ is an increasing function of λ, uniformly bounded on any compact interval.

Writing

$$(1-\mathcal{T}_{\lambda 2})^{-1} = \{1-[1-\mathcal{T}_{\lambda 1}]^{-1} (\mathcal{T}_{\lambda 2} - \mathcal{T}_{\lambda 1})\}^{-1} (1-\mathcal{T}_{\lambda 1})^{-1} \quad (11)$$

we obtain, provided

$$(\operatorname{diam}\Omega)\, D_1\, C(\lambda_1) \left| \frac{1}{\lambda_2} - \frac{1}{\lambda_1} \right| < 1 \quad (12)$$

$$C(\lambda_2) \leqslant C(\lambda_1) \{1-(\operatorname{diam}\Omega)\, D_1 \left| \frac{1}{\lambda_2} - \frac{1}{\lambda_1} \right| C(\lambda_1)\}^{-1}\,. \quad (13)$$

Now, from (13) there is evidently some closed interval contained in $]0,+\infty[$ on which $C(\lambda)$ is uniformly bounded. More precisely for each $\lambda_o \in]0,+\infty[$, $C(\lambda_o)$ is bounded and, from (13) with $\lambda_1 = \lambda_0$ one can find some open interval $I(\lambda_o))$, containing (λ_o) over which $C(\lambda)$ is uniformly bounded. Let $[\alpha,\beta]$ be any closed interval contained in $]0,+\infty[$, we may cover it with a finite number of $I(\lambda_o)$ and, consequently $C(\lambda)$ is uniformly bounded on $[\alpha,\beta]$. On the other hand $\mathcal{T}_\infty$ is well defined and we may apply (13) with $\lambda_1 = \infty$. As a consequence there is some γ for which $C(\lambda)$ is uniformly bounded on $[\gamma,+\infty[$.

2.2.1.8.1. We state the result of the analysis of the previous paragraphs: for any hard potential angular cut-off or rigid sphere or radial cut-off, under the assumption that $\partial\Omega$ satisfies a Liapounov condition, Ω being bounded, the operator $\mathcal{C}_\lambda$ defined in 2.2.1.3.(3) is invertible in $L^2(\partial\Omega)$ for any $\lambda \in]0, \infty[$ and whatever be $a > 0$ we have

(1) $$\|(1-\mathcal{C}_\lambda)^{-1}\|_{L^2(\partial\Omega)} \leqslant C(a) \ , \quad \text{for} \quad \lambda \in [a, +\infty[\ ,$$

with a constant $C(a)$ which does not depend on $\lambda \in [a, +\infty[$.

2.2.1.9. We come back to 2.2.1.2.(1) in which we take account of 2.2.1.3.(6) and 2.2.1.8.(1), getting, for $\lambda \in [a, +\infty[$,

(1) $$[\![f^+]\!] \leqslant (1 + (2\pi)^{-1/2}\, C(a))\{[\![\phi^+]\!] + |G|_{\tilde{H},\tilde{H}}\, [\![\Pi J^- Bf]\!]\} .$$

On the other hand, using 2.2.1.3.(6) and 2.2.1.8.(1) in 2.2.1.2.(3) we obtain

(2) $$\begin{aligned} k[\![\Pi J^- Bf]\!]^2 \leqslant \frac{1}{2}[\![\phi^+]\!]^2 + |G|_{\tilde{H},\tilde{H}}\, [\![\Pi J^- Bf]\!]\, [\![\phi^+]\!] + \\ (2\pi)^{-1/2} C(a)\, [\![\phi^+]\!]\,([\![\phi^+]\!] + |G|_{\tilde{H},\tilde{H}}\, [\![\Pi J^- Bf]\!]) . \end{aligned}$$

Using the obvious inequality

(3) $$ab \leqslant \epsilon\, a^2 + (4\epsilon)^{-1}\, b^2$$

we derive

(4) $$\begin{aligned} (k - \epsilon_1 |G|_{\tilde{H},\tilde{H}} - (2\pi)^{-1/2} C(a)\, \epsilon_2 |G|_{\tilde{H},\tilde{H}})\, [\![\Pi J^- Bf]\!]^2 \leqslant \\ \{\frac{1}{2} + (2\pi)^{-1/2} C(a) + [(4\epsilon_1)^{-1} + C(a)(\sqrt{2\pi}\, 4\epsilon_2)^{-1}]\, |G|_{\tilde{H},\tilde{H}}\}\, [\![\phi^+]\!]^2 . \end{aligned}$$

we choose

$$4\epsilon_1 |G|_{\tilde{H},\tilde{H}} = k \qquad 4\epsilon_2 |G|_{\tilde{H},\tilde{H}} \, C(a)\,(2\pi)^{-1/2} = k \tag{5}$$

and we obtain

$$[\![\Pi\, J^- Bf]\!] \leqslant k^{-1/2} \{ 1 + 2(2\pi)^{-1/2} C(a) + 2k^{-1} [1 + (2\pi)^{-3/2} C^2(a)] \}^{-1/2} [\![\phi^+]\!] \tag{6}$$

then, from (1),

$$[\![f^+]\!] \leqslant (1 + (2\pi)^{-1/2} C(a)) \cdot \tag{7}$$

$$\cdot \{ (1 + 2(2\pi)^{-1/2} C(a) + 2k^{-1} [1 + (2\pi)^{-3/2} C^2(a)]) K^{-1/2} |G|_{\tilde{H}\tilde{H}} + 1 \} [\![\phi^+]\!]$$

2.2.1.9.1. We state the result that we have just obtained. Under the assumption of a hard potential angular cut-off, or rigid spheres, or radial cut-off, and a boundary $\partial\Omega$ which satisfies a Liapounov condition, Ω being bounded, and G continuous from $\tilde{H}$, to $\tilde{H}$, satisfying 1.4.5.4.(1), we have that for any $\lambda \in [a, +\infty]$ with $a > 0$

$$[\![f^+]\!] \leqslant C(a, G, \nu, \partial\Omega) [\![(1 - \vartheta_\lambda) f^+]\!] \tag{1}$$

for any $[\![f^+]\!] \in L^2(\partial\Omega, H)$ and $\mathcal{V}_\lambda$ as defined in 2.2.1.1.(2). The constant C depends on a, the operator G, the function ν the boundary $\partial\Omega$ but not on $\lambda \in [a, +\infty[$. We state (1) otherwise as

$$\|T_\lambda f^+\|_{L^2(\partial\Omega,\tilde{H})} \leqslant |T_\lambda|_{L^2(\partial\Omega,\tilde{H})} \|f^+\|_{L^2(\partial\Omega,\tilde{H})} , \tag{2}$$

where $T_\lambda = (1 - \mathcal{V}_\lambda)^{-1}$ (see 2.1.2.4).

2.3. Existence Theory for the Linear Interior Boundary Value Problem. Second Step

2.3.1. Invertibility of $1-\mathfrak{V}$ *in* $L^2(\partial\Omega, H)$

2.3.1.1. We come back to 2.1.3.8.(4) we change ν to ν/λ and let $\partial\Omega^+(\xi)$ for that part of $\partial\Omega$ along which ξ is emerging

(1) $$[\![J^- BE_\lambda \phi^+]\!]^2 \leqslant \int_{R^3} |\xi|\, \omega\, d\xi \int_{(\partial\Omega)^+(\xi)} e^{-2\lambda^{-1}\nu d(y,\xi)|\xi|^{-1}} |\phi^+(y,\xi)|^2 \cos\theta_y\, dS_y$$

with

$$d(y,\xi) = |x-y|\ ,\quad x\in\partial\Omega,\ y\in\partial\Omega,\ \xi\wedge(x-y) = 0$$

$n_x \cdot (x-y) > 0$, $\rho x + (1-\rho) y \in \Omega$ for $\rho \in]0,1[$, $d = \infty$ otherwise,
We assume that the interactions are rigid spheres like
or that we have a radial cut-off so that

(2) $$\operatorname*{Inf}_{\xi} \frac{\nu(|\xi|)}{|\xi|} = \beta > 0\,.$$

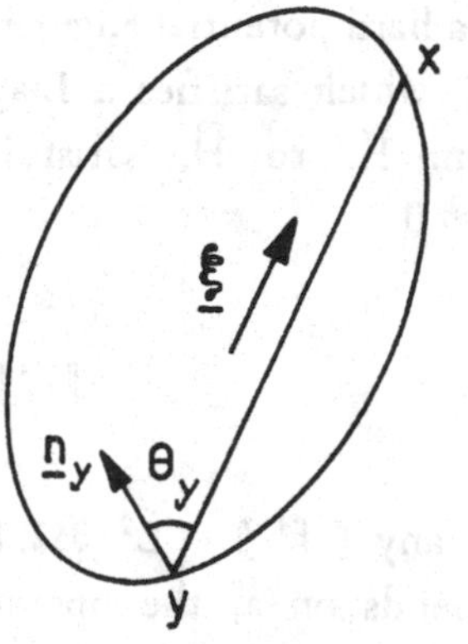

we state the existence of some $r^*(\lambda)$ such that

(3) $$e^{-2\beta\lambda^{-1} d} \cos\theta \leqslant e^{-2\beta\lambda^{-1} r^*(\lambda)}\ ,$$

and

(4) $$\lambda^{-1} r^*(\lambda) \to \infty\ ,\quad \text{when } \lambda \to 0\,.$$

It is sufficient to prove (3) for $|x-y| < r_1 < r_0$ with $\sqrt{2}\,E\,r_0^\alpha \leqslant 1$, where we refer to 2.2.1.7. assuming that 2.2.1.5. (L_α) holds. With the notation

(5) $$x = (0,0,0) \qquad y = (x, y, f(x,y)) \qquad f(x,y) = z$$

we have, from 2.2.1.7.(9)

$$r < r_1 \Rightarrow z \leqslant \sqrt{2}\, E\, r^{\alpha+1} \tag{6}$$

and, from $z = r\cos\theta$ we have

$$\cos\theta \leqslant \sqrt{2}\, E\, r^{\alpha} \tag{7}$$

so that

$$e^{-2\beta\lambda^{-1}d}\cos\vartheta \leqslant \sup\ \{e^{-\beta r/\lambda}\ , \sqrt{2}\, E\, r^{\alpha}\} \tag{8}$$

and $r^*(\lambda)$ is defined by

$$e^{-2\beta\lambda^{-1} r^*(\lambda)} = \sqrt{2}\, E[r^*(\lambda)\,]^{\alpha}\ , \tag{9}$$

from which (4) clearly holds. As a consequence we have

$$[\![J^- B E_\lambda\, \phi^+]\!] \leqslant e^{-\beta\lambda^{-1} r^*(\lambda)}\, \|\phi^+\|_{L^2(\partial\Omega,\, H)}\ , \tag{10}$$

and assuming that 2.2.1.1. $(\mathcal{H}_{\tilde{H},H})$ holds, we get

$$\|\mathcal{V}_\lambda\, \phi^+\|_{L^2(\partial\Omega,\, H)} \leqslant e^{-\beta\lambda^{-1} r^*(\lambda)}\, |G|_{\tilde{H},H}\, \|\phi^+\|_{L^2(\partial\Omega, H)}\ . \tag{11}$$

Let us choose λ_0 so that

$$e^{-\beta\lambda_0^{-1} r^*(\lambda_0)}\, |G|_{\tilde{H},H} < 1\ , \tag{12}$$

then, for any $\lambda \in]0, \lambda_0]$, $1 - \mathfrak{V}_\lambda$ is invertible by a bounded operator in $L^2(\partial\Omega, H)$. We observe that $L^2(\partial\Omega, H)$ is dense in $L^2(\partial\Omega, \tilde{H})$. As a matter of fact, let $\phi^+ \in L^2(\partial\Omega, H)$, we define $\phi^+_{\delta,\bar{\epsilon}}$ by

(13) $$\begin{cases} \phi^+_{\delta,\epsilon}(x, \xi) = \phi^+(x, \xi), & \text{if } |\xi| \geqslant \epsilon \text{ and } |\xi \cdot n_x| \geqslant \delta |\xi|, \\ \phi^+_{\delta,\epsilon}(x, \xi) = 0, & \text{if } |\xi| \leqslant \epsilon \text{ or } |\xi \cdot n_x| < \delta |\xi| \end{cases}$$

and, from

(14) $$\|\phi^+_{\delta,\epsilon}\|_{L^2(\partial\Omega,H)} \leqslant \epsilon^{-1}\, \delta^{-1}\, \|\phi^+\|_{L^2(\partial\Omega,\tilde{H})}$$

we see that $\phi^+_{\delta,\epsilon} \in L^2(\partial\Omega, H)$. On the other hand

(15) $$\|\phi^+ - \phi^+_{\delta,\epsilon}\|^2_{L^2(\partial\Omega,\tilde{H})} \leqslant \int_{\partial\Omega} dS_x \int_{V(x,\delta,\epsilon)} \omega\, |\xi \cdot n_x|\, |\phi^+(x, \xi)|^2\, d\xi$$

where

(16) $$V(x, \delta, \epsilon) = \{\, \xi \mid |\xi| < \epsilon \text{ or } |\xi \cdot n_x| < \delta |\xi| \,\}$$

and, from

(17) $$\int_{\partial\Omega} dS_x \int_{V(x,\delta,\epsilon)} \omega\, |\xi \cdot n_x|\, d\xi \to 0 \quad \text{when } \delta + \epsilon \to 0$$

a known result of measure theory (Dunford-Schwartz[22] vol. 1 page 114 theorem 20) implies that

(18) $$\|\phi^+ - \phi^+_{\epsilon,\delta}\|_{L^2(\partial\Omega,\tilde{H})} \to 0, \quad \text{when } \delta + \epsilon \to 0 .$$

Let

$$f^+_{\delta,\epsilon} = (1-\mathcal{V}_\lambda)^{-1}\ \phi^+_{\delta,\epsilon}\ , \qquad \lambda \in]0,\lambda_o] \tag{19}$$

we know, from our analysis, that $f^+_{\epsilon,\delta}$ exists and belongs to $L^2(\partial\Omega, H)$. Now

$$f^+_{\delta,\epsilon} = \phi^+_{\delta,\epsilon} + \mathcal{V}_\lambda\, f^+_{\delta,\epsilon} \tag{20}$$

and, if we assume that 2.1.1.1. $(\mathcal{H}_{H,\tilde{H}})$ holds we have that $f^+_{\delta,\epsilon} \in L^2(\partial\Omega, \tilde{H})$ and

$$\begin{aligned} \|f^+_{\delta,\epsilon}\|_{L^2(\partial\Omega,\tilde{H})} &\leqslant \|\phi^+_{\delta,\epsilon}\|_{L^2(\partial\Omega,\tilde{H})} + |G|_{H,\tilde{H}}\ |G|_{\tilde{H},H}\ e^{-\beta\lambda^{-1} r^*(\lambda)}\, \|f^+_{\delta,\epsilon}\|_{L^2(\partial\Omega,\tilde{H})} \\ &\leqslant \|\phi^+_{\delta,\epsilon}\|_{L^2(\partial\Omega,H)} + |G|_{H,\tilde{H}} \|f^+_{\delta,\epsilon}\|_{L^2(\partial\Omega,H)} \end{aligned} \tag{21}$$

taking (12) into account. From 2.2.1.9.1. we have

$$\|f^+_{\delta,\epsilon}\|_{L^2(\partial\Omega,\tilde{H})} \leqslant C\ \|\phi^+_{\delta,\epsilon}\|_{L^2(\partial\Omega,\tilde{H})} \tag{22}$$

and under $\delta + \epsilon \to 0$, $f^+_{\delta,\epsilon} \to f^+$ in $L^2(\partial\Omega, \tilde{H})$ such that

$$(1-\mathcal{V}_\lambda)\, f^+ = \phi^+ . \tag{23}$$

From this analysis we conclude that, for $\lambda \in]\,0, \lambda_o]$, $1-\mathcal{V}_\lambda$ is invertible by a bounded operator in $L^2(\partial\Omega, \tilde{H})$.

2.3.1.2. We are going to extend by continuity the invertibility of $1-\mathcal{V}_\lambda$ to all values of $\lambda > 0$. We have analogously to 2.3.1.1.(1).

$$(1)\quad [\![J^- BE_{\lambda_2}\phi^+ - J^- BE_{\lambda_1}\phi^+]\!]^2 \leqslant \int_{R^3} |\xi|\omega \, d\xi \int_{(\partial\Omega)^+(\xi)} e^{-2\lambda_o^{-1}\nu d(x,\xi)|\xi|^{-1}} (1 - e^{-2\nu|\lambda_1^{-1}-\lambda_2^{-1}|\nu|\xi|^{-1}})^2 \cos\theta_y \, |\phi^+(y,\xi)|^2 \, dS_y$$

with

$$(2)\qquad \lambda_o = \sup(\lambda_1, \lambda_2) \; , \qquad \lambda_m = \mathrm{Inf}(\lambda_1, \lambda_2) \, .$$

From 2.3.1.1.(2) and boundedness of Ω, we have*

$$(3)\quad (1 - \exp[-2\nu \, |\lambda_2^{-1} - \lambda_1^{-1}| \, |\xi|^{-1} d]) \exp(-\nu \lambda_o^{-1} |\xi|^{-1} d) \leqslant \frac{2|\lambda_2 - \lambda_1|}{\lambda_m}$$

and substituting this into (1) we obtain

$$(4)\quad [\![J^- BE_{\lambda_2}\phi^+ - J^- BE_{\lambda_1}\phi^+]\!]^2 \leqslant 4\left(\frac{\lambda_2 - \lambda_1}{\lambda_m}\right)^2 \int_{R^3} |\xi|\omega d\xi \int_{(\partial\Omega)^+(\xi)} \cos\theta_y \, |\phi^+(y,\xi)|^2 \, dS_y$$

and, then, from the argument of 2.1.3.8.,

$$(5)\qquad [\![J^- BE_{\lambda_2}\phi^+ - J^- BE_{\lambda_1}\phi^+]\!]^2 \leqslant 4\left(\frac{\lambda_2 - \lambda_1}{\lambda_m}\right)^2 [\![\phi^+]\!] \, .$$

Using 2.2.1.1. $(\mathcal{H}_{H,H})$ we have

$$(6)\qquad |\mathfrak{V}_{\lambda_2} - \mathfrak{V}_{\lambda_1}|_{\mathcal{L}(L^2(\partial\Omega,\tilde{H}))} \leqslant 2|G|_{\tilde{H},\tilde{H}} \, |\lambda_m|^{-1} \, |\lambda_2 - \lambda_1| \, .$$

From 2.2.1.8.1. and the previous estimate (6) we extend, by continuity, the invertibility of $1 - \mathfrak{V}_\lambda$ to $\lambda \in \,]0, +\infty[$. The argument is simple and uses the formula

*) This estimate arises from $\sup_{\alpha > 0} e^{-\alpha}(1 - e^{-\alpha x}) = x(1+x)^{-(x+1)/x} \leqslant x$

$$(1-\mathfrak{V}_\lambda)^{-1} = \{1-(1-\mathfrak{V}_{\lambda_1})^{-1}(\mathfrak{V}_{\lambda_2}-\mathfrak{V}_{\lambda_1})\}^{-1}(1-\mathfrak{V}_{\lambda_1})^{-1} \tag{7}$$

to prove that if $1-\mathfrak{V}_\lambda$ is invertible, then $1-\mathfrak{V}_\lambda$ is invertible for $\lambda \in [a, +\infty[\cap]\lambda_1-\delta, \lambda_1+\delta[$ with $\delta = 1/2\, a(C(a)|G|_{\tilde{H},\tilde{H}})^{-1}$, and we observe that a may be chosen arbitrarily in $]0, +\infty[$.

2.3.1.3. We state the result that we have obtained. We assume that $\partial\Omega$ satisfies the Liapounov condition, Ω being bounded; we assume that $\nu(|\xi|)$ satisfies

$$\operatorname*{Inf}_{\xi} \frac{\mathfrak{V}(|\xi|)}{|\xi|} = \beta > 0\,, \qquad \nu \geqslant \mathfrak{V}_0 > 0\,, \tag{1}$$

and that G satisfies 1.4.4.5.(1)(2)(3) and 1.4.5.4.(1), and is bounded from $L^2(\partial\Omega,X)$ into $L^2(\partial\Omega,Y)$ where X and Y are any of H and $\tilde{H}$, then $(1-\mathfrak{V})$ is invertible in $L^2(\partial\Omega,\tilde{H})$ by a bounded operator T

$$T(1-\mathfrak{V})^{-1} = (1-\mathfrak{V})T = 1\,, \quad \text{in} \quad L^2(\partial\Omega, J^+\bar{H}) \tag{2}$$

If the complementary hypothesis that G is bounded in $L^2(\partial\Omega,\tilde{H}_\nu)$, where

$$\bar{H}_\nu = \{f(\xi) \mid f\in\bar{H}\,,\ \nu^{1/2} f\in\bar{H}\}\,; \quad \|f\|_{\bar{H}_\nu} = \|\nu^{1/2} f\|_{\bar{H}} \tag{3}$$

then T is bounded in $L^2(\partial\Omega, J^+H_\nu)$. We set

$$|T| = \|T\|_{\mathcal{L}(L^2(\partial\Omega,J^+\tilde{H}))}\,, \quad |T|_\nu = \|T\|_{\mathcal{L}(L^2(\partial\Omega,J^+H_\nu))} \tag{4}$$

2.3.1.4. The statements of 2.3.1.2. hold if G is replaced by G^*, $\mathfrak{V}$ by $\mathfrak{V}^*$ and T by T^*, while 1.4.5.4.(1) is replaced by 1.4.6.3.(8).

2.3.2. Properties of the Operators W and W. Some Function Spaces**

2.3.2.1. From 2.3.1.2. we see that the operators W and $\mathcal{E}$ of 2.1.2.4.(10) are well defined, the first from $L^2(\Omega, H)$ into itself and the second from $L^2(\partial\Omega, \tilde{H})$ or $L^2(\partial\Omega, \tilde{H}_s)$ into $L^2(\Omega, H)$ furthermore they are bounded operators. This last statement results from 2.3.1.3. and 2.1.3.3.(10), 2.1.3.5.(7), 2.1.3.6.(5), which yield

(1) $$|||W\varphi||| \leq (\nu_0^{-1} + (2\nu_0)^{-1}|T|\,|G|_{\tilde{H},\tilde{H}})\,|||\varphi|||\,,$$

(2) $$|||\mathcal{E}\phi^+||| \leq (2\nu_0)^{-1/2}|T|[\![\phi^+]\!]\,.$$

These statements do not need that G be bounded from $L^2(\partial\Omega, \tilde{H})$ into $L^2(\partial\Omega, \tilde{H}_\nu)$. When this complementary hypothesis holds, $TGJ^- BU_\varphi \in L^2(\partial\Omega, \tilde{H}_\nu)$. For $\varphi \in L^2(\Omega, H)$ and $T\phi^+ \in L^2(\partial\Omega, \tilde{H}_\nu)$ for $\phi^+ \in L^2(\partial\Omega, H_\nu)$. If we assume furthermore that G is bounded from $L^2(\partial\Omega, \tilde{H}_\nu)$ into itself, then we may apply separately $\xi\cdot\nabla$ and ν and we have, using in particular 2.1.3.5.(4)

(3) $$|||\nu W\varphi||| \leq (1 + 2^{-1}|T|_\nu\,|G|_{\tilde{H}_\nu,\tilde{H}_\nu})\,|||\varphi|||\,,$$

(4) $$|||\xi\cdot\nabla W\varphi||| \leq (1 + 2^{-1}|T|_\nu|G|_{\tilde{H}_\nu,\tilde{H}_\nu} + \nu_0^{-1} + (2\nu_0)^{-1}|T|\,|G|_{H,H})\,|||\varphi|||\,,$$

(5) $$|||\nu\,\mathcal{E}\phi^+||| \leq 2^{-1/2}|T|_\nu\,\|\phi^+\|_{L^2(\partial\Omega,\tilde{H}_\nu)}\,,$$

(6) $$|||\xi\cdot\nabla\,\mathcal{E}\phi^+||| \leq 2^{-1/2}|T|_\nu\,\|\phi^+\|_{L^2(\partial\Omega,\tilde{H}_\nu)}\,.$$

*) The hypothesis for the whole of 2.3.2. are that of 2.3.1.2. and 2.3.1.3.

Even without the two complementary hypothesis, which have been stated, we may apply the operator $\mathscr{M}$, as a whole, and

$$\mathscr{M} W \varphi = \varphi \,, \quad \mathscr{M} \mathscr{E} \phi^+ = 0 \,. \tag{7}$$

Again without the complementary hypothesis we have

$$[\![J^- \mathrm{B} W \varphi]\!] \leqslant (2\nu_0)^{-1/2} (1 + |\mathrm{T}| \, |\mathrm{G}|_{\tilde{\mathrm{H}},\tilde{\mathrm{H}}}) \, |\!|\!| \varphi |\!|\!| \,, \tag{8}$$

$$[\![J^- \mathrm{B} \mathscr{E} \phi^+]\!] \leqslant |\mathrm{T}| \, [\![\phi^+]\!] \,, \tag{9}$$

$$[\![J^+ \mathrm{B} W \varphi]\!] \leqslant (2\nu_0)^{-1/2} \, \nu_0^{-1/2} \, |\mathrm{T}| \, |\mathrm{G}|_{\tilde{\mathrm{H}},\tilde{\mathrm{H}}} \, |\!|\!| \varphi |\!|\!| \,, \tag{10}$$

$$[\![J^+ \mathrm{B} \mathscr{E} \phi^+]\!] \leqslant |\mathrm{T}| \, [\![\phi^+]\!] \,, \tag{11}$$

from 2.1.2.3.(3) to 2.1.2.4.(9) (10); on the other hand

$$J^- \mathrm{B} W \varphi = J^- \mathrm{B} \mathrm{U} \varphi + J^- \mathrm{B} \mathrm{E} \mathrm{T} \mathrm{G} J^- \mathrm{B} \mathrm{U} \varphi \,, \tag{12}$$

$$J^+ \mathrm{B} W \varphi = \mathrm{T} \mathrm{G} J^- \mathrm{B} \mathrm{U} \varphi \,, \tag{13}$$

$$J^- \mathrm{B} \mathscr{E} \phi^+ = J^- \mathrm{B} \mathrm{E} \mathrm{T} \phi^+ \,, \tag{14}$$

$$J^+ \mathrm{B} \mathscr{E} \phi^+ = \mathrm{T} \phi^+ \,, \tag{15}$$

(16) $$J^+ B \mathcal{W}\varphi - G J^- B \mathcal{W}\varphi = (1-\mathfrak{V}) T\ G J^- B \cup\varphi - G J^- B \cup\varphi = 0$$

(17) $$J^+ B \mathcal{E}\phi^+ - G J^- B\mathcal{E}\phi^+ = (1-\mathfrak{V})\, T\phi^+ = \phi^+ .$$

From all the statements of this paragraph, we conclude that

(18) $$f = \mathcal{W}\varphi + \epsilon\,\phi^+$$

solves, in a strong sense in $L^2(\partial\Omega, H)$ and $L^2(\partial\Omega, \tilde{H})$, the problem

(19) $$\mathcal{M} f = \varphi , \quad J^+ Bf - G J^- Bf = \phi^+ .$$

2.3.2.1.1. Under 2.3.1.3., all the statements of 2.3.2.1. hold if G, T, $\mathcal{W}$, $\mathcal{E}$, $\cup$, E are replaced by G^*, T^*, $\mathcal{W}^*$, $\mathcal{E}^*$, $\cup^*$, E^* and if J^+ and J^- are interchanged in 2.3.2.1.(8) to (17) and (19).

2.3.2.2. We define the spaces V and V^* as

(1) $$V : \{ f \mid \exists \varphi \in L^2(\Omega, H) \text{ such that }, \ f = \mathcal{W}\varphi \}$$

(2) $$V^* : \{ f \mid \exists \varphi \in L^2(\Omega, H) \text{ such that }, \ f = \mathcal{W}^*\varphi \}$$

and we make them Banach spaces with the norms

(3) $$\| f \|_V = \{ ||| f |||^2 + |||\mathcal{M} f |||^2 \}^{1/2} ,$$

(4) $$\| f \|_{V^*} = \{ ||| f |||^2 + |||\mathcal{M}^* f |||^2 \}^{1/2} .$$

We need only prove that V and V* are complete. We give the proof for V. Let $\{ f_n \}$ be a Chauchy sequence in V, we have $f_n = W\varphi_n$ and

$$|||\varphi_n - \varphi_p||| = |||\mathcal{M}W\varphi_n - \mathcal{M}W\varphi_p||| \leqslant \|f_n - f_p\|_V \tag{5}$$

from which we conclude that $\varphi_n \to \varphi \in L^2(\Omega, H)$ and if we set $f = W\varphi$ we have $\| f - f_n \|_V \to 0$.

From 2.3.2.1.(1) and 2.3.2.2.(3) we have

$$f \in V \Rightarrow |||f||| \leqslant \nu_0^{-1}(1 + 2^{-1}\,|T|\,|G|_{\tilde{H},\tilde{H}})\ |||\mathcal{M}f|||\ , \tag{6}$$

and, consequently

$$f \in V \Rightarrow \|f\|_V \leqslant \{1 + \nu_0^{-2}(1 + 2^{-1}\,|T|\,|G|_{\tilde{H},\tilde{H}})^2\}^{1/2}\ |||\mathcal{M}f|||\ , \tag{7}$$

and in a similar way

$$f \in V^* \Rightarrow \|f\|_{V^*} \leqslant \{1 + \nu_0^{-2}(1 + 2^{-1}\,|T|\,|G^*|_{\tilde{H},\tilde{H}})^2\}^{1/2}\ |||\mathcal{M}^*f|||\ . \tag{8}$$

From 2.3.2.1.(8) (10) and the analogue we have

$$(2\nu_0)^{-1/2}(1 + 2\,|T|\,|G|_{\tilde{H},\tilde{H}})\ \|f\|_v$$

$$f \in V \Rightarrow [\![Bf]\!] \leqslant \nu_0^{-1/2}(2^{-1/2} + (1 + 2^{-1/2})\,|T|\,|G|_{\tilde{H},\tilde{H}}\ \|f\|_V\ ,\ J^+Bf = GJ^-Bf\ , \tag{9}$$

$$f \in V^* \Rightarrow [\![Bf]\!] \leqslant \nu_0^{-1/2}(2^{-1/2} + (1 + 2^{-1/2})|T|\,|G|_{\tilde{H},\tilde{H}})\ \|f\|_{V^*},\ J^-Bf = G^*J^+Bf \tag{10}$$

If the two complementary hypothesis of 2.3.2.1. hold, then

$$f \in V \Rightarrow |||\nu f||| \leqslant (1 + 2^{-1}\,|T|_\nu\,|G|_{\tilde{H}_\nu,\tilde{H}_\nu})\ \|f\|_V\ , \tag{11}$$

$$f \in V \Rightarrow |||\xi \cdot \nabla f||| \leqslant (1 + 2^{-1} |T|_\nu |G|_{\tilde{H}_\nu \tilde{H}_\nu} + \nu_0^{-1} + (2\nu_0)^{-1} |T| |G|_{\tilde{H},\tilde{H}}) \|f\|_V \ , \tag{12}$$

while, if the two complementary hypothesis hold for G^* we have

$$f \in V^* \Rightarrow |||\nu f||| \leqslant (1 + 2^{-1} |T^*|_\nu |G^*|_{\tilde{H}_\nu,\tilde{H}_\nu}) \|f\|_{V^*} \tag{13}$$

$$f \in V^* \Rightarrow |||\xi \cdot \nabla ||| \leqslant (1 + 2^{-1} |T^*|_\nu |G^*|_{\tilde{H}_\nu,\tilde{H}_\nu} + \nu_0^{-1} + (2\nu_0)^{-1} |T| |G|_{\tilde{H},\tilde{H}}) \|f\|_{V^*} \ . \tag{14}$$

2.3.2.3. We define two further spaces W and W^* as

(1) $W : \{ f \mid \exists \varphi \in L^2(\Omega, \tilde{H}) \ , \ \exists \phi^+ \in L^2(\partial\Omega, \tilde{H}) \text{ such that } f = \mathcal{W}\varphi + \mathcal{E}\phi^+ \}$

(2) $W^* : \{ f \mid \exists \varphi \in L^2(\Omega, \tilde{H}) \ , \ \exists \phi^- \in L^2(\partial\Omega, \tilde{H}) \text{ such that } f = \mathcal{W}^\phi \varphi + \mathcal{E}^\phi \phi^+ \}$

and we make of them Banach spaces with norms

$$\|f\|_W = \{ |||f|||^2 + |||\mathcal{L}\mathcal{M} f|||^2 + [\![\mathcal{J}^+ Bf - G\mathcal{J}^- Bf]\!]^2 \}^{1/2} \ , \tag{3}$$

$$\|f\|_{W^*} = \{ |||f|||^2 + |||\mathcal{L}\mathcal{M}^* f|||^2 + [\![\mathcal{J}^- Bf - G^* \mathcal{J}^+ Bf]\!]^2 \}^{1/2} \ . \tag{4}$$

Let $\{f_n\}$ be a Cauchy sequence in W with

$$f_n = \mathcal{W}\varphi_n + \mathcal{E}\phi_n^+ \tag{5}$$

from

$$|||\varphi_n - \varphi_p||| \leq |||f_n - f_p|||_W \ , \ [\![\phi_n^+ - \phi_p^+]\!] \leq \|f_n - f_p\|_W \tag{6}$$

we see that $\{\varphi_n\}$ and $\{\phi_n^+\}$ are Cauchy sequences in $L^2(\Omega, H)$ and $L^2(\partial\Omega, \tilde{H})$. As a consequence there exist, unique, f and ϕ^+ such that

$$|||\varphi_n - \varphi||| \to 0 \ , \quad [\![\phi_n^+ - \phi^+]\!] \to 0 \ , \quad \text{as } n \to \infty \tag{7}$$

and, with

$$f = \mathcal{W}\varphi + \mathcal{E}\phi^+ \ , \tag{8}$$

we have

$$\|f_n - f\|_W \to 0 \ , \qquad \text{as } n \to \infty \ . \tag{9}$$

2.3.2.4. For $f \in \mathcal{W}$ (or $f \in W^*$) we may state a set of inequalities analogous to the ones in 2.3.2.2.

$$f \in W \Rightarrow \|f\|_W \leq C(|||\mathcal{M}f||| + [\![\mathcal{J}^+Bf - G\mathcal{J}^-Bf]\!]) \tag{1}$$

With some constant that we do not state explicitly. In the same way, with different constant

$$f \in W \Rightarrow [\![Bf]\!] \leq C(|||\mathcal{M}f||| + [\![\mathcal{J}^+Bf - G\mathcal{J}^-Bf]\!]) \ , \tag{2}$$

(3) $$f \in W^* \Rightarrow \|f\|_{W^*} \leqslant C(|||\mathcal{LM}f||| + [\![\mathcal{J}^- Bf - G^* \mathcal{J}^+ Bf]\!]) ,$$

(4) $$f \in W^* \Rightarrow [\![Bf]\!] \leqslant C(|||\mathcal{LM}f||| + [\![\mathcal{J}^- Bf - G^* \mathcal{J}^+ Bf]\!] .$$

If the two complementary hypothesis of 2.3.2.1. hold, then

(5) $$f \in W \Rightarrow |||\nu f||| \leqslant C(|||\mathcal{LM}f||| + [\![\mathcal{J}^+ Bf - G\mathcal{J}^- Bf]\!]) ,$$

(6) $$f \in W \Rightarrow |||\xi \cdot \nabla f||| \leqslant C(|||\mathcal{LM}f||| + [\![\mathcal{J}^+ Bf - G\mathcal{J}^- Bf]\!]) ,$$

while if the two complementary hypothesis hold for G^*, then

(7) $$f \in W^* \Rightarrow |||\nu f||| \leqslant C(|||\mathcal{LM}^* f||| + [\![\mathcal{J}^- Bf - G^* \mathcal{J}^+ Bf]\!]) ,$$

(8) $$f \in W^* \Rightarrow |||\xi \cdot \nabla f||| \leqslant C(|||\mathcal{LM}^* f||| + [\![\mathcal{J}^- Bf - G^* \mathcal{J}^+ Bf]\!]) .$$

2.3.2.5. Let $f \in W$, φ, ϕ^+ be as in 2.3.2.3.(8) then we have

(1) $$f = U\varphi + E f^+ ,$$

(2) $$f^+ = TG\mathcal{J}^- BU\varphi + T\phi^+ .$$

We can find a sequence $\{\varphi_n\}$ of functions, continuous on $\Omega \times R^3$ vanishing in a vicinity of $|\xi| = \infty$ such that

$$|||\varphi_n - \varphi||| \to 0\,, \quad \text{as } n \to \infty\,. \tag{3}$$

We set

$$f_n = U\varphi_n + E f_n^+ = \mathcal{W}\varphi_n + \mathcal{E}\phi^+\,, \tag{4}$$

$$f_n^+ = T G \mathcal{J}^- B U\varphi_n + T\phi^+\,, \tag{5}$$

and, for each fixed n, we can find a sequence $\{f_{n,p}^+\}$ of functions continuous on the set $\{(x, \xi) \mid x \in \partial\Omega,\ \xi \in R_+^3(\xi)\}$, vanishing in a vicinity of $|\xi| = \infty$, such that

$$[\![f_{n,p}^+ - f_n^+]\!] \to 0\,, \quad \text{as } p \to \infty,\ n \text{ fixed}\,. \tag{6}$$

Each $f_{n,p}^+$ allows to define an $f_{n,p}$

$$f_{n,p} = U\varphi_n + E f_{n,p}^+ \tag{7}$$

which is piecewise continuous* on $\Omega \times R^3$, vanishes in a vicinity of infinity in velocity space and continuously differentiable, along rays, with respect to position. Each $f_{n,p}$ belongs to W, namely

$$f_{n,p} = \mathcal{W}\varphi_n + \mathcal{E}\phi_{n,p}^+\,, \tag{8}$$

$$\phi_{n,p}^+ = (1 - \mathfrak{V})\, f_{n,p}^+ - G \mathcal{J}^- B U \varphi_n \tag{9}$$

---

*Continuous for convex Ω.

as is shown by the following computation

$$W\varphi_n + \mathcal{E}\phi^+_{n,p} = \cup\varphi_n + ETGJ^- B\cup\varphi_n$$

(10) $$+ ET(1-\mathfrak{V})\, f^+_{n,p} - ETGJ^- B\cup\varphi_n$$

$$= \cup\varphi_n + Ef^+_{n,p} .$$

Now

(11) $$\| f - f_n \|_W \leqslant ||| W(\varphi_n - \varphi) ||| + ||| \varphi_n - \varphi ||| \leqslant C ||| \varphi_n - \varphi |||$$

(12) $$\| f_{n,p} - f_n \|_W \leqslant ||| E(f^+_{n,p} - f_n) ||| + [\![\phi^+_{n,p} - \phi^+]\!]$$

while

(13) $$\phi^+_{n,p} - \phi^+ = (1-\mathfrak{V})\,(f^+_{n,p} - f^+_n)$$

so that

(14) $$\| f_{n,p} - f_n \|_W \leqslant C\, [\![f^+_{n,p} - f^+_n]\!] .$$

The two constants C in (11) and (14) are different but they do not depend on n nor p. From (3) and (6) taking into account (11) and (13) we see that, to each we may associate some p(n) such that

(15) $$\| f_{n,p(n)} - f \|_W \to 0 \quad \text{as } n \to \infty$$

2.3.2.5.1. We state the result that we have just obtained*. The set of functions $f(x, \xi)$ which are piecewise continuous on $\Omega \times R^3$ continuously differentiable along rays with respect to position and which vanish in a vicinity of $|\xi| = \infty$ is dense in W (resp W*). The same result holds for V (resp V*) when it is considered as a topological subspace of W (resp W*) with the same Banach structure.

2.3.2.6. Assume that the two complementary hypothesis of 2.3.2.1. hold and let $f \in W$ with

$$f = W\varphi + \mathcal{E}\phi^+ . \tag{1}$$

Define (refer to 2.3.2.1.)

$$\chi = 2\nu W\varphi + 2\nu \mathcal{E}\phi^+ - \varphi , \tag{2}$$

$$\psi^- = J^- B W\varphi - G^* T G J^- B U\varphi + J^- BET\phi^+ - G^* T\phi^+ , \tag{3}$$

we obtain

$$f = W^*\chi + \mathcal{E}^*\psi^- , \tag{4}$$

and $f \in W^*$. We have

$$\| f \|_W = \{ |||f|||^2 + |||\varphi|||^2 + [\![\phi^+]\!]^2 \}^{1/2} \tag{5}$$

$$\| f \|_{W^*} = \{ |||f|||^2 + |||\chi|||^2 + [\![\psi^-]\!]^2 \}^{1/2} \tag{6}$$

---

*) The hypothesis are the ones of 2.3.1.3 and 2.3.1.4.

and we leave to the reader the proof that there exist two constants C_1 and C_2, such that

$$\|f\|_{W*} \leq C_1 \|f\|_W , \quad \|f\|_W \leq C_2 \|f\|_{W*} . \tag{7}$$

The two space W and W* are identical as topological vector spaces but they are endowed with different norms.

2.3.3. The Basic Inequality

2.3.3.1. Let $f \in W$ and consider the sequence $\{f_n\}$ with the properties stated in 2.3.2.5.1. From 2.1.1.7. we may apply 2.1.1.7.(3) to each f_n namely

$$\mu |||Pf_n|||^2 + k [\![\Pi J^- Bf_n]\!]^2 \leq |||\mathcal{L} f_n||| \, |||f_n||| + \frac{1}{2} [\![J^+Bf_n - GJ^- Bf_n]\!]^2 +$$

$$+ \operatorname{Re}\{ J^+ Bf_n - GJ^- Bf_n , G\Pi J^- Bf_n \} + \tag{1}$$

$$+ \sqrt{2\pi} \int_{\partial\Omega} \operatorname{Re}(J^+Bf_n - GJ^- Bf_n , 1^+)_{\tilde{H}} \overline{(J^- Bf_n , 1^-)_{\tilde{H}}} \, ds .$$

Let

$$f_n = W\varphi_n + \mathcal{E}\phi_n^+ , \tag{2}$$

$$f = W\varphi + \mathcal{E}\phi^+ , \tag{3}$$

we know that

$$|||\varphi_n - \varphi||| \to 0, \quad [\![\phi_n^+ - \phi^+]\!] \to 0 , \quad |||f_n - f||| \to 0 \tag{4}$$

and, from 2.3.2.1.(8)(9)(10)(11) we conclude that

$$[\![J^- Bf_n - J^- Bf]\!] \to 0 \ , \ [\![J^+ Bf_n - J^+ Bf]\!] \to 0 \ , \tag{5}$$

then, trivially

$$[\![\Pi J^- Bf_n - \Pi J^- Bf]\!] \to 0 \tag{6}$$

and, thanks to 2.2.1.1. $(\mathcal{H}_{\tilde{H},\tilde{H}})$ we conclude that

$$[\![G J^- Bf_n - G J^- Bf]\!] \to 0 \ , \ [\![G\Pi J^- Bf_n - G\Pi J^- Bf]\!] = 0 \tag{7}$$

while, provided A is bounded

$$|||\mathcal{L}f_n - \mathcal{L}f||| \leqslant |||\varphi_n - \varphi||| + |A| \, |||f_n - f||| \to 0 \, . \tag{8}$$

From all of these we conclude that it is legitimate to take the limit of each term of (1).

2.3.3.2. We state the result that we have just proven. The hypothesis are the ones of 2.3.1.3., such boundedness of A in H, then for any $f \in W$ the following inequality holds

$$\mu \, |||Pf|||^2 + k \, [\![\Pi J^- Bf]\!]^2 \leqslant |||\mathcal{L}f||| \ |||f||| +$$

$$\frac{1}{2} [\![J^+ Bf - G J^- Bf]\!]^2 + \mathrm{Re}\{J^+ Bf - G J^- Bf, G\Pi J^- Bf\} \tag{1}$$

$$+ \sqrt{2\pi} \int_{\partial\Omega} \mathrm{Re}(J^+ Bf - G J^- Bf, 1^+)_{\tilde{H}} \, \overline{(J^- Bf, 1^-)_{\tilde{H}}} \, ds$$

2.3.3.3.* For any $f \in V$ the last three terms in 2.3.3.2.(1) are zero and we have

(1) $$\mu |||Pf|||^2 + k [\![\Pi J^- Bf]\!]^2 \leqslant |||\mathcal{L} f|||\ |||f||| .$$

2.3.3.4.* For any $f \in W^*$ we have

(1) $$\begin{aligned} \mu |||Pf|||^2 + k^* [\![\Pi J^+ Bf]\!]^2 \leqslant\ & |||\mathcal{L}^* f|||\ |||f||| + \\ & + \frac{1}{2} [\![J^- Bf - G^* J^+ Bf]\!]^2 + \\ & + \operatorname{Re} \{ J^- Bf - G^* J^+ Bf , G^* \Pi J^+ Bf \} \\ & + \sqrt{2\pi} \int_{\partial\Omega} \operatorname{Re}(J^+ Bf - G^* J^+ Bf, 1^+)_{\tilde{H}} \overline{(J^+ Bf, 1^+)_{\tilde{H}}}\, ds . \end{aligned}$$

2.3.3.5.* For any $f \in V^*$ we have

(1) $$\mu |||Pf|||^2 + k^* [\![\Pi J^+ Bf]\!]^2 \leqslant |||\mathcal{L}^* f|||\ |||f|||$$

2.3.4. The Generalized Identity of Moments

2.3.4.1. Consider an $f \in W$ and a $g \in W^*$, we have

(1) $$f = W\varphi + \mathcal{E} \phi^+ , \quad \varphi \in L^2(\Omega\, H) , \quad \phi^+ \in L^2(\partial\Omega, \tilde{H})$$

(2) $$g = W^* \chi + \mathcal{E}^* \psi^- , \quad \chi \in L^2(\Omega, H) , \quad \psi^- \in L^2(\partial\Omega, \tilde{H})$$

*) The hypothesis for 2.3.3.3., 2.3.3.4. and 2.3.3.5. are the same as for 2.3.3.2., namely the ones of 2.3.1.3. (or, are as precedingly, 2.3.1.4.) and boundedness of A in H.

and, then

$$\varphi = \mathscr{M} \mathrm{f}, \; \chi = \mathscr{M}^* \mathrm{g}, \tag{3}$$

$$\phi^+ = J^+ \mathrm{Bf} - \mathrm{G} J^- \mathrm{Bf}, \quad \psi^- = J^- \mathrm{Bg} - \mathrm{G}^* J^+ \mathrm{Bg}. \tag{4}$$

We may find sequences $\{\varphi_n\}$, $\{\chi_n\}$, $\{\phi_n^+\}$, $\{\psi_n^-\}$ of continuous functions such that

$$|||\varphi_n - \varphi||| \to 0, \quad |||\chi_n - \chi||| \to 0, \tag{5}$$

$$[\![\phi_n^+ - \phi^+]\!] \to 0, \quad [\![\psi_n^- - \psi^-]\!] \to 0. \tag{6}$$

To each set $\varphi_n, \chi_n, \phi_n^+, \psi_n^-$ we may apply 2.1.2.6.(5), namely

$$\begin{aligned} &((\varphi_n, W^* \chi_n + \mathscr{E}^* \psi_n^-)) - ((W\varphi_n + \mathscr{E} \phi_m^+, \chi_n)) \\ &+ \{\phi_n^+, J^+ \mathrm{B} W^* \chi_n + J^+ \mathrm{B} \mathscr{E}^* \psi_n^-\} - \{J^- \mathrm{B} W \varphi_n + J^- \mathrm{B} \mathscr{E} \phi_n^+, \psi_n^-\} = 0 \end{aligned} \tag{7}$$

according to (5) (6) and

$$|||W\varphi_n - W\varphi||| \to 0, \quad |||\mathscr{E}\phi_n^+ - \mathscr{E}\phi^+||| \to 0, \tag{8}$$

$$|||W^*\chi_n - W^*\chi||| \to 0, \quad |||\mathscr{E}^*\psi_n^- - \mathscr{E}^*\psi^-||| \to 0, \tag{9}$$

$$[\![J^- \mathrm{B} W\varphi_n - J^- \mathrm{B} W\varphi]\!] \to 0, \quad [\![J^- \mathrm{B}\mathscr{E}\phi_n^+ - J^- \mathrm{B}\mathscr{E}\phi^+]\!] \to 0, \tag{10}$$

$$(11) \qquad [\![J^+ B W^* \chi_n - J^+ B W^* \chi]\!] \to 0 \,, \quad [\![J^+ B \mathcal{E}^* \psi_n^- - J^+ B \mathcal{E}^* \psi^-]\!] \to 0 \,,$$

which result from 2.3.2.1.(1) (2) (8) (9) (10) (11) and 2.3.2.1.1. (observe that we do not need the complementary hypothesis of 2.3.1.3. and 2.3.1.4.) we can take the limit of each term in (7) and conclude that, for any $\varphi \in L^2(\Omega, H)$, $\chi \in L^2(\Omega, H)$, $\phi^+ \in L^2(\partial\Omega, \tilde{H})$, $\psi \in L^2(\partial\Omega, \tilde{H})$ we have

$$(12) \qquad \begin{aligned} &((\varphi, W^*\chi + \mathcal{E}^*\psi^-)) - ((W\varphi + \mathcal{E}\phi^+, \chi)) + \\ &+ \{\phi^+; J^+BW^*\chi + J^+B\mathcal{E}^*\psi^-\} - \{J^- BW\varphi + J^- B\mathcal{E}\phi^+, \psi^-\} = 0 \,. \end{aligned}$$

From (1) (2) (3) (4) we may state that, for any $f \in W$, $g \in W^*$ we have

$$(13) \qquad ((\mathcal{M}f, g)) - ((f, \mathcal{M}^* g)) + \{J^+Bf - GJ^- Bf, J^+Bg\} - \{J^- Bf, J^- Bg - G^* J^+Bg\} = 0 \,.$$

2.3.4.2. Provided A is bounded in H, which occurs for a power law angular cut-off* or rigid spheres we have that, for any $f \in W$, $g \in W^*$ the following generalized identity of moments holds

$$(1) \qquad ((\mathcal{L}f, g)) - ((f, \mathcal{L}^*g)) + \{J^+Bf - GJ^- Bf, J^+Bg\} - \{J^- Bf, J^- Bg - G^* J^+Bg\} = 0 \,.$$

This relation holds equally well if the domain of A is included in the one of ν and if, concurrently, the two complementary hypothesis of 2.3.2.1. and 2.3.2.2. hold.

2.3.4.3. For $f \in V$, $g \in V^*$ we have

$$(1) \qquad ((\mathcal{L}f, g)) - ((f, \mathcal{L}^* g)) = 0 \,.$$

---

*) But, then, 2.3.1.2.(1) does not hold and our analysis fills.

For $f \in W$, $g \in V^*$ we have

$$((\mathcal{L}f, g)) - ((f, \mathcal{L}^* g)) + \{J^+ Bf - G J^- Bf, J^+ Bg\} = 0 \,, \tag{2}$$

while, for $f \in V$, $g \in W^*$ we obtain

$$((\mathcal{L}f, g)) - ((f, \mathcal{L}^* g)) - \{J^- Bf, J^- Bg - G^* J^+ Bg\} = 0 \,. \tag{3}$$

2.3.4.4. Assume that for some f and $\varphi \in L^2(\Omega, H)$ and $\phi^+ \in L^2(\partial\Omega, \tilde{H})$ there exists $\theta^- \in L^2(\partial\Omega, J^- \tilde{H})$ such that, for any $h \in W^*$

$$h = \mathcal{W}^* \chi + \mathcal{E}^* \psi^- \tag{1}$$

one has

$$((f, \mathcal{M}^* h)) - \{\phi^+, J^+ Bh\} = ((\varphi, h)) - \{\theta^-, \psi^-\}. \tag{2}$$

From (1) we have

$$((f, \chi)) - \{\phi^+, J^+ B \mathcal{W}^* \chi + J^+ B \mathcal{E}^* \psi^-\} = ((\varphi, \mathcal{W}^* \chi + \mathcal{E}^* \psi^-)) - \{\theta^-, \psi^-\}, \tag{3}$$

and this relation, taking 2.3.4.1.(12) into account, may be rewritten as

$$((f - \mathcal{W}\varphi - \mathcal{E}\phi^+, \chi)) + \{\theta^- - J^- B \mathcal{W}\varphi - J^- B \mathcal{E}\phi^+, \psi^-\} = 0 \,. \tag{4}$$

This relation holds for any $\chi \in L^2(\Omega, H)$ and $\psi^- \in L^2(\partial\Omega, \tilde{H})$, and, consequently, we may choose $\gamma = f - \mathcal{W}\varphi - \mathcal{E}\phi^+$ and $\psi^- = \theta^- - J^- B \mathcal{W}\varphi - J^- B \mathcal{E}\phi^+$ and conclude that

(5) $$f = \mathcal{W}\varphi + \mathcal{E}\phi^+ , \quad \theta^- = \mathcal{J}^- Bf ,$$

but this implies that $f \in W$ and $\phi^+ = J^+Bf - G\mathcal{J}^- Bf$.

2.3.4.4.1. We state the result which we have obtained. The hypothesis are the ones of 2.3.1.3. and 2.3.1.4. (boundedness of Ω, Liapunov condition for $\partial\Omega$, 2.3.1.3.(1) for ν 1.4.4.5.(1) (2) (3) and 1.4.5.4.(1) for G 1.4.6.3.(8) for G*, and finally boundedness of G and G* from $L^2(\partial\Omega, X)$ into $L^2(\partial\Omega, Y)$ where X and Y are any of H and $\tilde{H}$). Given any f and $\varphi \in L^2(\Omega, H)$ and $\phi^+ \in L^2(\partial\Omega, \mathcal{J}^+ \tilde{H})$, the necessary and sufficient conditions in order that the following holds

(1) $$f \in W , \quad \mathcal{M}f = \varphi , \quad \mathcal{J}^+Bf - G\,\mathcal{J}^- Bf = \phi^+ ,$$

is that there exists a $\theta^- \in L^2(\partial\Omega, \tilde{H})$ such that for any $h \in W^*$ with $h = \mathcal{W}^* \chi + \mathcal{E}^* \psi^-$ the relation

(2) $$((f, \mathcal{M}^* h)) - \{\phi^+, \mathcal{J}^+Bh\} - ((\varphi, h)) = -\{\theta^-, \psi^-\} ;$$

holds; then, this $\theta^- = \mathcal{J}^- Bf$.

2.3.4.4.2. Under the hypothesis of 2.3.4.4.1. given g and $\chi \in L^2(\Omega, H)$ and $\psi^- \in L^2(\partial\Omega, \tilde{H})$ the necessary and sufficient condition in order that the following holds

(1) $$g \in W^* , \mathcal{M}^* g = \chi , \quad \mathcal{J}^- Bg - G^* \mathcal{J}^+ Bg = \psi^- ,$$

is that there exists a $\theta^+ \in L^2(\partial\Omega, H)$ such that, for any $f \in W$, $f = W\varphi + \mathcal{E}\phi^+$, the relation

$$((\mathcal{M}f,g)) - \{J^- Bf, \psi^-\} - ((f,\chi)) - \{\theta^+, \phi^+\}. \quad (2)$$

holds; then, this $\theta^+ = J^+ Bg$.

2.3.4.4.3. With the hypothesis that A is bounded in H or that the domain of A in H is contained in that one of ν, and furthermore (for A not bounded) that the complelentary hypothesis and 2.3.2.1. holds, the statements of 2.3.4.4.2. hold with $\mathcal{L}^*$ standing for $\mathcal{M}^*$, with the convention that the complementary hypothesis of 2.2.1.3. replaces the one of 2.2.1.2. for A not bounded.

2.3.4.5. Let us come back to the argument of 2.3.4.4. and assume that, in 2.3.4.4.(2), h is restricted to V^*, then we obtain (4) with $\psi^- = 0$ and χ arbitrary. Choosing $\chi = f - W\varphi - \mathcal{E}\phi^+$ we obtain

$$f = W\varphi - \mathcal{E}\phi^+ \quad (1)$$

2.3.4.5.1. The statements of 2.3.4.4.1., 2.3.4.4.2. and 2.3.4.4.3. hold if W^* and W are replaced by V and V^* respectively and if the right hand side of 2.3.4.1.(2) and 2.3.4.4.2(2) are replaced to zero.

2.3.4.5.2. For further reference we state explicitely a corollary of one of the statements of 2.3.4.5.1. The hypothesis are that of 2.3.4.4.1. and 2.3.4.5.1. The hypothesis are that of 2.3.4.4.1. and 2.3.4.4.3., then the necessary and sufficient condition in order that given $\varphi \in L^2(\Omega, H)$, some of $f \in L^2(\Omega, H)$ solves the problem

$$f \in V, \quad \mathcal{L}f = \varphi, \quad (1)$$

is that, for any $h \in V^*$ we have

$$((f, \mathcal{L}^* h)) = ((\varphi, h)). \quad (2)$$

We observe that (1) $\Rightarrow \mathcal{J}^+ Bf = G\mathcal{J}^- Bf$.

*2.3.5. The a-Priori Estimate**

2.3.5.1. We consider an f such that

$$f \in V \ , \ ((f, 1)) = 0 \tag{1}$$

and we set

$$\mathcal{L} f = \varphi \ . \tag{2}$$

From 2.3.3.2. we have

$$\mu \, |||Pf|||^2 + k \, [\![\Pi \mathcal{J}^- Bf]\!]^2 \leqslant |||\varphi||| \ |||f||| \ , \tag{3}$$

but this inequality does not allow to estimate $|||f|||$. From the definition of **P** (1.5.4.4.(8)) we state that

$$f = \sum_{\alpha=0}^{4} \rho_\alpha(x) \, \psi_\alpha(\xi) + Pf \ ; \ (\mathbf{P}f, \psi_\alpha) = 0 \ , \ \alpha = 0, \ldots 4, \tag{4}$$

and, from the second of (1) we have

$$(\rho_0, 1)_{L^2(\Omega)} = 0 \ . \tag{5}$$

From the last of (4) and 1.4.3.3.(1) we find

$$|||f|||^2 = \sum_{\alpha=0}^{4} \|\rho_\alpha\|^2_{L^2(\Omega)} + |||Pf|||^2 \tag{6}$$

and, using (3) and 2.2.1.9.(3) we get

*) Through all of 2.3.5. we assume that the hypothesis of 2.3.2.1. hold as well as boundedness of A.

$$(\mu-\epsilon_1) \, |||Pf|||^2 + k[\![\Pi J^- Bf]\!]^2 \leqslant \epsilon_1 \sum_{\alpha=0}^{4} |||\rho_\alpha|||^2_{L^2(\Omega)} + (4\epsilon_1)^{-1} |||\varphi|||^2 \tag{7}$$

and what we need is an estimate of each $|||\rho_\alpha|||_{L^2(\Omega)}$ in terms of $|||Pf|||$.

2.3.5.2. We use 2.3.4.5.2.(2) and choose

$$h = \theta(x) \; \phi(\xi) \tag{1}$$

with θ continuously derivable, vanishing on $\partial\Omega$ while $\phi, |\xi|\phi, \nu\phi$ belong to H. We have

$$((f, \mathcal{L}^* h)) = -\sum_{\alpha=0}^{4} (\psi_\alpha, \xi\phi)_H \int_\Omega \rho_\alpha \; \nabla\theta \; dx + ((Pf, \mathcal{L}^* \theta\phi)) \tag{2}$$

thanks to $(\psi_\alpha, L\phi)_H = 0$. From (2) and 2.3.4.5.2.(2) we get

$$\Big| \sum_{\alpha=0}^{4} (\psi_\alpha, \xi\phi)_H \int_\Omega \rho_\alpha \cdot \nabla\theta \, dx \Big| \leqslant |||\varphi||| \; |||\theta\phi||| + |||Pf||| \; |||\mathcal{L}^*\theta\phi||| . \tag{3}$$

Let us set

$$|\theta|_0 = \Big| \int_\Omega |\theta|^2 \, dx \Big|^{1/2} , \quad |\theta|_1 = \Big| \int_\Omega \sum_i \Big|\frac{\partial\theta}{\partial x_i}\Big|^2 dx \Big|^{1/2} \tag{4}$$

we have

$$|||\theta\phi||| = |\theta|_0 \; \|\phi\|_H \tag{5}$$

and

$$\begin{aligned}(6)\qquad |||\mathcal{L}^*\theta\phi||| &\leqslant |||\xi\cdot\nabla\theta\phi||| + |||\nu\theta\phi||| + |||\theta A\phi||| \\ &\leqslant |||\xi\cdot\nabla\theta\,\phi||| + |\theta|_0\,(|||\nu\phi||| + |||A\phi|||)\end{aligned}$$

on the other hand

$$\begin{aligned}|||\xi\cdot\nabla\,\theta\phi||| &= \int_\Omega dx\ \|\sum_{i=1}^{3}\xi_i\frac{\partial\theta}{\partial x_i}\ \phi\|_H^2 \\ (7)\qquad &= \int_\Omega\sum_{i=1}^{3}\left|\frac{\partial\theta}{\partial x_i}\right|^2\ \|\xi_i\phi\|_H^2\ dx \\ &\leqslant |||\xi|\phi\|_H\ |\theta|_1\ .\end{aligned}$$

Substituting (5) (6) (7) into (3) we obtain

$$\begin{aligned}(8)\qquad \left|\sum_{\alpha=0}^{4}(\psi_\alpha,\xi\phi)_H\int_\Omega\rho_\alpha\cdot\nabla\,\theta\,dx\ \right| &\leqslant |||\varphi|||\ \|\phi\|_H\ |\theta|_0 + \\ &+ |||Pf|||\,([\|\nu\phi\|_H + \|A\phi\|_H]\ |\theta|_0 + \|\,|\xi|\,\phi\|_H\ |\theta|_1)\end{aligned}$$

2.3.5.3. We observe that $\phi(\xi)$ in (8) may be chosen arbitrarily provided only that $\|\,\xi|\phi\|_H$, $\|\nu\phi\|_H$ and $\|A\phi\|_H$ be bounded. The set of ϕ in H such that these conditions hold is some subspace that we designate as *K*. The set of ϕ in *K* such that $(\xi\,\psi_\alpha,\phi)_H = 0$ is again a subspace of codimension 5. We set

$$X_\alpha : \{\phi \mid \phi \in H, (|\xi|\phi, \nu\phi, A\phi) \in H, (\xi\psi_\alpha, \phi)_H = 0\} . \tag{1}$$

Now

$$Y_\alpha = \bigcap_{\substack{\beta \in (0,1,2,3,4) \\ \beta \neq \alpha}} X_\beta \tag{2}$$

is again a linear subspace of codimension at most 12, and given a unit vector e we may find in Y_α some $\phi(\alpha, e)$ such that

$$(\xi\psi_\alpha, \phi(\alpha, e))_H = e. \tag{3}$$

From the previous analysis we conclude that with $\phi = \phi(\alpha, e)$ in 2.3.5.2.(8) we have

$$\left| \int_\Omega \rho_\alpha \, e \cdot \nabla\theta \, dx \right| \leq |||\varphi||| \, \|\phi(\alpha, e)\|_H \, |\theta|_0 + \\ + |||Pf||| \, \{ \|\nu\phi(\alpha, e)\|_H + \|A\phi(\alpha, e)\|_H) \, |\theta|_0 + \| \, |\xi|\phi(\alpha, \xi)\|_H \, |\theta|_1 \} \tag{4}$$

and this holds for each $\alpha = 0,1,2,3,4.$

Now we observe that, from the hypothesis that θ vanishes near Ω we may apply the Poincaré inequality which states that

$$|\theta|_0 \leq C(\Omega) \, |\theta|_1 , \tag{5}$$

where the constant $C(\Omega)$ depends only on Ω (boundedness of Ω is taken into account in this statement). From (4) and (5) we see that there exists some constant

$C(\alpha, \Omega, e, L)$ depending only on α, the domain Ω, the unit vector e and the operator L, such that

$$(6) \qquad \left| \int_\Omega \rho_\alpha \, e \cdot \nabla \theta \, dx \right| \leqslant C(\alpha, \Omega, e, L) \, (|||\varphi||| + |||Pf|||) \, |\theta|_1 \, .$$

Let e_1, e_2, e_3 be an orthonormal basis, from (3), we see that

$$(7) \qquad \phi(\alpha, e) = \sum_{i=1}^{3} \lambda_i \, \phi(\alpha, e_i) \quad \text{if} \quad e = \sum_{i=1}^{3} \lambda_i \, e \, , \quad |e| = 1$$

and

$$(8) \qquad \begin{aligned} C(\alpha, \Omega, e, L) &\leqslant \sum_{i=1}^{3} |\lambda_i| \, C(\alpha, \Omega, e_i, L) \\ &\leqslant \sum_{i=1}^{3} C(\alpha, \Omega, e_i, L) \, . \end{aligned}$$

From this argument we conclude that whatever be the unit vector e and θ belonging to the Sobolev space $H_1^o(\Omega)$ we have

$$(9) \qquad \left| \int_\Omega \rho_\alpha \, e \cdot \nabla \theta \, dx \right| \leqslant C(\alpha, \Omega, L) \, (|||\varphi||| + |||Pf|||) \, |\theta|_1$$

with a constant C which no longer depends on e, nor, of course, on the choice of θ in H_1^o.

2.3.5.4. Let us consider the Sobolev space $H_1^o(\Omega)$ with the Hilbert space structure defined by

$$(1) \qquad (\theta, \omega)_1 = \sum_{i=1}^{3} \left(\frac{\partial \theta}{\partial x_i} \, , \, \frac{\partial \omega}{\partial x_i} \right)_0 = \sum_{i=1}^{3} \int_R \frac{\partial \theta}{\partial x_i} \, \frac{\partial \omega}{\partial x_i} \, dx \, .$$

From the Poincaré inequality 2.3.5.3.(5) we see that $H_1^o(\Omega)$ is indeed endowed, in such a way, with an Hilbert space structure. Given ρ_α and e, the application

$$\theta \rightarrow \int_\Omega \rho_\alpha \, e \cdot \nabla \bar{\theta} \, dx \tag{2}$$

is a semi-linear form, which from 2.3.5.3.(9) is continuous on $H_1^o(\Omega)$ with its Hilbert space structure as was just defined. From the projection theorem* in Hilbert space, there exists an $F(\alpha, e) \in H_1^o(\Omega)$ such that

$$\int_\Omega (\nabla F(\alpha, e) - \rho_\alpha e) \cdot \nabla \bar{\theta} \, dx = 0 \tag{3}$$

whatever be $\theta \in H_1^o(\Omega)$. We observe that $F(\alpha, e)$ is unique when ρ_α, e are given. As a matter of fact, if there were two of them F and F′ the difference $F'' = F - F'$ would satisfy

$$\int_\Omega \nabla F'' \cdot \nabla \theta \, dx = 0 \tag{4}$$

and, as θ is arbitrary in $H_1^o(\Omega)$ we could put $\theta = F''$ in (4) which implies that $\nabla F'' = 0$. Now, let e_1, e_2, e_3 be an orthonormal basis it is evident that

$$F(\alpha, e) = \sum_{i=1}^{3} \lambda_i \, F(\alpha, e_i) \quad \text{with} \quad e = \sum_i \lambda_i e_i \tag{5}$$

is a solution of (3) and from the preceding argument it is the (unique) solution. We conclude from this that

$$F(\alpha, e) = e \cdot \chi_\alpha \tag{6}$$

and, from (3)

---

*) See Dunford-Schwartz[12] Vol. 1 p. 249 theorem 5.

(7) $$\int_{\Omega} \left(e_i \frac{\partial \chi_{\alpha,i}}{\partial x_j} - \rho_\alpha e_j\right) \frac{\partial \bar{\theta}}{\partial x_j} \, dx = 0$$

with summation over 1,2,3 for repeated indices. Any smooth θ which vanishes in an arbitrary small neighbourhood of $\partial\Omega$ belongs to $H_1^o(\Omega)$ and consequently, (7) holds for such a θ. This means that as distributions, we have

(8) $$e \cdot \Delta \chi_\alpha = e \cdot \nabla \rho_\alpha \Rightarrow \Delta \chi_\alpha = \nabla \rho_\alpha, \; \chi_\alpha \in \{H_1^0(\Omega)\}^3 \, .$$

Now from (3) and 2.3.5.3.(9) we conclude that

(9) $$|\chi_\alpha|_1 \leqslant C(\alpha, \Omega, L) \, (|||\varphi||| + |||Pf|||)$$

where, for any $v \in (H_1^o(\Omega))^3$ we set

(10) $$|v|_1 = \{\sum_{i=1}^{3} |v_i|_1^2\}^{1/2}$$

and, consequently

(11) $$(v, w)_1 = \sum_{i=1}^{3} (v_i, w_i)_1 \, .$$

From (8) we have first for any smooth vector v with compact support, then for any $v \in (H_1^o(\Omega))^3$,

(12) $$(\chi_\alpha, v)_1 = (\rho_\alpha, \operatorname{div}(v))_0 = \int_{\Omega} \varkappa \rho_\alpha \operatorname{div}(v) \, dx$$

where

$$\varkappa\rho_\alpha = \rho_\alpha - |\Omega|^{-1}(\rho_\alpha, 1)_0 \;, \quad |\Omega| = \int_\Omega dx \;. \tag{13}$$

We observe that

$$(1, \operatorname{div}(V))_0 = 0 \;. \tag{14}$$

Now we intend to prove the following statement: given any $g \in L^2(\Omega)$ such that $(g, 1)_0 = 0$, there exists an $u \in \{ H_1^0(\Omega) \}^3$ such that

$$\operatorname{div}(u) = g \,, \quad \text{and} \quad |u|_1 \leq k(\Omega)\, |g|_0 \;. \tag{15}$$

Deferring the proof of this statement we deduce the consequence which is of interest for us. From (13) we have $(\varkappa\rho_\alpha, 1)_0 = 0$ and we may choose $g = \varkappa\rho_\alpha$ in (15), then we substitute to v in (12) the solution u of (15) and we have

$$|\varkappa\rho_\alpha|_0^2 = (\chi_\alpha, u)_1 \leq k(\Omega)\, C(\alpha, \Omega, L)\, (|||\varphi||| + |||Pf|||)\, |\varkappa\rho_\alpha|_0 \;. \tag{16}$$

As a consequence, provided we give a proof of the statement in (15) we have proven the existence of a constant $C(\Omega, L)$ depending only on the domain Ω and on the linear collision operator L, such that

$$\|\varkappa\rho_\alpha\|_{L^2(\Omega)} \leq C(\Omega, L)\, (|||Pf||| + |||\varphi|||) \,, \tag{17}$$

where $\varkappa\rho_\alpha$ is related to ρ_α according to (13), and we observe that the argument of 2.3.5.5. will show that we need to put on $\partial\Omega$ the requirement that the unit normal n to $\partial\Omega$ depends in a piecewise continuously differentiable way on position on $\partial\Omega$.

2.3.5.5. Now we give a proof of the statement in 2.3.5.4.(1). We seek u in the form

(1) $$u = \nabla\varphi + \chi$$

with

(2) $$\Delta\varphi = g, \quad \operatorname{div}\chi = 0$$

(3) $$\frac{d\varphi}{dn} = 0, \quad \nabla_T\varphi + \chi_T = 0, \quad \text{and } \Omega$$

where the subscript T stands for the projection on the tangent plane to $\partial\Omega$. From

(4) $$\Delta\varphi = g, \quad \frac{d\varphi}{dn} = 0,$$

we conclude that

(5) $$|\nabla\varphi|_1 \leqslant C_1(\Omega)\ |g|_0,$$

(6) $$\|\nabla_T\varphi\|_{H_{1/2}}(\partial\Omega) \leqslant C_2(\Omega)\ |g|_0.$$

For the definition of $H_{1/2}(\partial\Omega)$ we refer to Lions-Magenes[24] Vol. 1 page 39-40 and for (6) to p. 47 theorem 9.4 of the same reference. We observe that in Lions-Magenes the boundary $\partial\Omega$ is assumed to be infinitely differentiable, but the property of $\partial\Omega$ which is actually used in the argument is the following one: in any neighbourhood of a point $P \in \partial\Omega$ the function $f(x, g)$ of 2.2.1.5.(1) is continuously differentiable, and from the Liapounov condition, we have even more than this. Now, we have to choose χ such that

$$\operatorname{div}(\chi) = 0\,, \quad \chi_T\big|_{\partial\Omega} = -\nabla_T\varphi\,\big|_{\partial\Omega}\,, \tag{7}$$

in order that u as defined by (1) satisfies 2.3.5.4.(15), and this will hold if

$$|\chi|_1 \leqslant C_3(\Omega)\,\|\nabla_T\varphi\|_{H_{1/2}(\Omega)}\,. \tag{8}$$

We set

$$\chi = \nabla \wedge \psi\,, \quad \psi_{|\partial\Omega} = 0\,. \tag{9}$$

From (9) we find that $\operatorname{div}(\chi) = 0$, and $\chi_{|\partial\Omega}\cdot \mathbf{n} = 0$ while

$$\chi_{|\partial\Omega} = \mathbf{n} \wedge \frac{\partial\psi}{\partial n}\,. \tag{10}$$

Finally we have to look for a χ such that

$$\psi \in \{H_2(\Omega)\}^3\,, \quad \psi_{|\partial\Omega} = 0\,, \quad \mathbf{n}\wedge\frac{\partial\psi}{\partial n}\Big|_{\partial\Omega} = -\nabla_T\varphi_{|\partial\Omega} \in (H_{1/2}(\partial\Omega)^2)\,, \tag{11}$$

and it is sufficient to find one such that

$$\psi \in \{H_2(\Omega)\}^3\,, \quad \psi\,\big|_{\partial\Omega} = 0\,, \quad \frac{\partial\psi}{\partial n}\Big|_{\partial\Omega} = \mathbf{n}\wedge\nabla_T\varphi \in (H_{1/2}(\partial\Omega)^2)\,. \tag{12}$$

From Lions-Magenes[24] Vol. 1 p. 47 theorem 9.4 we have an infinity of solutions for (12), but at least one is such that

(13) $$\|\psi\|_{H_2(\Omega)} \leqslant C_3(\Omega)\ \|\nabla_T \varphi\|_{H_{1/2}(\Omega)} ,$$

for some constant $C_3(\Omega)$, and from

(14) $$|\nabla \wedge \psi|_1 \leqslant \|\psi\|_{H_2(\Omega)} ,$$

We see that (8) holds. We observe that a condition for (12) to be solvable with the property (13) is that, for every sufficiently small neighbourhood of a point of $\partial\Omega$, the function $f(x, y)$ in 2.2.1.5.(1) be twice continuously differentiable, at least piecewise.

2.3.5.6. Referring to 2.3.5.4.(13) we set*

(1) $$\rho_\alpha = \varkappa\rho_\alpha + \rho_\alpha^{(o)} , \quad \rho_\alpha^{(o)} = |\Omega|^{-1} (\rho_\alpha, 1) , \quad \alpha = 0, 1, 2, 3, 4$$

and we try to estimate $\rho_\alpha^{(o)}$. In 2.3.4.5.2.(2) we substitute

(2) $$h_\alpha = \theta(x)\ \phi_\alpha(x, \xi)$$

and we choose θ and ϕ_α in the following manner. First of all θ is zero except in some neighbourhood of $\partial\Omega$ and it depends only on the distance of x to $\partial\Omega$; second $\phi_\alpha(x, \xi)$ depends only (observe that we need consider only an x in the previous neighbourhood) on the projection y of x on $\partial\Omega$.

As a consequence, setting

---

*) We observe that, from the second of 2.3.5.1.(1), $\rho_0^{(o)} = 0$.

$$x = y + z\ h(y)\ , \qquad 0 \leqslant z \leqslant \delta\ , \qquad (3)$$

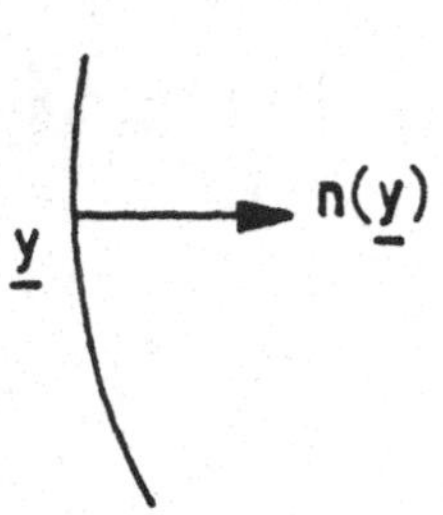

we have

$$h_\alpha = \theta(z)\ \phi_\alpha(y, \xi)\ , \quad \theta = 0 \ \text{ for } \ z > \delta \qquad (4)$$

In a manner completely analogous to the proof of 2.3.5.2.(4) we may prove

$$\Big| \sum_{\beta=0}^{4} \int_{\Omega} \rho_\beta\ \nabla\cdot[(\psi_\beta, \xi\phi_\alpha)_H\, \bar\theta]\ dx\ \Big| \leqslant |||\varphi|||\ \ |||\phi_\alpha\, \theta||| +$$

$$+\ |||Pf|||(\ |||\nu\phi_\alpha\theta||| + |||A\phi_\alpha\,\theta||| + |||(\xi\cdot\nabla\,\phi_\alpha)\,\theta||| + |||\xi\,\phi_\alpha\cdot\nabla\,\theta|||) \qquad (5)$$

We rearrange this observing that $\nabla\theta = d\theta/dz\ n$ and we have

$$\Big| \sum_{\beta=0}^{4} \int_{\Omega} \rho_\beta(\psi_\beta, \xi\cdot n\,\phi_\alpha)_H\ \frac{d\bar\theta}{dz}\ dx\ +$$

$$+ \sum_{\beta=0}^{4} \int \rho_\beta(\nabla\cdot(\psi_\beta, \xi\,\phi_\alpha))\ \bar\theta\ dx\ \Big| \leqslant |||\varphi|||\ |||\phi_\alpha|||\ |\theta|_0 + \qquad (6)$$

$$+\ |||Pf|||\,\{\,[\ |||\nu\phi_\alpha||| + |||A\phi_\alpha||| + |||\xi\cdot\nabla\,\phi_\alpha|||]\ |\theta|_0 + |||\,|\xi|\,\phi_\alpha|||\ |\theta|_1\}\ ,$$

then we substitute (1) into (6) and we obtain*

*) Remember that $\rho_0^{(o)} = 0$.

$$
\left| \sum_{\beta=1}^{4} \int_{\Omega} \rho_\beta^{(o)} (\psi_\beta, \xi \cdot n\, \phi_\alpha)_H \frac{d\bar{\theta}}{dz}\, dx \; + \right.
$$

$$
\left. + \sum_{\beta=1}^{4} \int_{\Omega} \rho_\beta^{(o)} (\nabla \cdot (\psi_\beta, \xi \phi_\alpha)_H\, \bar{\theta}\, dx \right| \leqslant
$$

(7)

$$
|||\varphi|||\; |||\phi_\alpha|||\; |\theta|_0 + |||\mathrm{Pf}|||\, \{ [\, |||\nu\phi_\alpha||| + |||A\phi_\alpha||| + |||\xi \cdot \nabla \phi_\alpha|||]\; |\theta|_0 \; +
$$

$$
+ \; |||\, |\xi|\phi_\alpha|||\; |\theta|_1 \} + \sum_{\beta=0}^{4} (|\theta|_1\; |||\xi\,|\phi_\alpha||| + |||\nabla\, (\xi\, \phi_\alpha)|||\; |\theta|_0)\; |\varkappa \rho_\beta|_0
$$

We observe that h_α in (4) has to be in V^* and this implies that

(8)
$$
\phi_\alpha^- = G^* \phi_\alpha^+ \; .
$$

Now we choose the $\phi_\alpha(y, \xi)$, $y \in \partial\Omega$ in such a way that

(9)
$$
(\psi_\beta, \xi \cdot n\, \phi_\alpha)_H = \delta_{\alpha,\beta} \qquad \beta \neq 0 \; ,
$$

and make the hypothesis

(10) It is possible to choose the ϕ_α according to (8)(9) such that $(\psi_\beta, \xi\, \phi_\alpha)_H$ is continuous and piecewise continuously differentiable on $\partial\Omega$.

The fact that β must be different from zero in (9) is a consequence of

(11)
$$
\phi_\alpha^- = G^* \phi_\alpha^+ \;\Rightarrow\; (\xi \cdot n,\; \phi_\alpha)_H = 0 \; .
$$

We observe that (10) is a very weak requirement. As a matter of fact there is a mathematical requirement and a physical one. The mathematical requirement is that in complement to the conditions already imposed to $\partial\Omega$ (n depends continuously and piecewise continuously differentiably on position) the operator G^* must depend continuously, piecewise continuously differentiably on position. The physical requirement is that there are distribution functions meeting the boundary condition on the wall for which there is a net momentum flux and energy flux to the wall at each point of it.

We choose a set of ϕ_α in agreement with (10), then the right hand side of (7) may be estimated using 2.3.5.4.(17) and we obtain

$$| \rho_\alpha^{(o)} \int_{\partial\Omega} dS \int_0^\delta J \frac{d\theta}{dz} dz | \leq \sum_{\beta=1}^{4} K_{\alpha\beta} | \rho_\beta^{(o)} | \int_{\partial\Omega} dS \int_0^\delta J |\theta| dz \tag{12}$$

$$+ C(\Omega, L, G^*) (|||Pf||| + |||\varphi|||)$$

where the constant $C(\Omega, L, G^*)$ differs from the one in 2.3.5.4.(17). We have set

$$K_{\alpha\beta} = \sup_{x \in \partial\Omega} | \nabla \cdot (\psi_\beta, \xi \phi_\alpha)_H | . \tag{13}$$

We observe that neither the constant $C(\Omega, L, G^*)$ nor the constant $K_{\alpha\beta}$ depend on the choice of θ. We designate as J the (modulus of) Jacobian of $x \rightarrow (y, z)$. From the hypothesis already made

$$J \leq (1 - M z)^{-1} , \quad M = M(\partial\Omega) \tag{14}$$

and choosing

$$\theta = 1 - \frac{z}{\delta} \tag{15}$$

we have

$$(16)\qquad \int_{\partial\Omega} dS \int_0^\delta J \frac{d\theta}{dz} dz = |\partial\Omega| (M\delta)^{-1} \operatorname{Log}(|1 - M\delta|^{-1})$$

$$(17)\qquad \int_{\partial\Omega} dS \int_0^\delta J |\theta| dz \leqslant \delta |\partial\Omega| (M\delta)^{-1} \operatorname{Log}(|1 - M\delta|^{-1})$$

and, from (12), with a constant D depending on Ω through the choice of θ

$$(18)\qquad |\rho_\alpha^{(o)}| - \delta \sum_{\beta=1}^{4} K_{\alpha\beta} |\rho_\beta^{(o)}| \leqslant \frac{C(\Omega, L, G^*)\, D(\Omega)}{|\partial\Omega|(M\delta)^{-1} \operatorname{Lg}(|1 - M\delta|^{-1})} (|||Pf||| + |||\varphi|||)$$

We choose δ such that

$$(19)\qquad \sup_\alpha \delta \sum_{\beta=1}^{4} K_{\alpha,\beta} = 1/2 , \quad \delta < \frac{1}{2M}$$

and we obtain, with a new constant

$$(20)\qquad |\rho_\alpha^{(o)}| \leqslant C(\Omega, L, G^*)(|||Pf||| + |||\varphi|||) , \quad \alpha = 1, 2, 3, 4.$$

Finally, from

$$(21)\qquad \|\rho_\alpha\|_{L^2(\Omega)} \leqslant \|\varkappa\rho_\alpha\|_{L^2(\Omega)} + |\rho_\alpha^{(o)}|\, |\Omega| ,$$

we obtain

$$(22)\qquad \|\rho_\alpha\|^2_{L^2(\Omega)} \leqslant C(\Omega, L, G^*) (|||Pf|||^2 + |||\varphi|||^2) .$$

Now substitute (22) into 2.3.5.1.(7), we find

$$\{\mu - \epsilon_1 - \epsilon_1 \, C(\Omega, L, G^*)\} \, |||Pf|||^2 + k[\![\Pi J^- Bf]\!]^2 \leq \{\epsilon_1 C(\Omega, L, G^*) + (4\,\epsilon_1)^{-1}\} \, |||\varphi|||^2 \tag{23}$$

2.3.5.7. We state our result. The hypothesis are: Ω is bounded, $\partial\Omega$ satisfies a Liapounov condition and moreover, the unit normal depends piecewise continuously differentiably on position. The linear collision operator is a cut-off one and the collision frequency satisfies

$$\text{Inf} \, \frac{\nu(|\xi|)}{|\xi|} = \beta > 0 \tag{1}$$

furthermore the dissipation inequality

$$(Lf, f)_H \geq \mu \, \|Pf\|_H^2 \tag{2}$$

holds and the operator A is bounded in H. The operator G satisfies 1.4.4.5.(1)(2)(3) and is continuous from $L^2(\partial\Omega, X)$ into $L^2(\partial\Omega, Y)$ where X and Y are any of H, $\tilde{H}$, the same holds for G^* while the kernels satisfy 1.4.5.4.(1) and 1.4.6.3.(8). Finally G^* satisfies 2.3.5.6.(10). Then for any $f \in V$ such that $((f, 1)) = 0$, we have

$$|||f||| \leq C(\Omega, L, G) \, |||\mathcal{L}f||| \, . \tag{3}$$

If the last hypothesis on G^* is replaced by the analogous one on G, then for $f \in V^*$, $((f, 1)) = 0$ we have

$$|||f||| \leq C(\Omega, L, G) \, |||\mathcal{L}^* f||| \tag{4}$$

2.3.6. Existence and Uniqueness

2.3.6.1. Consider problem 2.1.1.3.(1), we know from 2.1.2.7.(4)(5) that it is sufficient to consider the solution when $\phi^+ = 0$. From 2.3.4.5.2. we know that in order that f be a solution of this problem it is necessary and sufficient that the following relation

(1) $$((f, \mathcal{L}^* h)) = ((\varphi, h)) ,$$

holds for every $h \in V^*$. We know that, in order that f exists we must have

(2) $$((\varphi, 1)) = 0 .$$

2.3.6.2. Let us call C the set of almost everywhere constant functions. Obviously, from boundedness of Ω, $C \subset L^2(\Omega, H)$ We set C' for the orthogonal complement to C in $L^2(\Omega, H)$ so that 2.3.6.1.(2) means $\varphi \in C'$. We set

(1) $$x = V \cap C'$$

(2) $$x^* = V^* \cap C'$$

and we make x and x^* Banach spaces by using the norms of V and V^* respectively.

We observe that, in 2.3.6.1.(1) we may restrict h to x^*. Let $\mathcal{L}^*(x^*)$ be the range of $\mathcal{L}^*$ in $L^2(\Omega, H)$, from 2.3.5.7.(4), $\mathcal{L}^*$ has a well defined inverse $\mathcal{T}^*$ which is bounded from $\mathcal{L}^*(x^*)$ with the norm of $L^2(\Omega, H)$ to x^* and for $h \in x^*$ we may set

(3) $$h = \mathcal{T}^* \mathcal{L}^* h = \mathcal{T}^* \psi .$$

Arbitrariness of h in x^* implies arbitrariness of ψ in $\mathcal{L}^*(x^*)$. As a consequence, if f is solution of 2.1.1.3.(1) we must have from (1)

(4) $$((f, \psi)) = ((\varphi, \mathcal{T}^* \psi))$$

for any $\psi \in \mathcal{L}^*(x^*)$.

We observe that $\mathcal{L}^*(x^*)$ is a closed subspace of $L^2(\Omega, H)$ As a matter of fact let $\{\psi_n\}$ be a Cauchy sequence in $L^2(\Omega, H)$ of elements $\psi_n \in \mathcal{L}^*(x^*)$ we have $h_n = \mathcal{T}^*_{\psi n}$, $\psi_n = \mathcal{L}^* h_n$ and $\psi_n \to \psi$ in $L^2(\Omega, H)$. From boundedness of $\mathcal{T}^*$ or, directly from 2.3.5.7.(4)

$$
\begin{aligned}
|h_n - h_p|_{V^*} &\leqslant |||\mathcal{T}^* \psi_n - \mathcal{T}^* \psi_p||| + \\
&\quad + |||\mathcal{M}^* \mathcal{T}^* \psi_n - \mathcal{M}^* \mathcal{T}^* \psi_p||| \\
&\leqslant |||\mathcal{T}^* \psi_n - \mathcal{T}^* \psi_p||| + |||\psi_n - \psi_p||| \\
&\quad + |||A\, \mathcal{T}^* \psi_n - A\, \mathcal{T}^* \psi_p|||
\end{aligned}
\tag{5}
$$

that is

$$
|h_n - h_p|_{V^*} \leqslant \text{const}\; |||\psi_n - \psi_p||| \,. \tag{6}
$$

We conclude that h_n is a Cauchy sequence in V^* and, from 2.3.2.2., h_n converges to $h \in V^*$. Now we have

$$
\begin{aligned}
|||\mathcal{L}^* h_n - \mathcal{L}^* h||| &\leqslant |||A\, h_n - A h||| + |||\mathcal{M}^* h_n - \mathcal{M}^* h||| \\
&\leqslant (1 + |A|) \sqrt{2} |A|\; |h_n - h|_V
\end{aligned}
\tag{7}
$$

so that $\mathcal{L}^* h_n = \psi_n$. converges to $\mathcal{L}^* h$ in $L^2(\Omega, H)$ and we must have $\mathcal{L}^* h = \psi$. As a consequence the limit ψ of the ψ_n belongs to $\mathcal{L}^*(x^*)$ and we have proven that $\mathcal{L}^*(x^*)$ is closed.

To $\mathcal{L}^*(x^*)$ considered as a closed subspace of $L^2(\Omega, H)$ we may associate a projection operator $\mathcal{R}^*$ which is the orthogonal projection from $L^2(\Omega, H)$ to $\mathcal{L}^*(x^*)$. Obviously $\mathcal{R}^*$ is bounded with the norm of $L^2(\Omega, H)$. We do not change anything of (4) as it stands by rewritting it as

(8) $$((f, \psi)) = ((\varphi, \mathcal{T}^* \mathcal{R}^* \psi)) .$$

Now the right hand side of (δ) is well defined for any $\psi \in L^2(\Omega, H)$ and $\mathcal{T}^* \mathcal{R}^*$ may be considered as a bounded operator from C' into itself if we restrict ψ to C'. This bounded operator has a bounded adjoint for which we set temporarily Λ. This means that, for any φ and ψ belonging to C' we have

(9) $$((\Lambda\varphi, \psi)) = ((\varphi, \mathcal{T}^* \mathcal{R}^* \psi)) .$$

Setting $f = \Lambda\varphi$, we see that (8) holds for any $\psi \in C'$. Now we restrict ψ to $\mathcal{L}^*(x^*)$ by observing that $\mathcal{L}^*(x^*) \subset C'$. As a matter of fact, for $\psi = \mathcal{L}^* h$ with $h \in V^*$ we have, from 2.3.4.5.2., observing that $1 \in V$, $L\,1 = 0$

(10) $$((1, \psi)) = 0 .$$

We conclude that, with $f = \Lambda\varphi$, (4) holds for any $\psi \in \mathcal{L}^*(x^*)$ and then, that (1) holds for any $h \in x^*$. Now, any $h \in V^*$ may be written as $h = g + c$ with C a constant and $g \in x^*$ and

(11) $$((f, \mathcal{L}^* h)) = ((f, \mathcal{L}^* g)) = ((\varphi, g)) = ((\varphi, h))$$

from (2).

This argument ends the proof of existence of a solution 2.1.1.3.(1) for $\phi^+ = 0$ under the condition 2.1.1.6.(1), again for $\phi^+ = 0$.

2.3.6.3. Now we prove uniqueness. For a given φ and ϕ^+ let f_1 and f_2 be two solutions to 2.1.1.3.(1) then $f = f_2 - f_1$ belongs to V and $\mathcal{L} f = 0$. If we require furthermore that $((f, 1)) = 0$ we may use 2.3.5.7.(3) to conclude that $f = 0$ in $L^2(\Omega, H)$.

2.3.6.4. We state our main result. The hypothesis are as follows. For Ω it is bounded and meets the Liapounov requirement. Concerning the reflection operator we require that $\mathcal{G}$ and $\mathcal{G}^*$ satisfy 1.4.4.5.(1)(2)(3) and 1.4.5.4.(1) and are bounded from $L^2(\partial\Omega, x)$ and $L^2(\partial\Omega, y)$ where x and y are any of H and $\tilde{H}$. We require furthermore that 2.3.5.6.(10) and the corresponding hypothesis with $\mathcal{G}^*$ replaced by $\mathcal{G}$ holds. Concerning the collision operator we assume that it is applicable as $L = \nu - A$, with $\nu \leqslant \nu_0$, 2.3.1.2.(1) and A bounded in H while the dissipation inequality 1.5.4.4.(1) holds.

Then there exist two operators* $\mathcal{T}$ and $\mathcal{T}'$ the first from C' to x and the second from C' to x^*. Both are bounded if C' is considered with the norm of $L^2(\Omega, H)$ and x and x^* with the norms of V and V^* respectively. We recall that

$$C' : \{\varphi \mid \varphi \in L^2(\Omega, H), \ ((\varphi, 1)) = 0\} \tag{1}$$

$$x = V \cap C' ; \quad x^* = V^* \cap C' . \tag{2}$$

for any $\varphi \in C'$,

$$f = \mathcal{T}\varphi \text{ solves 2.1.1.3.(1) with } \phi^+ = 0 , \tag{3}$$

$$f = \mathcal{T}^*\varphi \text{ solves 2.1.1.4.(1) with } \phi^- = 0 \tag{4}$$

and by solves we mean, for $\mathcal{T}_\varphi$, that it belongs to V and that $\mathcal{L}f = \varphi$.

Boundedness of $\mathcal{T}$, for example, means that

$$\{ |||\mathcal{T}\varphi|||^2 + |||\mathcal{M}\mathcal{T}\varphi|||^2 \} \leqslant |\mathcal{T}| \ |||\varphi||| , \tag{5}$$

* The argument for $\mathcal{T}^*$ is analogous to the one for $\mathcal{T}$.

and, for $\mathcal{T}^*$

(6) $$\{ |||\mathcal{T}^*\varphi|||^2 + |||\mathcal{M}^*\mathcal{T}^*\varphi|||^2 \}^{1/2} \leqslant |\mathcal{T}^*| \; |||\varphi||| \, .$$

We observe that $\mathcal{T}_\varphi$ + const is also a solution 2.1.1.3.(1) and that the same is true with $\mathcal{T}^*\varphi$ + const and 2.1.1.4.(1).

On the other hand $\mathcal{T}_\varphi$ is the unique solution of 2.1.1.3.(1) in x and $\mathcal{T}^*_\varphi$ the unique solution of 2.1.1.4.(1) in x^*.

2.3.7. Some Further Comments

2.3.7.1. We have chosen to explain in great detail the work of reference[18] rather than to present a review of current work but we must make reference at this point to the important work of Cercignani[25] [26]. In reference[25] Cercignani considers the problem

(1) $$\xi \cdot \nabla f = \varphi \ , \quad \text{within } \Omega \ ,$$

(2) $$f^+ = \mathcal{G} f^- \ , \quad \text{on } \partial\Omega \ .$$

and assuming that $\mathcal{G}$ satisfies

(3) $$\|\sigma^{1/2} \mathcal{G} f^- \|^2_{\hat{H}} \leqslant C(\sigma) \|\sigma^{1/2} f^- \|^2_{\tilde{H}}$$

for any positive function $\sigma(|\xi|)$ such that

(4) $$\| \, |\xi| \, \sigma^{1/2} \|_H < +\infty \ , \quad \| |\xi| \sigma^{-1/2} \|_H < +\infty \ ,$$

he proves that the following a priori inequality holds

$$ |||\,|\xi|\ f\ |||^2 \leqslant [1 + K\, C(\sigma)]\ D^2\ |||\varphi|||^2\ , \tag{5} $$

where K is some constant, depending on σ, Ω and G and whose meaning is as follows. Referring to the notation of 2.2.1.1.(1) we set $\mathcal{V}_\infty$ for the operator $\mathcal{V}$ which is formed by assuming that $\nu \equiv 0$, then the constant K is the norm of $(1-\ _\infty)^{-1}$ according to $L^2(\partial\Omega, H_{\sigma 1/2})$, that is, if

$$ (1-\mathcal{V}_\infty)\ f^+ = h^+\ , \tag{6} $$

we have

$$ \| \sigma^{1/2}\ f^+ \|_{L^2(\partial\Omega,\, \tilde{H})} \leqslant K\ \| \sigma^{1/2}\ h^+ \|_{L^2(\partial\Omega,\, \tilde{H})}\ . \tag{7} $$

There is a technical point here, namely that 1^+ is always solution of (6), and, as a consequence, in order to ensure uniqueness one has to require

$$ (f^+,\ 1^+)_{L^2(\partial\Omega,\, \tilde{H})} = 0 \tag{8} $$

and for solubility

$$ (h^+, 1^+)_{L^2(\partial\Omega,\, \tilde{H})} = 0\ . \tag{9} $$

Cercignani gives a proof of the invertibility of $1-\mathcal{V}_\infty$ and of the estimate (7) in the case of 1.2.2.1.(1) with perfect accommodation, namely $a_r = 1$. This proof bears some resemblance to the proof we have given of the a priori estimate of 2.1.1. As a matter of fact for the model of reflection under consideration we have

$$ G = \alpha\, G_d + (1-\alpha)\, G_s \tag{10} $$

and as G_s is of norm unity in $L^2(\partial\Omega, H)$ if we write

$$\mathcal{V}_\infty = \alpha \mathcal{V}_{\infty,d} + (1-\alpha)\mathcal{V}_{\infty,s} \qquad 0 < \alpha < 1 \tag{11}$$

we obtain

$$\| \mathcal{V}_{\infty,s} f^+ \|_{L^2(\partial\Omega, \bar{H})} \leqslant \| f^+ \|_{L^2(\partial\Omega, \bar{H})} \tag{12}$$

by using 2.1.3.8.(5) for E_∞ instead of E where E_∞ corresponds to $\nu = 0$. Setting

$$f^+ = \Pi f^+ + \sqrt{2\pi}\,(f^+, 1^+)_{\bar{H}}\, 1^+ \tag{13}$$

$$h^+ = \Pi h^+ + \sqrt{2\pi}\,(h^+, 1^+)_{\bar{H}}\, 1^+ \tag{14}$$

Cercignani solves (6) in two steps, first

$$(1 - (1-\alpha)\mathcal{V}_{\infty,s})\, \Pi f^+ = \Pi h^+ \tag{15}$$

which according to (12) and $0 < \alpha < 1$ may be solved by a contracting iteration, then an ordinary integral equation is formed for $(f^+, 1^+)_{\bar{H}}$ and this integral equation is treated in a way much similar to the one in which we have treated 2.2.1.3.(2).

The second step in Cercignani's work amounts to trying to solve

$$(1 - \mathcal{W}A)\, f = \mathcal{E}\, \phi^+ , \tag{16}$$

by a proper use of the contracting mapping principle in $L^2(\Omega, H)$. This cannot be worked out as it stands and Cercignani uses an elegant process. Instead of starting from the obvious splitting

$$L = \nu - A \ , \tag{17}$$

of the collision operator, he starts from

$$L = (\alpha + 1)\nu - A_\alpha \ , \qquad \alpha \geqslant 0 \ , \tag{18}$$

and he sets instead of (16), the equivalent equation

$$(1 - W^{(\alpha)} A_\alpha) f = \mathcal{E}^{(\alpha)} \phi^+ \ , \tag{19}$$

where $W^{(\alpha)}$ and $\mathcal{E}^{(\alpha)}$ and analogous to W and $\mathcal{E}$ but with ν replaced by $(\alpha + 1)\nu$. We observe that

$$W^{(\alpha)} = W_{(\alpha+1)^{-1}} \ , \tag{20}$$

where W_λ is the inverse of 2.2.1.1.(1) with homogeneous boundary conditions. Cercignani proves that, with

$$\rho = ([(\alpha + 1)\nu]^2 + \zeta^2 D^{-2} |\xi|^2)^{1/2} \tag{21}$$

where ζ is some constant, the following holds

$$|||\rho^{1/2} W^{(\alpha)} f||| \leqslant |||\rho^{-1/2} f||| \tag{22}$$

while, for radial as well as for angular cut-off, we have

$$|||\rho^{-1/2} A^{(\alpha)} f||| \leqslant k \ |||\rho^{1/2} f||| \ , \qquad 0 \leqslant k \leqslant 1 \tag{23}$$

for a suitably chosen α.

We observe that this argument, if (7) is provable for a general reflection operator, avoids our analysis of 2.3.1.; and that it avoids the analysis of 2.3.5.

2.3.7.2. We have not considered at all unbounded domains. We mention here the work at Rigolot-Turbat which is a first step towards a theory of the exterior boundary value problem.

CHAPTER 3

THE NON LINEAR BOUNDARY VALUE PROBLEM

3.1. Statement of the Problem. The Method of Approach

3.1.1. Statement of the Problem

As for the linear problem of chapter 2 we consider a bounded domain Ω that, throughout the chapter, we assume to be convex with a smooth boundary in order that we may apply the existence theorem of 2.3.6.4. We assume that $\partial\Omega$ is a wall independent of time. At each point of $\partial\Omega$ the temperature parameter assumes a constant value a_w and the material velocity assumes a given value u_w, constant in time with a zero normal component so that $u_w \cdot n = 0$. We assume that a_w and u_w depends continuously on position and that

$$\sup_{x \in \partial\Omega} \{ |a_w(x) - 1| + |u_w(x)| \} < \delta \tag{1}$$

so that the whole of the wall is close to rest and thermal equilibrium.

Referring to 1.4.4.3. we know that the reflection condition on the wall for the distribution function

$$F = \omega(1 + f), \tag{2}$$

assumes the form

$$f^+ = G f^- + G_1 \{(1 + f)^-\}. \tag{3}$$

We know that G_1 is zero for $a_w = 1$ and $|u_w| = 0$ and we assume that, in some convenient sense, G_1 is small when (1) holds and δ is small. To define smallness of G_1, we use the space B_r of Grad introduced in 1.5.3.6. We restate that B_r is a space of continuous functions of ξ with the maximum norm

$$(4) \qquad \langle f \rangle_r = \sup_{\xi \in R^3} \omega^{1/2}(|\xi|)\,(1 + |\xi|^2)^{r/2}\,|f(\xi)|\,.$$

Our basic assumption concerning G_1 is that it carries continuous functions of ξ into continuous functions of ξ and that, given $\epsilon > 0$ there exists, at each point on the wall, some $\delta(x, \epsilon)$ such that

$$(5) \qquad |a_w(x) - 1| + |u_w(x)| \leqslant \delta(x,\epsilon) \Rightarrow \langle G_1 f^- \rangle_r \leqslant \epsilon \langle f^- \rangle_r\,,$$

for any $r \in [0, r_0]$ with some r_0 which may be taken, for later convenience, to be 3. The property (5) will be considered as a mathematical definition of the smallness of G_1, if $\sup \delta(x, \epsilon) \leqslant \delta(\epsilon)$.

Our task will be to try to find a solution to the following boundary value problem

$$(6) \qquad \xi \cdot \nabla F = J(F,F) \quad \text{in } \Omega$$

$$(7) \qquad F^+ = K F^- \quad \text{on } \partial\Omega$$

under the condition of weak nonlinearity (1).

We use the formalism of linearisation according to (1) (3) and 1.4.1.1. to translate (6) (7) into

$$(8) \qquad \begin{aligned} &\xi \cdot \nabla f + Lf = Q(f,f)\,, \quad \text{in } \Omega\,,\\ &f^+ = Gf^- + G_1(1+f)^-\,, \quad \text{on } \partial\Omega\,. \end{aligned}$$

and our task, throughout this chapter, will be to solve (8) for f according to the work of Guiraud [28].

3.1.2. The Method of Approach

3.1.2.1. We shall rely heavily on the result of chapter 2 (refer to 2.3.6.4.) and setting

$$\varphi = Q(f,f) \qquad \phi^{+} = G_1\,(1 + f)^{-} \tag{1}$$

we write formally, using 2.1.2.7.(5)

$$f = \mathfrak{T}\{\, Q(f,f) + A\mathfrak{E}\; G_1\,(1 + f)^{-}\} + \mathfrak{E}\; G_1\,(1 + f)^{-} \;. \tag{2}$$

We have set the constant in 2.1.2.7.(5) equal to zero by reference to the following argument. Given a solution F of 3.1.1.(6)(7) and n an arbitrary constant we may write

$$F = \omega(1 + f) = n\;\omega\;(1 + f) \tag{3}$$

this changes slightly the process of linearisation leaving L and G unchanged. Then we have

$$\int_{\Omega} dx \int_{R^3} F\; d\xi = n(\,|\Omega| + ((1,g)) \qquad , \tag{4}$$

and we may always choose n in such a way that

$$((1,g)) = ((1, \mathfrak{E}\; G_1\,(1 + g)^{-}))\;, \tag{5}$$

then, through a proper choice of units we may restrict ourselves to n = 1. As a matter, setting the constant equal to zero as in (2) amounts to select, out of a one parameter family of solutions, the particular one for which the relation (5) holds.

Referring to gas dynamics we cannot expect the flow within Ω, to be given by conditions on temperature and velocity at the wall if we do not specify the (global) level of density. On the other hand, the way in which the solution depends on this factor is far from being simple, as we may expect from gas dynamics.

3.1.2.2. There is a minor point that we must elucidate. From 2.1.1.6. we know that in order that 3.1.2.1.(2) makes sense we must have

$$(1) \qquad ((1, Q(f,f))) + \{ 1^+, G_1(1+f)^- \} = 0 .$$

From 1.4.1.1. we know that

$$(2) \qquad J\{ \omega(1+f), \ \omega(1+f)\} = -\omega L f + \omega Q(f,f)$$

and from 1.1.3.1.(2), for any collisional invariant ψ_α,

$$(3) \qquad (\omega^{-1} J \{\omega(1+f), \ \omega(1+f)\}, \ \psi_2) = 0 ,$$

independently of f and, as such

$$(4) \qquad (Lf, \psi_\alpha)_H = 0 ,$$

$$(5) \qquad (Q(f,f)) \ \psi_\alpha)_H = 0 .$$

Taking $\psi_0 = 1$, we see that (1) reduces to

$$(6) \qquad \{ G_1(1+f)^- , \ 1^+\} = 0 .$$

We detail (6) as

$$\int_{\partial\Omega} dS \int_{R^3} \omega\ \xi\cdot \mathbf{n}\ G_1[(1+f)^-]\ d\xi = 0\ , \tag{7}$$

and to prove that (7) holds whatever be f we refer 1.2.3.2.(3) and observe that, thanks to $\mathbf{u}_w \cdot \mathbf{n}$ we have $\xi_r \cdot \mathbf{n} = \xi \cdot \mathbf{n}$, then

$$\{G(1+f)^- + G_1(1+f)^-, 1^+\} - \{(1+f)^-, 1^-\} = 0\ . \tag{8}$$

Now we observe that (8) holds whatever the distribution of a_w and $\mathbf{u}_w$ (with $\mathbf{u}_w \cdot \mathbf{n} = 0$) on $\partial\Omega$ and, in particular for $a_w = 1$ and $\mathbf{u}_w = 0$ (observe that f^- is independent of a_w and $\mathbf{u}_w$) so that

$$\{G(1+f)^-, 1^+\} - \{(1+f)^-, 1^-\} = 0 \tag{9}$$

and finally

$$\{G_1(1+f)^-, 1^+\} = 0 \tag{10}$$

which is (6). We present the argument in another way, we have locally

$$\int_{R^3_+} \omega\, \xi\cdot\mathbf{n}\ G(1+f)^- + G_1(1+f)^-\ d\xi + \int_{R^3_-} \omega\ \xi\cdot\mathbf{n}\,(1+f)^-\ d\xi = 0 \tag{11}$$

and this, for fixed f^- must be identically satisfied when a_w and $\mathbf{u}_w$ are varied. For $a_w = 1,\ \mathbf{u}_w = 0$ we have

$$\int_{R^3_+} \omega\, \xi\cdot\mathbf{n}\ G(1+f)^-\ d\xi + \int_{R^3_-} \omega\, \xi\cdot\mathbf{n}\ (1+f)^-\ d\xi = 0 \tag{12}$$

then we achieve as previously.

3.1.2.3. From the argument of 3.1.2.2. we may separate the terms in 3.1.2.1.(2) and write

$$f = \mathfrak{T}\, Q(f,f) + \mathfrak{T}\, A\mathfrak{E}\; G_1(1+f)^- + \mathfrak{E}G_1(1+f)^- \; . \tag{1}$$

our method of approach will be to look for f as a fixed point of equation (1).

3.1.3. Function Spaces

3.1.3.1. We extend the spaces B_r of Grad to spaces of functions of **x** and ξ. As $B_r(\Omega)$ we consider the space of functions $f(\mathbf{x}, \xi)$ continuous on $\Omega \times R^3$ with the maximum norm

$$< f >_r \; = \sup_{\substack{x \in \partial\Omega \\ \xi \in R^3}} \; \omega^{1/2}(|\xi|)\,(1 + |\xi|^2)^{r/2}\; |f(\mathbf{x}, \xi)| \tag{1}$$

Throughout the chapter we consider only spaces of real functions. We observe that $< \cdot >_r$ stands for the norm in B_r or in $B_r(\Omega)$. It will prove that in each case it will be clear which of these two norms matter and we do not bother to make a notational distinction.

3.1.3.2. For functions defined on $\partial\Omega$ we use spaces of functions continuous on $\{(\mathbf{x}, \xi)\,|\,\mathbf{x} \in \partial\Omega, \xi \in R^3_+\}$ or on $\{(\mathbf{x}, \xi)\,|\,\mathbf{x}\; \partial\Omega, \xi \in R^3\}$ with the maximum norm

$$< f >_r \; = \sup_{\substack{x \in \Omega \\ \xi \in R^3}} \; \omega^{1/2}(|\xi|)\,(1 + |\xi|^2)^{r/2}\; |Bf(\mathbf{x}, \xi)|\,. \tag{1}$$

and we set $B_r(\partial\Omega)$ for such a space.

3.1.4. The Method of Solution

3.1.4.1. We shall prove, and this will be the most lengthy part of our work, that the operators $\mathfrak{T}Q$ and $\mathfrak{E}$ and $\mathfrak{T}A\mathfrak{E}$ convey functions of $B_r(\Omega)$ or $B_r(\partial\Omega)$ into functions of $B_r(\Omega)$; moreover we shall prove the following estimates

$$< \mathfrak{T}Q(f,g) >_r \leqslant \gamma < f>_r < g>_r \,, \tag{1}$$

$$< (1 + \mathfrak{T}A)\, \mathcal{E}\, \psi^+ >_r \leqslant \beta < \psi^+ >_r \,, \tag{2}$$

with some constants β and γ which depend on the domain Ω, the operator G and the collision operator. We observe that *we shall restrict our considerations to rigid spheres* and then the collision operator depends only on the diameter σ of the spheres and on the level of density as explained 3.1.2.1.

3.1.4.2. We use the most standard fixed point argument that we repeat here for the benefit of the reader. We consider in $B_r(\Omega)$ the ball

$$S_{r,R} : \{ f \mid f \in B_r(\Omega) \,,\quad < f>_r \leqslant R \} \tag{1}$$

and consider the nonlinear functional

$$f \to F(f) = \mathfrak{T}Q(f,f) + (1 + \mathfrak{T}A)\, \mathcal{E}\, G_1 (1 + f)^- \,; \tag{2}$$

from 3.1.4.1.(1) and (2) this functional conveys $S_{r,R}$ into itself provided

$$\gamma R^2 + \beta \epsilon R + \beta \epsilon < 1^- >_r \leqslant R \,, \tag{3}$$

where we have used the argument of 3.1.1., namely

$$\sup_{x \in \partial\Omega} \{ |a_w(x) - 1| + |u_w(x)| \} \leqslant \delta(\epsilon) \Rightarrow < G_1 f^- >_r \leqslant \epsilon < f^- >_r \,. \tag{4}$$

Referring to the figure we set R_1 and R_2 for the two roots of

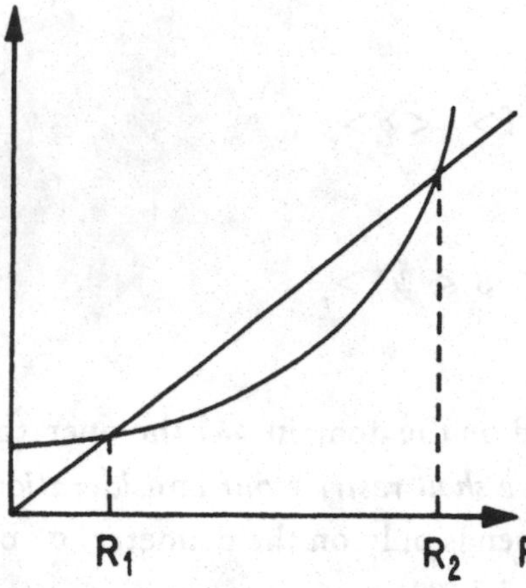

$$(5) \qquad \gamma x^2 + \beta\epsilon x + \beta\epsilon < 1^- >_r = x$$

provided they are real, which occurs if

$$(6) \quad 4\beta\gamma\epsilon < 1^- >_r < (1-\beta\epsilon)^2 .$$

If (6), holds for any $R \epsilon [R_1, R_2]$ the functional F will convey the ball $S_{r,R}$ into itself.

From

$$(7) \qquad F(f) - F(g) = \mathcal{T}Q(f-g,\, f+g) + (1+\mathcal{T}A)\mathcal{E}G_1(f^- - g^-) ,$$

we see that

$$(8) \quad < F(f) - F(g) >_r \leqslant \gamma(<f>_r + <g>_r) <f-g>_r + \beta\epsilon <f-g>_r ,$$

and provided

$$(9) \qquad \epsilon\beta + 2\gamma R \leqslant k < 1 ,$$

we shall have

$$(10) \qquad f \text{ and } g \epsilon S_{r,R} \Rightarrow <F(f) - F(g)>_r \leqslant k <f-g>_r .$$

We observe that

$$R_x = \frac{R_1 + R_2}{2} \,, \tag{11}$$

is such that

$$\epsilon\beta + 2\gamma\, R_x = 1 \,, \tag{12}$$

and, as a consequence, if (6) holds and we choose any $R \in [R_1, R_*]$ the non-linear functional F will apply the ball $S_{r,R}$ into itself and will be contracting according to (12)(11) and (10); then the simplest iteration

$$f_o = 0 \;; \quad f_n = F(f_{n-1}) \quad n \geqslant 1 \tag{13}$$

will converge in $B_r(\Omega)$ to a solution of 3.1.2.3.(1) and such is unique in the ball $S_{r,R}$.

We shall investigate later in what sense the preceding solution of 3.1.2.3.(1) is also a solution of 3.1.1.(8).

3.1.5. Regularisation of the Linear Solution

3.1.5.1. The basic problem is to prove the properties stated in 3.1.4.1. We observe that Grad[29] has proven that

$$f, g \in B_r \Rightarrow \nu^{-1} Q(f,g) \in B_r \text{ and } < \nu^{-1} Q(f,g) >_r \leqslant \alpha_2 < f >_r < g >_r \tag{1}$$

for power law angular cut-off including rigid spheres. It is a trivial matter to extend (1) to $B_r(\Omega)$ and what we have to prove is that

$$\nu^{-1} f \in B_r(\bar{\Omega}) \Rightarrow \quad \mathfrak{T} f \in B_r(\bar{\Omega}) \quad \text{and } < \mathfrak{T} f >_r \leqslant \alpha_2^{-1} \gamma < \nu^{-1} f >_r \,, \tag{2}$$

in order that 3.1.4.1.(1) hold. For 3.1.4.1.(2) we need to prove that

(3) $\phi^+ \in B_r(\partial\Omega) \Rightarrow \mathcal{E}\phi^+ \in B_r(\Omega)$ and $< \mathcal{E}\phi^+>_r \leqslant (1+\gamma b\alpha_2^{-1} k_{r+1})^{-1}\beta < \phi^+>_r$

where we make reference to 1.5.3.6.(3) and 1.5.2.3.(7).

We observe that according to boundedness of Ω and smoothness of $\partial\Omega$ we have

(4) $$r < 3/2 \Rightarrow B_r(\Omega) \in L^2(\Omega, H) \text{ and } B_r(\partial\Omega) \in L^2(\partial\Omega, H)$$

(5) $$r > 5/2 \Rightarrow B_r(\partial\Omega) \in L^2(\partial\Omega, \tilde{H})$$

(6) $$r > 7/2 \Rightarrow B_r(\partial\Omega) \in L^2(\partial\Omega, \bar{H}_\nu)$$

and choosing $r > 7/2$, we know from 2.3.6. and 2.1.2.7 that $\mathcal{G}$ may be applied to any f such that $\nu^{-1} f \in B_r(\Omega)$ and $\mathcal{E}$ to any $\phi^+ \in B_r(\partial\Omega)$. The point is to prove that $\mathcal{G}f$ and $\mathcal{E}\phi^+$ then, belong to $B_r(\Omega)$ and not merely to $L^2(\Omega, H)$.

3.1.5.2. Referring to 2.1.2., we restate the basic relation

(1) $$\mathcal{G} = W + WA\mathcal{G},$$

which we iterate as

(2) $$\mathcal{G} = \sum_{n=0}^{N-1} (WA)^n W + (WA)^N \mathcal{G}.$$

We use (2) on the basis of the observation that A has a smoothing effect in velocity space (refer to 1.5.3.6.) while W has a smoothing effect along rays. As AWA mixes rays we expect that $(WA)^2$ has a smoothing effect in both position and velocity space. We are going to prove that, with a sufficiently high N, $\nu(W\,A)^N$ conveys $L^2(\Omega,H)$ into $B_r(\bar{\Omega})$, any r (N will depend on r). Then we prove separately that νW applies $B_r(\Omega)$ into itself.

3.1.5.3. As a matter of fact the preceding argument would be simple if W were replaced by U. A lot of complications arise from the fact that W is related to U in a complicated way

$$W = U + E\,T\,G\,J^{-}\,B\,U \tag{1}$$

according to 2.1.2.4.(7), and we recall that

$$T = (1-\mathcal{V})^{-1}\ , \qquad \mathcal{V} = G\,J^{-}\,BE \tag{2}$$

The complications arise from the fact that T is not at all explicit and that its existence and boudedness is proven only in $L^2(\partial\Omega,\tilde{H})$ or $L^2(\partial\Omega,\tilde{H}_\nu)$. The same difficulty precludes any direct treatment of

$$\mathcal{E} = E\,T\,. \tag{3}$$

As a matter of fact, the smoothing property of W proceeds from smoothing properties of both U and G. The smoothing property of U arises from it being a convolution along rays. The smoothing property of G arises from hypothesis on the kernel Γ that we shall formulate later. To take advantage of the smoothing property of G we use

$$T = 1 + \mathcal{V}T\ , \tag{4}$$

and we iterate on it as

$$T = \sum_{n=0}^{N-1} \mathcal{V}^n + \mathcal{V}^N\,T\,. \tag{5}$$

From the stated hypothesis on G it will be easy to prove that $\mathcal{V}$ is continuous in $B_r(\partial\Omega)$, then, through a somewhat lengthy technical process we shall iterate on the smoothing property of $\mathcal{V}$ to end with the result that, for a sufficiently great N, $\mathcal{V}^N$ will convey continuously $L^2(\partial\Omega, \bar{H})$ into $B_r(\partial\Omega)$, any r, N depending on r.

3.1.5.4. The first application of the previous scheme may be found in Pao[30]. He works with the Couette flow problem and instead of applying a reflection operator on each wall he assumes that f is given on it. This eliminates all the technical difficulties associated with the operator T. Furthermore, in the Couette (plane) flow problem smoothness along rays implies smoothness in position space and it is not necessary to use the fact that AVA mixes rays.The technical work is consequently far more simple.

3.1.6. Hypothesis

3.1.6.1. We have already stated the hypothesis needed in order to obtain an existence theorem in $L^2(\Omega, H)$. We state here the hypothesis that we will need in order to work out the previously sketched scheme concerning $\mathcal{W}A$ and $\mathcal{E}$. We need more but not all the hypothesis of existence theory. Of course, for the final argument we need a set of hypothesis which include both. We may tentatively conjecture that existence theory in $L^2(\Omega, H)$ might be improved.

3.1.6.2. As far as Ω is concerned we assume boundedness, convexity and smoothness of $\partial\Omega$ in Liapounov sense (refer to 2.2.1.5.), we set α for the corresponding exponent. We observe that, without convexity we should have to replace $B_r(\Omega)$ by a set of functions continuous only on rays.

Concerning $\partial\Omega$ we need two more hypothesis that we state in 3.1.6.2.1. and 3.1.6.2.2.

3.1.6.2.1. Let $\mathbf{n_x}$ be the unit normal to $\partial\Omega$ at the point $\mathbf{x}$, directed towards the interior. Let $\mathbf{x}_1$ and $\mathbf{x}_2$ be any pair of points on $\partial\Omega$. Set $\mathbf{n}_1 = \mathbf{n}_{x1}$, $\mathbf{n}_2 = \mathbf{n}_{x2}$ and

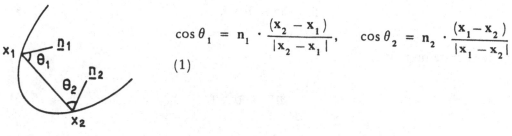

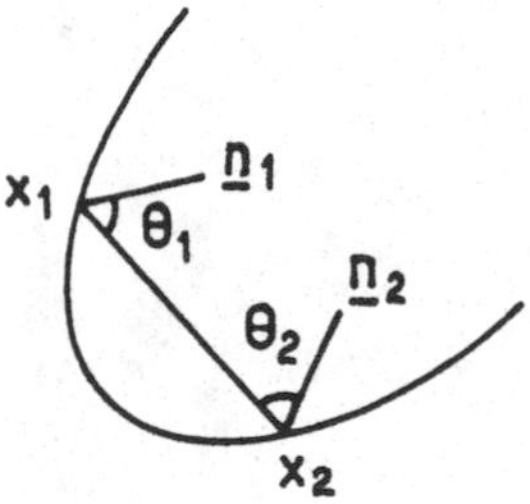

$$\cos\theta_1 = \mathbf{n}_1 \cdot \frac{(\mathbf{x}_2 - \mathbf{x}_1)}{|\mathbf{x}_2 - \mathbf{x}_1|}, \qquad \cos\theta_2 = \mathbf{n}_2 \cdot \frac{(\mathbf{x}_1 - \mathbf{x}_2)}{|\mathbf{x}_1 - \mathbf{x}_2|} \tag{1}$$

We state our hypothesis as

$$\cos\theta_1 > 0\,,\ \cos\theta_2 > 0 \Rightarrow \frac{\cos\theta_1}{\cos\theta_2} \leqslant P < +\infty \tag{2}$$

with a constant **P** independent of both $\mathbf{x}_1$ and $\mathbf{x}_2$.

3.1.6.2.2. We refer to the notations of 3.1.6.2.1. and we state our hypothesis as implying the relation

$$|x_1 - x_2| \leqslant R\ \mathrm{Inf}\,(|\cos\theta_1|, |\cos\theta_2|)\,, \quad R > 0\,. \tag{1}$$

3.1.6.2.3. We observe that convexity implies that $\cos\theta \geqslant 0$ and 3.1.6.2.2. that $\cos\theta_1 > 0$ and $\cos\theta_2 > 0$ whenever $|x_2 - x_1| > 0$.

3.1.6.3. Our basic requirement on the operator G is that its kernel (refer to 1.4.4.4.) $\Gamma(x, \xi, \zeta)$ is continuous on the set

$$\{\,(x, \xi, \zeta) \mid x \in \partial\Omega, \xi \in R^3_+(x)\,,\ \ \zeta \in R^3_-(x)\,\} \tag{1}$$

which is a rather mild requirement, and that it satisfies the following estimate

$$0 \leqslant \Gamma(x, \xi, \zeta) \leqslant \Gamma_1(x, \xi)\ \ \Gamma_2(x, \zeta) \tag{2}$$

with conditions imposed on Γ_1 and Γ_2 that will be stated presently. We observe that (2) is rather strong a requirement with respect to any realistic kernel like Cercignani's (1.2.3.4.) which does not satisfy (2). The reflection operator of 1.2.2.1. does not meet the requirement of continuity for the kernel unless $\alpha = 0$ which corresponds to purely diffuse reflection. Then we observe that the corresponding operator is nonlinear unless there is perfect accommodation $\alpha_E = 1$, so that only diffuse reflection with perfect accommodation may be considered within the framework of our linear theory of the reflection on the wall. The corresponding kernel is easily picked out

(3) $$\Gamma(x,\xi,\zeta) = \sqrt{2\pi}\ |\zeta \cdot n|\ \omega(|\zeta|)$$

and it is seen that (2) holds with $\Gamma_1 \equiv 1$, $\Gamma_2 = \sqrt{2\pi}\,\omega(|\zeta|)\,|\zeta \cdot n|$. In order to demonstrate the potentialities of our method we restrict ourselves to the consideration of kernels which satisfy (2) and we make the conjecture that with a substantial amount of technical work the process could be carried out with Cercignani's kernel. We observe also that it seems likely that, with a small amount of complementary work, the analysis to follow could be worked out in the case of specular diffuse reflection with complete accommodation. Of course the necessary work could include a reworking of chapter 2.

3.1.6.3.1. Concerning the functions of 3.1.6.3. we require that they meet one of two sets of conditions.

A) The following inequalities hold

(i) $$\Gamma_1(x,\xi) \leqslant \Gamma_1(|\xi|),\quad \Gamma_2(x,\zeta) \leqslant \Gamma_2(|\zeta|)\ ,$$

(ii) $$\int_0^\infty |\zeta|^2\ \Gamma_1(|\zeta|)\ \Gamma_2(|\zeta|)\ d|\zeta| \leqslant M_1\ ,$$

(iii) $$\int_0^\infty |\xi|^2\ \omega(|\xi|)\ (\rho(|\xi|))^{2p}\ \Gamma_1^2(|\xi|)\ d|\xi| \leqslant M_2(p)\ ,\ p = 0,1,2,3,4$$

(iv) $$\int_0^\infty |\zeta|^2\ \omega^{-1}(|\zeta|)\ \Gamma_2^2(|\zeta|)\ d|\zeta| \leqslant M_3\ ,$$

(v)$_r$ $$\sup_{\substack{x \in \partial\Omega \\ \xi \in R^3_+(x)}} \omega^{1/2}(|\xi|)\ (1 + |\xi|^2)^{r/2}\ \Gamma_1(x,\xi) \leqslant M_{4,r}\ ,$$

(vi)$_r$ $$\sup_{x \in \partial\Omega} \int_{R^3} \omega^{-1/2}(|\zeta|)\ (1 + |\zeta|^2)^{-r/2}\ \Gamma_2(x,\zeta)\ d\zeta \leqslant M_{5r}\ ,$$

where $\rho(|\xi|)$ stands for

$$\rho(|\xi|) = (1 + \mathrm{Log}\ \frac{1 + 2|\xi|}{|\xi|})^{1/2} . \tag{1}$$

B) We assume that all the inequalities of A) are satisfied but, furthermore, we require $(i)_B$ $(ii)_B$ $(iv)_B$ to hold. As to $(i)_B$ it is obtained from (i) by replacing the second of (i) by

$$\Gamma_2(x, \zeta) \leqslant \frac{|\zeta \cdot n|}{|\zeta|} \Gamma_{2,B}(|\zeta|) \tag{2}$$

on the other hand $(ii)_B$ and $(iv)_B$ have the same form as (ii) and (iv) but with $\Gamma_{2,B}$ in place of Γ_2.

3.1.6.3.2. We state a set of inequalities that we require for the functions Γ_1 and Γ_2 in complement to the ones of 3.1.6.3.1.

$$\sup_{x \in \partial\Omega} \int_{R^3_+} \omega\, |\xi \cdot n|\, (1 + \nu(|\xi|)\,)\, \Gamma_1^2(x, \xi)\, d\xi \leqslant M_6^2 , \tag{i}$$

$$\sup_{x \in \partial\Omega} \int_{R^3_+} \omega\, \Gamma_1^2(x, \xi)\, d\xi \leqslant M_7^2 , \tag{ii}$$

$$\sup_{y \in \partial\Omega} \int_{R^3_-} \omega^{-1} (1 + |\zeta \cdot n|^{-1})\, \Gamma_2^2(y, \zeta)\, d\zeta \leqslant M_8^2 , \tag{iii}$$

$$\sup_{y \in \partial\Omega} \int_{R^3_-} \omega^{-1} |\zeta \cdot n|^{-1}\, \Gamma_2^2(y, \zeta)\, d\zeta \leqslant M_9^2 . \tag{iv}$$

3.1.6.4. Concerning A we assume that it corresponds to rigid sphere and, correspondingly we have (refer to 1.5.2.3.)

$$a(1 + |\xi|^2)^{1/2} \leqslant \nu(|\xi|) \leqslant b(1 + |\xi|^2)^{1/2} \tag{1}$$

3.1.7. A Few Preliminary Lemmas

3.1.7.1. We first consider a consequence of 3.1.6.3.2. Consider the following norms

(1) $$|f^-|^2_X = \int_{\partial\Omega} dS \int_{R^3_-} \lambda(x,\xi)\ \omega(|\xi|)\ |f^-(x,\xi)|^2\ d\xi$$

(2) $$|f^+|^2_Y = \int_{\partial\Omega} dS \int_{R^3_+} \mu(x,\xi)\ \omega(|\xi|)\ |f^+(x,\xi)|^2\ d\xi$$

with $\lambda \geqslant 0$, $\mu \geqslant 0$. We have

(3) $$|Gf^-|^2_Y = \int_{\partial\Omega} dS \int_{R^3_+} \mu(x,\xi)\ \omega(|\xi|)\ \{ \int_{R^3_-} \Gamma(x,\xi,\zeta)\ \omega^{-1/2}(|\zeta|)\ \lambda^{-1/2}(x,\zeta)\ \omega^{1/2}(|\zeta|)\ \lambda^{1/2}(x,\zeta)\ |f^-(x,\zeta)|\ d\zeta\ \}^2\ d\xi$$

and we use 3.1.6.3.1.(1) and 3.1.6.3.2. with the Schwarz inequality to obtain

(4) $$|Gf^-|^2_Y \leqslant \int_{\partial\Omega} dS \int_{R^3_+} \mu(x,\xi)\ \Gamma_1^2(x,\xi)\ \omega(|\xi|)\ d\xi \cdot \int_{R^3} \Gamma_2^2(x,\zeta)\ \omega^{-1}(|\zeta|)\ \lambda^{-1}(x,\zeta)\ d\zeta \int_{R^3} \omega(|\zeta|)\ \lambda(x,\zeta)\,|f(x,\zeta)|^2\ d\zeta$$

$$\leqslant \left(\sup_{x\in\partial\Omega} \int_{R^3_+} \mu\,\omega\,\Gamma_1^2\ d\xi\right)\left(\sup_{x\in\partial\Omega} \int_{R^3_-} \omega^{-1}\lambda^{-1}\,\Gamma_2^2\ d\zeta\right)\ |f^-|^2_X$$

We apply (4) with

$$X = L^2(\partial\Omega, X) \Rightarrow \lambda = 1, \qquad \lambda^{-1} = 1 \tag{5}$$
$$X = L^2(\partial\Omega, \tilde{H}) \Rightarrow \lambda = |\zeta\cdot n|, \ \lambda^{-1} = |\zeta\cdot n|^{-1},$$
$$X = L^2(\partial\Omega, \tilde{H}_\nu) \Rightarrow \lambda = |\zeta\cdot n|\nu, \lambda^{-1} = |\zeta\cdot\zeta|^{-1}\nu^{-1} \leqslant a^{-1}(1+|\zeta|^2)^{-1/2}|\zeta\cdot n|^{-1}$$

$$Y = L^2(\partial\Omega, H) \Rightarrow \mu = 1,$$
$$Y = L^2(\partial\Omega, \tilde{H}) \Rightarrow \mu = |\xi\cdot n| \leqslant |\xi\cdot n|\ (1+\nu), \tag{6}$$
$$Y = L^2(\partial\Omega, \tilde{H}_\nu) \Rightarrow \mu = |\xi\cdot n|\nu \leqslant |\xi\cdot n|\ (1+\nu),$$

and we conclude that 3.1.6.3.2. implies that G is continuous from X to Y with any of (5) as X and any of (6) as Y. The norm $|G|_{X\to Y}$ is bounded by a constant given in the following table

Y \ X	$L^2(\partial\Omega, H)$	$L^2(\partial\Omega, \tilde{H})$	$L^2(\partial\Omega, \tilde{H}_\nu)$
$L^2(\partial\Omega, H)$	$M_7\ M_8$	$M_7\ M_8$	$a^{-1} M_7\ M_8$
$L^2(\partial\Omega, \tilde{H})$	$M_6\ M_8$	$M_6\ M_8$	$a^{-1} M_6\ M_8$
$L^2(\partial\Omega, \tilde{H}_\nu)$	$M_6\ M_8$	$M_6\ M_8$	$a^{-1} M_6\ M_8$

(7)

3.1.7.2. We state a result concerning the operator A that will be one of our basic tools in the later work. The proof of this statement will be found in 3.1.7.3. The operator $f \to Af$ is an integral operator, namely

$$A f(\xi) = \int_{R^3} \omega^{-1/2}(|\xi|)\ B(\xi,\zeta)\ \omega^{1/2}(|\zeta|)\ f(\zeta)\ d\zeta \tag{1}$$

and the kernel is continuous and bounded except in $\zeta = \xi$ where it has an integrable singularity. Furthermore, for rigid spheres the following inequalities hold

$$(2) \qquad \sup_{\xi \in R^3} \int_{R^3} | B(\xi, \zeta) | \, d\zeta \leqslant \sigma^2 C_1 ,$$

$$(3) \qquad \sup_{\zeta \in R^3} \int_{R^3} (1 + |\xi|^{-2+\alpha}) (1 + \mathrm{Log} \frac{1 + 2|\xi|}{|\xi|})^p \, | B(\xi, \zeta) | \, d\xi \leqslant \sigma^2 C_1' (\alpha, p) ,$$

$$(4) \qquad \left\{ \begin{array}{l} \int_0^\infty (1 + \mathrm{Log} \dfrac{1 + 2|\xi|}{|\xi|})^p \, | (B(|\xi| u , |\zeta| v) | \, d|\xi| \leqslant \\ \sigma^2 C_2(p) \, (1 + \mathrm{Log} \dfrac{(1 + 2|\zeta|)}{|\zeta|})^{p+1} (1 + \mathrm{Log} | \mathrm{Sin} \dfrac{\theta}{2} |^{-1}) . \end{array} \right.$$

In these formulas $0 < \alpha \leqslant 2$, $p > 0$ and θ is the angle between the unit vectors u and v

$$(5) \qquad | u \wedge v | = \mathrm{Sin}\, \theta , \quad u \cdot v = \mathrm{Cos}\, \theta .$$

3.1.7.3.1. We refer to 1.5.3.2. for the value of B in the case of rigid spheres and we restate the result

$$(1) \qquad |B| \leqslant B_1 + B_2 ,$$

$$(2) \qquad B_1 = \frac{\sigma^2}{2\sqrt{2\pi}} |\xi - \zeta| \exp \{ - \frac{1}{4}(|\xi|^2 + |\zeta|^2) \} ,$$

$$(3) \qquad B_2 = \frac{\sigma^2}{2\sqrt{2\pi}} | \xi - \zeta|^{-1} \exp \{ - \frac{1}{8} | \xi - \zeta|^2 - \frac{1}{8} \frac{(|\xi|^2 - |\zeta|^2)^2}{|\xi - \zeta|^2} \} .$$

We prove first 3.1.7.2.(2). We have

$$\int_{R^3} B_1(\xi,\zeta)\,d\zeta \leqslant \frac{\sigma^2}{2\sqrt{2\pi}} \int_{R^3} |u|\, e^{-\frac{1}{4}(|\xi|^2+|\xi+u|^2)}\,du \tag{4}$$

then, from

$$|\xi+u|^2+|\xi|^2 \geqslant 2|\xi|^2 - 2|\xi|\,|u| + |u|^2 = (|\xi|-|u|)^2 + |\xi|^2 \tag{5}$$

we get

$$\int_{R^3} B_1(\xi,\zeta)\,d\zeta \leqslant \sigma^2\sqrt{2\pi}\int_0^\infty u^3\, e^{-1/4|\xi|^2 - 1/4(|\xi|-u)^2}\,du$$

$$= \sigma^2\sqrt{2\pi}\, e^{-1/4|\xi|^2}\left\{\int_0^{|\xi|}(|\xi|-V)^3 e^{-1/4V^2}\,dv + \int_{|\xi|}^\infty (|\xi|+v)^3 e^{-1/4V^2}\,dv\right\} \tag{6}$$

$$\leqslant \sigma^2 C\,(1+|\xi|)^3\, e^{-1/4|\xi|^2}$$

where C is some numerical constant. On the other hand

$$\int_{R^3} B_2(\xi,\zeta)\,d\zeta \leqslant \frac{\sigma^2}{2\sqrt{2\pi}}\int_{R^3}|u|^{-1} e^{-1/8|u|^2}\,du \leqslant C\,\sigma^2 \tag{7}$$

with another numerical constant C. From (6) and (7) we conclude that 3.1.7.2.(2) holds.

3.1.7.3.2. Now we prove 3.1.7.2.(3). We consider separately the contributions to the integral from $|\xi| \leqslant 1$ and from $|\xi| > 1$. For $|\xi| \leqslant 1$ we start from

$$B_1 \leqslant \frac{\sigma^2}{2\sqrt{2\pi}}(1+|\zeta|)\,e^{-1/4|\zeta|^2} \leqslant C\,\sigma^2\,, \tag{1}$$

$$B_2 \leqslant \frac{\sigma^2}{2\sqrt{2\pi}}|\xi-\zeta|^{-1}\,, \tag{2}$$

and we find

$$\int_{|\xi|\leqslant 1} (1+|\xi|^{-2+\alpha})(\rho(|\xi|))^{2p}\,|\,B(\xi,\zeta)\,|\,d\xi \leqslant \tag{3}$$

$$C\sigma^2 \int_{|\xi|\leqslant 1} (1+|\xi-\zeta|^{-1})(1+|\xi|^{-2+\alpha})(\rho(|\xi|))^{2p}\,d\xi \leqslant C\sigma^2$$

with ρ as in 3.1.6.3.1.(1).

We use the convention, now and later, that C stands for some numerical constant which changes from estimate to estimate. We observe that the constant C in (3) depends on p and α and that it grows indefinitely when $\alpha \to 0$. For $|\xi| > 1$ we have

$$(1+|\xi|^{-2+\alpha})\,\rho^{2p}(|\xi|) \leqslant (1+\mathrm{Log}\,3)^p\,(1+|\xi|^{-2+\alpha}) \tag{4}$$

on the other hand

$$\int_{|\xi|>1} (1+|\xi|^{-2+\alpha})\,B_1(\xi,\zeta)\,d\xi \leqslant \frac{\sigma^2}{2\sqrt{2\pi}}\Big(\int_{|\xi|>1} (1+|\xi|^{-2+\alpha})(|\xi|+1)e^{-1/4|\xi|^2}\,d\xi\Big). \cdot (1+|\zeta|)\,e^{-1/4|\zeta|^2}$$

$$\leqslant C\,\sigma^2 \tag{5}$$

and

$$\int_{|\xi|>1} (1+|\xi|^{-2+\alpha})\,B_2(\xi,\zeta)\,d\xi \leqslant \frac{\sigma^2}{2\sqrt{2\pi}} \int_{|\xi|>1} \frac{1+|\xi|^{-2+\alpha}}{|\xi-\zeta|}\,e^{-1/8|\xi-\zeta|^2}\,d\xi \tag{6}$$

$$\leqslant \frac{\sigma^2}{\sqrt{2\pi}} \int_{R^3} |u|^{-1}\,e^{-1/8|\mu|^2}\,du = C\,\sigma^2$$

taking into account $\alpha \leqslant 2$.

3.1.7.3.3. We prove 3.1.7.2.(4). We start from

$$\int_0^\infty \rho^{2p}(\xi)\,|B(\xi u,\zeta v)|\,d\xi \leqslant \int_0^\infty \rho^{2p} B_1\, d\xi + \int_0^\infty \rho^{2p} B_2\, d\xi \tag{1}$$

and, first

$$\int_0^\infty \rho^{2p} B_1\, d\xi \leqslant \frac{\sigma^2}{2\sqrt{2\pi}} \int_0^\infty (\xi+\zeta)\, \rho^{2p}(\xi)\, e^{-1/4(\xi^2+\zeta^2)}\, d\xi \leqslant C\sigma^2 \tag{2}$$

$$\int_0^\infty \rho^{2p} B_2\, d\xi \leqslant \frac{\sigma^2}{2\sqrt{2\pi}} \int_0^\infty \rho^{2p}\, |\xi u - \zeta v|^{-1} \exp\Big\{-\frac{1}{8}\,|\xi u - \zeta v|^2 - \frac{(\xi^2-\zeta^2)^2}{|\zeta u - \zeta v|^2}\Big\}\, d\xi = \frac{\sigma^2}{2\sqrt{2\pi}}\, I\,. \tag{3}$$

In order to estimate the integral I we set $\xi = \tau\zeta$ and observe that

$$|\xi u - \zeta v| = \zeta(\tau^2 + 1 - 2\tau\cos\theta)^{1/2} \tag{4}$$

then

$$\begin{aligned} I \leqslant & \int_0^2 \Big(1 + \mathrm{Log}\,\frac{1+2\tau\zeta}{\tau\zeta}\Big)^p (\tau^2+1-2\tau\cos\theta)^{-1/2} \exp\Big\{-\frac{1}{8}(\tau^2+1-2\tau\cos\theta)\Big\}\, d\tau \,+ \\ & + \int_2^\infty \Big(1 + \mathrm{Log}\,\frac{1+2\tau\zeta}{\tau\zeta}\Big)^p (\tau^2+1-2\tau\cos\theta)^{-1/2} \exp\Big\{-\frac{1}{8}\,\frac{(\tau^2-1)^2\,\zeta^2}{\tau^2+1-2\tau\cos\theta}\Big\}\, d\tau \\ = & \; J_1 + J_2\,. \end{aligned} \tag{5}$$

With some constant depending only on p

$$J_1 \leqslant C \int_0^2 \{(1+\mathrm{Log}\,\frac{1+4\zeta}{\zeta})^p + (\mathrm{Log}\,\frac{1}{\tau})^p\}\frac{d\tau}{\sqrt{(\tau-1)^2+4\tau\ \sin^2\frac{\theta}{2}}}$$

(6)

$$= C(1+\mathrm{Log}\,\frac{1+4\zeta}{\zeta})^p\ K_1 + C\ K_2$$

and

(7)
$$J_2 \leqslant \int_2^\infty (1+\mathrm{Log}\,\frac{(1+2\tau\zeta)}{\tau\zeta})^p\ \frac{e^{-1/8(\tau-1)^2\zeta^2}}{\tau-1}\,d\tau = K_3\ .$$

We estimate K_1 through

$$K_1 = \int_0^{1-\epsilon} + \int_{1-\epsilon}^{1+\epsilon} + \int_{1+\epsilon}^{2} \leqslant 2\int_0^{1-\epsilon}\frac{d\tau}{1-\tau} + \int_{1-\epsilon}^{1+\epsilon}\frac{d\tau}{[(\tau-1)^2+4(1-\epsilon)\ \sin^2\frac{\theta}{2}\ ]^{1/2}}$$

(8)

$$\leqslant 2\ \mathrm{Log}\,\frac{1}{\epsilon} + \int_{-\epsilon/2(1-\epsilon)^{1/2}\sin\theta/2}^{\epsilon/2(1-\epsilon)^{1/2}\sin\theta/2}\frac{du}{\sqrt{1+u^2}} \leqslant \mathrm{const} + 2\ \mathrm{Log}\,|\sin\frac{\theta}{2}|^{-1}$$

then

(9)
$$K_2 = \int_0^2 (\mathrm{Log}\,\frac{1}{\tau})^p\ \frac{d\tau}{[(\tau-1)^2+4\tau\sin^2\frac{\theta}{2}]^{1/2}} \leqslant C(p) \qquad p>0$$

(10)
$$K_2 \leqslant \mathrm{const} + 2\ \mathrm{Log}\,|\sin\theta/2|^{-1}\ , \qquad p=0$$

and

(11)
$$K_3 = \int_\zeta^\infty (1+\mathrm{Log}\,\frac{1+2\zeta+2x}{\zeta+x})^p\ e^{-1/8x^2}\ \frac{dx}{x}$$

and, from

$$\frac{1 + 2\zeta + 2x}{\zeta + x} \leqslant 2 + \frac{1}{\zeta} , \tag{12}$$

we have

$$K_3 \leqslant (1 + \mathrm{Log}\,\frac{1 + 2\zeta}{\zeta})^p \int_\zeta^\infty e^{-1/8\,x^2}\,\frac{dx}{x} \leqslant \mathrm{const}\,(1 + \mathrm{Log}\,\frac{1 + 2\zeta}{\zeta})^{p+1} . \tag{13}$$

summing up we obtain 3.1.7.2.(3).

3.1.8. Some Further Lemmas

3.1.8.1. Let us consider the notations of the figure, where e stands for a fixed direction, we state that

$$I = \int_\Omega \frac{(1 + \mathrm{Log}\,|\sin\theta/2|^{-1})\,(1 + \mathrm{Log}\,|\sin\varphi/2|^{-1})}{|x-z|^2\ |z-y|^2}\,dv_z$$

$$\leqslant C_3(\Omega)\,|x-y|^{-1} . \tag{1}$$

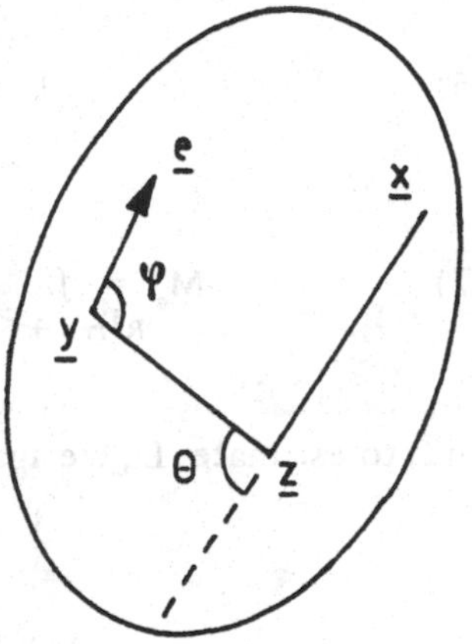

Let J be the same integral as in (1) but extended over the whole three dimensional space instead of only Ω . We have

$$J = \frac{1}{|x-y|} \int_{R^3} \frac{(1 + \mathrm{Log}|\sin\theta/2|^{-1})\,(1 + \mathrm{Log}|\sin\varphi/2|^{-1})}{|x-z'|^2\ |z'-y|^2}\,dv_{z'} = \frac{K}{|x-y|} , \tag{2}$$

where

$$y' = x + |x-y|^{-1}\,(y-x) \ ; \quad z' = x + |y-x|^{-1}\,(z-x) , \tag{3}$$

(4) $$K = \int_{R^3} \frac{(1 + \mathrm{Log}|\sin \theta/2|^{-1})\,(1 + \mathrm{Log}|\sin \varphi/2|^{-1})}{|x-z|^2 \;\; |z-y|^2}\, dv_z\,, \qquad |x-y| = 1\,.$$

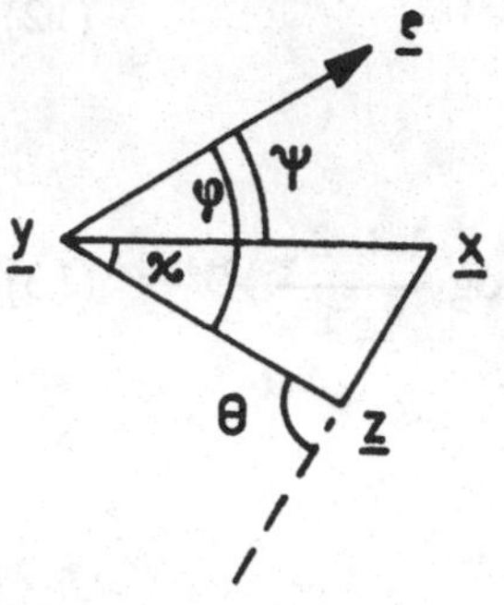

We have, dividing the range of integration in two sets,

(5) $$K \leqslant (1 + \mathrm{Log}|\sin \frac{\epsilon}{2}\,|^{-1})\; L + M_\epsilon$$

with

(6) $$L = \int_{R^3} \frac{1 + \mathrm{Log}/\sin \varphi/2\,|^{-1}}{|x-z|^2 \;\; |z-y|^2}\, dv_z\,,$$

(7) $$M_\epsilon = \int_{R^3 \cap \{\theta \leqslant \epsilon\}} \frac{(1 + \mathrm{Log}|\sin \theta/2|^{-1})\,(1 + \mathrm{Log}|\sin \varphi/2)^{-1})}{|x-z|^2 \;\; |z-y|^2}\, dv_z\,,$$

and, to estimate L, we again divide the range of integration

(8) $$L \leqslant \delta^{-2} \int_{R^3} \frac{1 + \mathrm{Log}|\sin \varphi/2|^{-1}}{|y-z|^2}\, d\nu_z + \int_{R^3 \cap \{|x-z| \leqslant \delta\}} \frac{1 + \mathrm{Log}|\sin \varphi/2|^{-1}}{|x-z|^2 \; |z-y|^2}\, dv_z$$

$$\leqslant C\,\delta^{-2} + N_\delta\,.$$

From $|x-z| \leqslant \delta \Rightarrow |y-z| \geqslant 1-\delta$ we have

(9) $$N_\delta \leqslant (1-\delta)^{-2} \int_{R^3 \cap \{|x-z| \leqslant \delta\}} \frac{1 + \mathrm{Log}\,|\sin \varphi/2|^{-1}}{|x-z|^2}\, dv_z = (1-\delta)^{-2}\, P$$

Let $x - y = w$, $|w| = 1$, we observe that N_δ depends only on ψ with $\sin\psi = |w \wedge e|$, $\cos\psi = w \cdot e$, with $|e| = 1$, and N_δ is bounded except possibly when ψ goes to zero. As a consequence, in order to estimate P we may restrict ourselves to $0 \leqslant \psi \leqslant \gamma$. Let $z - x = \rho u$, $|u| = 1$ we have

$$P = C \int_0^\delta d\rho \int_{|u|=1} \{1 + \operatorname{Log}|w \wedge e + \rho(u \wedge e)|^{-1}\}\, du = C(Q + 4\pi\delta) \qquad (10)$$

where we have used

$$\sin\varphi/2 = \frac{\sin\varphi}{2\cos\varphi/2} \geqslant C \sin\varphi \qquad (11)$$

for $\rho < \delta$, $\psi < \gamma$ with δ and γ sufficiently small. Changing the notations we obtain

$$Q = \int_0^\delta d\rho \int_{|u|=1} \operatorname{Log}|a + \rho u|^{-1}\, du$$

$$= 2\pi \int_0^\delta d\rho \int_0^\pi \{\operatorname{Log}\,|\,|a|^2 + \rho^2 + 2|a|\rho\cos\theta\,|^{-1}\} \sin\theta\, d\theta$$

$$= 2\pi \int_0^\delta d\rho \int_{-1}^{+1} \operatorname{Log}\,|\,|a|^2 + \rho^2 + 2|a|\rho t\,|^{-1}\, dt$$

$$= \pi |a|^{-1} \int_0^\delta \rho^{-1}\, d\rho \int_{(|a|-\rho)^2}^{(|a|+\rho)^2} (\operatorname{Log}|x|^{-1})\, dx \qquad (12)$$

$$= \pi |a|^{-1} \int_0^\delta \rho^{-1} \{x\,[\operatorname{Log}(x^{-1}) + 1]\}_{(|a|-\rho)^2}^{(|a|+\rho)^2}\, d\rho$$

$$= \pi |a|^{-1} \int^{\delta|a|} u^{-1} \{|a|^2(1+u)^2\,[1 - \operatorname{Log}|a|^2 - \operatorname{Log}(1+u)^2]$$

$$- |a|^2(1-u)^2\,[1 - \operatorname{Log}|a|^2 - \operatorname{Log}(1-u)^2]\}\, du$$

$$= 4\pi\delta(1 - \operatorname{Log}|a|^2) + \pi|a| \int_0^{\delta|a|} \{(1-u)^2 \operatorname{Log}(1-u)^2 - (1+u)^2 \operatorname{Log}(1+u)^2\} \frac{du}{u}$$

we may choose δ and then, the only thing we want to prove is that Q remains bounded when $|a| \to 0$ because $|a| < 1$ and Q is continuous on $]0,1]$. We have, for $u \to \infty$

$$(13) \quad (1-u)^2 \operatorname{Log}(1-u)^2 - (1+u)^2 \operatorname{Log}(1+u)^2 = -4u \operatorname{Log} u^2 - 4u + 0(\frac{1}{u})$$

and, for $|a| \to 0$, Q behaves as

$$Q = 4\pi\,\delta(1-\operatorname{Log}|a|^2) - 4\pi\,|a| \int_0^{\delta/|a|} \operatorname{Log}|u|^2\, du + 0(|a| \operatorname{Log}|a|)$$

$$(14) \quad = 4\pi\,\delta(1-\operatorname{Log}|a|^2) - 8\pi\,|a|\;\delta\,|a|^{-1}\,(\operatorname{Log}[\delta\,|a|^{-1}] - 1) + 0(|a| \operatorname{Log}|a|)$$

$$= 0(1) + 0(|a| \operatorname{Log}|a|)\;.$$

Summing up (8)(9)(12)(14) we conclude that L is bounded and from (5) we have now to prove that M_ϵ is bounded. In the range of integration occurring in M_ϵ we have

$$(15) \qquad \sin\frac{\theta}{2} \geqslant \frac{\sin\theta}{2\cos\epsilon/2} = \frac{\sin\chi}{2\,|x-z|\cos\epsilon/2}$$

and it is sufficient to prove that

$$(16) \quad R = \int_{R^3 \cap \{\theta < \epsilon\}} \frac{(1 + \operatorname{Log}|\sin\chi|^{-1} + \operatorname{Log}|x-z|)\,(1 + \operatorname{Log}|\sin\frac{\varphi}{2}|^{-1})}{|x-z|^2\,|z-y|^2}\, dv_z$$

is bounded. From

$$(1 + \operatorname{Log}|\sin\chi|^{-1} + \operatorname{Log}|x-z|)\,(1 + \operatorname{Log}|\sin\varphi/2|^{-1}) \leqslant 1 + \operatorname{Log}|\sin\chi|^{-1}$$

$$(17) \quad + \operatorname{Log}|\sin\varphi/2|^{-1} + \operatorname{Log}|x-z|^{-1} + \frac{1}{2}(\operatorname{Log}|\sin\chi|^{-1})^2 + (\operatorname{Log}|\sin\varphi/2|^{-1})^2$$

$$+ \frac{1}{2}(\operatorname{Log}|x-z|^{-1})^2$$

we see that it is sufficient to prove boundedness of

$$R_1 = \int_{R^3} \frac{dv_z}{|x-z|^2 \, |z-y|^2} \quad , \tag{18}$$

$$R_2 = \int_{R^3 \cap \{|\theta| \leq \epsilon\}} \frac{(\mathrm{Log}\,|x-z|^{-1})^2}{|x-z|^2 \, |z-y|^2} \, dv_z \quad , \tag{19}$$

$$R_3 = \int_{R^3 \cap \{|\theta| < \epsilon\}} \frac{(\mathrm{Log}\,|\sin \chi|^{-1})^2}{|x-z|^2 \, |z-y|^2} \, dv_z \quad , \tag{20}$$

$$R_4 = \int_{R^3 \cap \{|\theta| < \epsilon\}} \frac{(\mathrm{Log}\,|\sin \varphi/2|^{-1})^2}{|x-z|^2 \, |z-y|^2} \, dv_z \quad . \tag{21}$$

We observe that R_1 is a pure number, defined by a convergent integral (remember that $|x-y| = 1$) and, as such, it is obviously bounded. The same argument shows that R_2 is bounded. Referring to the figure showing the meaning of ψ we see that

$$R_4 \leqslant (\mathrm{Log}\,|\sin \epsilon/2|^{-1})^2 \; R_1 \qquad \text{when} \quad \psi > 2\epsilon \; , \tag{22}$$

so that it is sufficient to prove boundedness of R_4 under the assumption of $\psi < 2\epsilon$. Regarding R_3 we have

$$|\sin \chi|^{-1} \leqslant (2\cos\frac{\epsilon}{2})^{-1} |\sin \chi/2|^{-1} \tag{23}$$

and R_3 is bounded by a constant times R_1 plus a particular value of R_4 corresponding to $\psi = 0$. We need only concentrate on

$$R_5 = \int_{R^3} \frac{(\mathrm{Log}\,|\sin \varphi/2|^{-1})^2}{|x-z|^2 \, |z-y|^2} \, d\nu_z \quad . \tag{24}$$

Referring to the evaluation of L in (6) (8) (9) (10) (11) it is sufficient to prove boundedness of

$$S = \int_0^\delta d\rho \int_{|u|=1} (\mathrm{Log}\,|a + \rho u|^{-1})^2 \, du$$

$$= \pi \, |a|^{-1} \int_0^\delta \rho^{-1} \, d\rho \int_{(|a|-\rho)^2}^{(|a|+\rho)^2} (\mathrm{Log}\,|x|^{-1})^2 \, dx$$

(25) $$= \pi \, |a|^{-1} \int_0^\delta \rho^{-1} \{ x(\mathrm{Log}\,x)^2 - 2\,x\,\mathrm{Log}\,x + 2\,x \}_{(|a|-\rho)^2}^{(|a|+\rho)^2} \, d\rho$$

$$= \pi|a| \int_0^{\delta/|a|} u^{-1} \{(1+u)^2 \, [(\mathrm{Log}\,|a|^2 + \mathrm{Log}(1+u)^2)^2 - 2\,\mathrm{Log}|a|^2 - 2\,\mathrm{Log}(1+u)^2 + 2]$$

$$- (1-u)^2 \, [(\mathrm{Log}\,|a|^2 + \mathrm{Log}(1-u)^2)^2 - 2\,\mathrm{Log}\,|a|^2 - 2\,\mathrm{Log}(1-u)^2 + 2]\} \, du$$

Developing we find

$$S = 4\pi \, |a| \, (\mathrm{Log}\,|a|^2)^2 \int_0^{\delta/|a|} du \, +$$

$$+ \, 2\pi \, |a| \, \mathrm{Log}|a|^2 \int_0^{\delta/a} u^{-1} \{(1+u)^2 \, \mathrm{Log}(1+u)^2 - (1-u)^2 \, \mathrm{Log}(1-u)^2 - 4u\} \, du$$

(26)

$$+ \, \pi|a| \int_0^{\delta|a|} u^{-1} \{(1+u)^2 [(\mathrm{Log}(1+u)^2)^2 - 2\,\mathrm{Log}(1+u)^2 + 2] -$$

$$(1-u)^2 \, [(\mathrm{Log}(1-u)^2)^2 - 2\,\mathrm{Log}(1-u)^2 + 2] \} \, du$$

and, as with a in (12), we need only prove boundedness of S when $|a| \to 0$. To investigate this behaviour we use the relations

$$u^{-1}\{(1+u)^2 \operatorname{Log}(1+u)^2 - (1-u)^2 \operatorname{Log}(1-u)^2 - 4u\} - 4 \operatorname{Log} u^2 + 0(1) \quad (27)$$

$$u^{-1}\{(1+u)^2 [(\operatorname{Log}(1+u)^2)^2 - 2\operatorname{Log}(1+u)^2 + 2] -$$

$$- (1-u)^2 [(\operatorname{Log}(1-u)^2)^2 - 2 \operatorname{Log}(1-u)^2 + 2]\} \quad (28)$$

$$= 4(\operatorname{Log} u^2)^2 + 0\{u^{-1} (\operatorname{Log} u^2)\}$$

which are valid when $u \to \infty$ and, consequently, the behaviour of S when $|a| \to 0$ is

$$S = 4\pi\delta(\operatorname{Log}|a|^2)^2 + 16\pi|a| \operatorname{Log}|a|^2 \int_0^{\delta|a|} \operatorname{Log} u \, du +$$

$$+ 16\pi|a| \int_0^{\delta/|a|} (\operatorname{Log} u)^2 \, du + 0(1) \quad (29)$$

and, evaluating the right hand side

$$S = 16\pi\delta(\operatorname{Log}|a|)^2 + 32\pi\delta \operatorname{Log}|a|)(\operatorname{Log}\delta - \operatorname{Log}|a| - 1)$$

$$+ 16\pi\delta\{(\operatorname{Log}\delta - \operatorname{Log}|a|)^2 - 2\operatorname{Log}\delta + 2\operatorname{Log}|a| + 2\} + 0\,1) \quad (30)$$

$$= 0(1)$$

we find, as required, that S is bounded when $|a| \to 0$. Summing up, this achieves the proof that M_e in (5) is bounded, as we have already proven that L is bounded this achieves the proof of (1).

3.1.8.2. Let $0 \leqslant p < 3$, $0 \leqslant q < 3$ be arbitrary, we have

$$(1)\quad \begin{cases} p+q>3 \Rightarrow \int_\Omega \dfrac{dv_z}{|x-z|^p\,|z-y|^q} \leqslant C_{p,q}(\Omega)\,|x-y|^{3-p-q} , \\ p+q=3 \quad \text{and } \Omega \text{ bounded} \\ \Rightarrow \int_\Omega \dfrac{dv_z}{|x-z|^p\,|z-y|^q} \leqslant C_{p,q}(\Omega)\,(1+|\mathrm{Log}\,|x-y||) , \\ p+q<3 \quad \text{and } \Omega \text{ bounded} \\ \Rightarrow \int_\Omega \dfrac{dv_z}{|x-z|^p\,|z-y|^q} \leqslant C_{p,q}(\Omega) . \end{cases}$$

When $p+q>3$ we may replace Ω by R^3 and then the result is evident from the same type of argument as in the beginning of 3.1.8.1., we may even state that when $p+q>3$, $C_{p,q}$ is uniformly bounded with respect to Ω . Let us consider the situation when $p+q<3$. We divide Ω in two subregions

$$(2)\qquad \Omega = \Omega_1 \cup \Omega_2$$

such that

$$(3)\qquad \begin{aligned} \Omega_1 &: \{\, z \mid |z-y| \geqslant |x-z| \,\} \\ \Omega_2 &: \{\, z \mid |x-z| > |z-y| \,\} \end{aligned}$$

we find

$$\int_{\Omega} \frac{dv_z}{|x-z|^p \, |z-y|^q} \leqslant \int_{\Omega_1} \frac{dv_z}{|x-z|^{p+q}} + \int_{\Omega_2} \frac{dv_z}{|z-y|^{p+q}} \leqslant C_{p,q}(\Omega) \,. \tag{4}$$

Considering the case $p+q=3$ we set D for the diameter of Ω, and $B_D(x)$ for the ball of radius D and center at x. We have

$$\int_{\Omega} \frac{dv_z}{|x-z|^p \, |z-y|^q} \leqslant \int_{B_D(x)} \frac{dv_z}{|x-z|^p \, |z-y|^q} = \int_0^D r^{2-p} \, dr \int_{|u|=1} \frac{du}{|Re+ru|^q} = I \tag{5}$$

where R stands for $|x-y|$ and e for a unit vector. Integrating in polar coordinates we find

$$I = 2\pi \int_0^D r^{2-p} \, dr \int_0^\pi \frac{\sin\theta \, d\theta}{(R^2 + r^2 + 2Rr\cos\theta)^{q/2}}$$

$$= \pi R^{-1} \int_0^D r^{1-p} \left(|R+r|^{2-q} - |R-r|^{2-q} \right) dr \tag{6}$$

$$= \frac{\pi}{2-q} \int_0^{D/R} u^{1-p} \left(|1+u|^{2-q} - |1-u|^{2-q} \right) du \,.$$

When $u \to \infty$ we have

$$u^{1-p}[(1+u)^{2-q} - |1-u|^{2-q}] \sim 2(2-q) \, u^{2-p-q} \sim 2(2-q) \, u^{-1} \tag{7}$$

and we conclude that, as $R \to 0$

(8) $$I = 2\pi \ \mathrm{Log}(D\,R^{-1}) + 0(1)\ .$$

This achieves the proof of (1).

3.1.8.3. Let $0 \leqslant p < 2$, $0 \leqslant q < 2$ be arbitrary and assume that $\partial\Omega$ satisfies a Liapounov condition and has bounded area, we have, for

(1) $$\begin{cases} p+q>2 \Rightarrow \displaystyle\int_{\partial\Omega} \frac{dS_z}{|x-z|^p\,|z-y|^q} \leqslant C'_{p,q}(\partial\Omega)\,|x-y|^{2-p-q}\ , \\ p+q=2 \Rightarrow \displaystyle\int_{\partial\Omega} \frac{dS_z}{|x-z|^p\,|z-y|^q} \leqslant C'_{p,q}(\partial\Omega)\,(1+\big|\mathrm{Log}|x-y|\big|)\ , \\ p+q<2 \Rightarrow \displaystyle\int_{\partial\Omega} \frac{dS_z}{|x-z|^p\,|z-y|^q} \leqslant C'_{p,q}(\partial\Omega)\ . \end{cases}$$

Set I for the integral under consideration and assume $|x-y| > \epsilon$ we divide the range of integration in two sets in accordance to

(2) $$\partial\Omega = \Sigma_1 \cup \Sigma_2$$

with

(3) $$\begin{aligned} \Sigma_1 &: \{ z \mid |x-z| \geqslant |z-y| \} \\ \Sigma_2 &: \{ z \mid |z-y| \geqslant |x-z| \} \end{aligned}$$

and we have

$$I \leqslant \left(\frac{2}{\varepsilon}\right)^p \int_{\Sigma_1} \frac{dS_z}{|y-z|^q} + \left(\frac{2}{\varepsilon}\right)^q \int_{\Sigma_2} \frac{dS_z}{|x-z|^q} < \frac{C}{\varepsilon^r} \tag{4}$$

with $r = \sup(p,q)$. As a consequence it is sufficient to restrict our attention to the case when $\mathbf{x}$ and $\mathbf{y}$ are close to each other. To be more specific, let r_0 be the radius of the Liapounov sphere as explained in 2.2.1.5. and assume $|\mathbf{x}-\mathbf{y}| \leqslant N^{-1} r_0$ with some conveniently chosen $N > 2$.

We choose a system of local coordinates x, y, z with origin at $\mathbf{x}$ and set

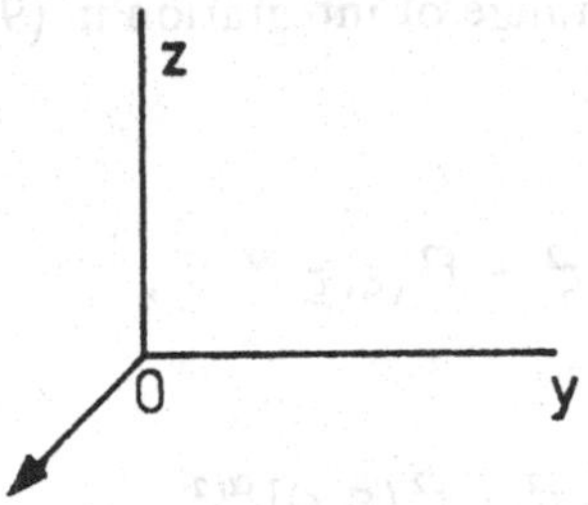

$$\mathbf{y} = (x, y, f(x,y)) \tag{5}$$

$$\mathbf{z} = (\xi, \zeta, f(\xi, \zeta)) \tag{6}$$

when $\mathbf{z}$ is on the part of Ω which is within the Liapounov sphere centered at $\mathbf{x}$. We set $B_{r_0}(\mathbf{x})$ for such a sphere and $\Sigma_{r_0}(\mathbf{x}) = \partial\Omega \cap B_{r_0}(\mathbf{x})$, while $\Delta_{r_0}(\mathbf{x})$ is the complement to $\Sigma_{r_0}(\mathbf{x})$ on $\partial\Omega$. We have

$$I = \int_{\Sigma_{r_0}(x)} + \int_{D_{r_0}(x)} \{\quad\} \, dS_z = I_1 + I_2 , \tag{7}$$

and

$$I_2 \leqslant \left(\frac{N}{N-1}\right)^q \frac{1}{r_0^{p+q}} \, |\partial\Omega| , \tag{8}$$

so that it is sufficient to prove boundedness of

(9) $$I_1 = \int_{\Sigma_{r_0}(x)} \frac{dS_z}{|x-z|^p \; |z-y|^q} \; .$$

We set, with a slight change of notation by reference to 2.2.1.5.,

(10) $$\frac{\partial f}{\partial x} = \lambda \qquad \frac{\partial f}{\partial y} = \mu$$

and we know, from 2.2.1.7.(8)(9) that, within the range of integration in (9),

(11) $$\lambda^2(\xi,\zeta) + \mu^2(\xi,\zeta) \leqslant 2\,E^2\,[\xi^2 + \zeta^2 + f^2(\xi,\zeta)]^{\alpha} \quad ,$$

(12) $$f(\xi,\zeta) \leqslant \sqrt{2}\;E\;(\xi^2+\zeta^2)^{1/2}\;[\xi^2+\zeta^2+f^2(\xi,\zeta)]^{\alpha/2} \; .$$

On the other hand

$$I_1 \leqslant \iint_{\xi^2+\zeta^2 \leqslant r_0^2} \frac{\{1+\lambda^2(\xi,\zeta)+\mu^2(f,\zeta)\}^{1/2}}{[\xi^2+\zeta^2+f^2(\xi,\zeta)]^{p/2}\;[(x-\xi)^2+(y-\zeta)^2+(f(x,y)-f(\xi,\zeta))^2]^{q/2}}\,d\xi\,d\zeta$$

(13) $$\leqslant (1+\sqrt{2}E\,r_0^{\alpha}) \iint_{\xi^2+\zeta^2 \leqslant r_0^2} \frac{d\xi\,d\zeta}{(\xi^2+\zeta^2)^{p/2}\,[(x-\xi)^2+(y-\zeta)^2]^{q/2}}$$

$$\leqslant \iint_{\xi^2+\zeta^2 \leqslant r_0^2} \frac{2\,d\xi d\zeta}{(\xi^2+\zeta^2)^{p/2}\,[(x-\xi)^2+(y-\zeta)^2]^{q/2}} = J$$

if we take 2.2.1.7.(7) into account. We compute J using polar coordinates and, setting $x^2 + y^2 = r$, we find

$$J = \int_0^{r_0} d\rho \int_0^{2\pi} d\theta \; \rho^{1-p} (\rho^2 + r^2 - 2r\rho\cos\theta)^{-q/2}$$

$$= r^{2-p-q} \int_0^{r_0/r} u^{1-p}\, du \int_0^{2\pi} (1 + u^2 - 2u\cos\theta)^{-q/2}\, d\theta \qquad (14)$$

$$= r^{2-p-q} F_{p,q}\left(\frac{r_0}{r}\right)$$

observing that r is equivalent to $|x-y|$ when $|x-y| \to 0$ we conclude that the proof of (1) amounts to prove that, when $t \to \infty$ we have

$$p + q > 2 \Rightarrow F_{p,q}(t) \text{ remains bounded,}$$

$$p + q = 2 \Rightarrow F_{p,q}(t) - C \operatorname{Log} t + D, \qquad (15)$$

$$p + q < 2 \Rightarrow F_{p,q}(t) - C\, t^{2-p-q} + o(t^{2-p-q}),$$

and inspection of (14) shows that this holds.

3.2. Regularisation of the Solution of the Linear Problem. First Step

3.2.1. The Method of Approach

3.2.1.1. Our technique is to show that WA has a smoothing property. The precise property is that for any integer p and with $f \in L^2(\Omega, H)$,$(WA)^{5+p} f \in B_{2p+1}(\Omega)$. The technique will be to prove the following estimate for any continuous f,

$$(1) \qquad < (WA)^{5+p}\ f >_r \ \leqslant \text{Const} \ ||| f ||| \ , \qquad r = 2p \ .$$

Then, given any $f \in L^2(\Omega, H)$ we can find a sequence of functions f_n continuous which converge to f in $L^2(\Omega, H)$. We shall prove that each of $(WA)^{p+5} f_n$ is continuous so that the sequence of $(WA)^{5+p} f_n$ will converge to $(WA)^{5+p} f \in B_{2p}(\Omega)$. As A conveys a continuous function into a continuous one the proof of the continuity statement will amount merely to a proof that W conveys a continuous function into a continuous one. From 2.1.2.4.(5) we see that the proof has to be given separately for U and ETG and as the continuity property holds for G and E it will be sufficient to prove the statement for T. As a matter of fact we shall prove that T is continuous in $B_r(\partial\Omega)$ and it will be sufficient to observe that a continuous f that belongs to $L^2(\Omega, H)$ is converted by $G J^- BU$ to a function of $B_r(\partial\Omega)$. The proof that T is continuous is $B_r(\partial\Omega)$ will be one of the lengthy parts of our work.

3.2.1.2. Our main task is to prove the estimate 3.2.1.1.(1). In order to do this we shall iterate on a weaker estimate concerning WA that we describe now and prove later. First we define

$$(1) \qquad \{f\}^2(x) = \int_{R^3} \omega(\xi) \ | f(x, \xi) |^2 \ d\xi$$

then we estimate $\{\rho^p \mu WAf\}(x)$, where ρ is a function of $|\xi|$ while μ is a function of $|\xi|^{-1} \xi$, that is of the orientation of ξ. Our estimate holds for any function ρ and for

$$(2) \qquad \mu_e = \left(1 + \text{Log} \left| \sin \frac{\psi}{2} \right|^{-1}\right)^{1/2}, \ |\xi \wedge e| = |\xi| \sin\psi \ , \ \xi \cdot e = |\xi| \cos\psi$$

where e is any unit vector. To state our estimate we make reference to the figure where the angles φ and θ are defined, and we state that, for any continuous f such that the right hand side is finite, we have

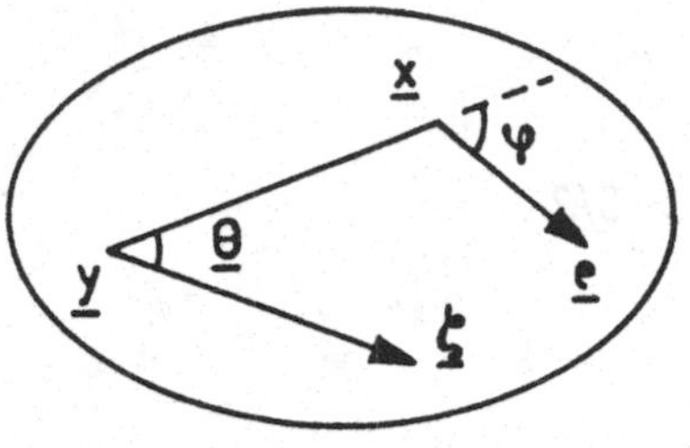

$$\{\rho^p \mu_e WAf\}^2(x) \leqslant \text{Const}\,\{\, |||\rho f|||^2 + \tag{3}$$

$$\int_\Omega \frac{dv_y}{|x-y|^{2-\alpha}} \int_{R^3} \omega(|\zeta|)\,\rho^2(|\zeta|)\,|f(y,\zeta)|^2\,d\zeta +$$

$$+ \int_\Omega \frac{1+\mathrm{Log}|\sin\varphi/2|^{-1}}{|x-y|^2}\,dv_y \int_{R^3} \omega(|\zeta|)\,(\rho(|\zeta|)^{2(p+1)}\,(1+\mathrm{Log}|\sin\frac{\theta}{2}|^{-1})\,|f(y,\zeta)|^2\,d\zeta$$

Iterating on (3) we shall be able to prove 3.2.1.1.(1).

3.2.1.3. Assume that 3.2.1.1.(1) has been proven, we are in a position to obtain an estimate concerning the operator $\mathcal{C}_0$. To this end we start from 3.1.5.2.(2) and observe that in the process of proving 3.2.1.1.(1) we shall prove that for $r > 5/2$, νW is continuous from $B_r(\Omega)$ into itself and $\nu \geqslant \nu_0 > 0$, then for any $\varphi \in B_r(\Omega) \cap L^2(\Omega, H)$ we deduce from 3.1.5.2.(2) and the estimate to be proven that

$$<\mathcal{C}_\varphi>_r \;\leqslant\; |W|_{Br} \sum_{n=0}^{N-1} (|WA|_{Br})^n <\varphi>_r + C\,|\mathcal{C}|_{L^2(\Omega,H)}\, |||\varphi||| \tag{1}$$

provided we choose $N-5 \geqslant 2r$. Now, for $r > 3/2$, $B_r(\Omega) \sqsubset L^2(\Omega, H)$ and

$$|||\varphi||| \leqslant A(r)\,|\Omega|\,<\varphi>_r\,, \quad r > 3/2 \tag{2}$$

with

$$A(r) = 4\pi \int_0^\infty \frac{u^2}{(1+u^2)^r}\,du\,. \tag{3}$$

From a standard continuity argument we see that, for any $r > 3/2$, $\mathcal{T}$ applies continuously $B_r(\Omega)$ into $B_r(\Omega)$, then, from 1.5.3.6.(3), and from continuity of νW as an operator from $B_r(\Omega)$ into itself and from 1.5.2.1.(6) we conclude that the following estimate holds

(4) $$< \mathcal{T}\varphi >_r \leqslant \text{Const} < \varphi >_{r-1}, \quad r > 5/2 .$$

which is 3.1.5.1.(2), thanks again to 1.5.2.1.(6)

3.2.2. An Estimate for the Operator UA

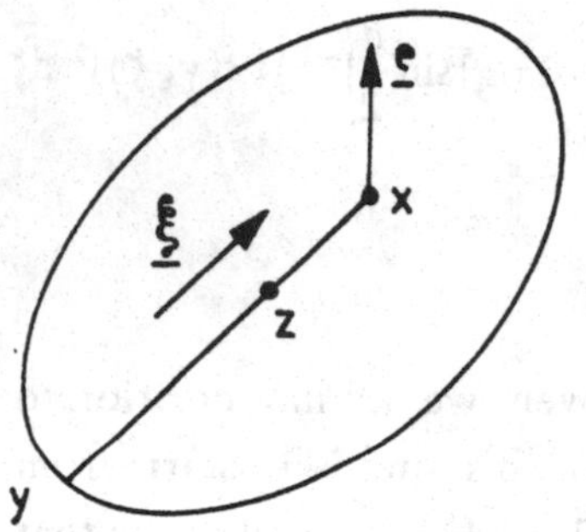

(1) $$\mathsf{U} f(x, \xi) = \int_y^x |\xi|^{-1} e^{-\nu|x-z||\xi|^{-1}} f(z, \xi)\, ds_z$$

where ds_z is the differential of length along the ray from $y \in \partial\Omega$ to x, parallel to ξ. We consider x as fixed and we set, in short

(2) $$|\xi|^{-1} \exp(-|\xi|^{-1} \nu |x-z|) = E(\xi, z)$$

then, we have

(3) $$\{\rho^p \mu_e \mathsf{U} f\}^2 (x) = \int_{R^3} \omega\, \rho^{2p} \mu_e^2\, d\xi \int_y^x E(\xi, z_1) f(z_1, \xi) ds_{z_1} \int_y^x E(\xi, z_2) f(z_2, \xi)\, ds_{z_2}$$

and, using, in an obvious notation,

(4) $$E_1 E_2 f_1 f_2 \leqslant \frac{1}{2} E_1 E_2 (f_1^2 + f_2^2)$$

we have

$$\{\rho^p \mu_e Uf\}^2(x) \leq \int_{R^3} \omega \rho^{2p} \mu_e^2 \, d\xi \int_y E(\xi, z) \, | f(z, \xi) |^2 \, ds_z \int_y E(\xi, z_1) \, ds_{z_1} \; . \tag{5}$$

On the other hand

$$\int_y^x E(\xi, z_1) \, ds_{z_1} = \frac{1}{\nu}(1 - e^{-\nu|x-y|\,|\xi|^{-1}}) \leq \frac{|x-y|}{|\xi|} \tag{6}$$

and

$$\{\rho^p \mu_e U f\}^2(x) \leq \int |\xi|^{-2} \rho^{2p} \mu_e^2 \, \omega \, d\xi \; \cdot$$
$$\cdot \int_y^x |x-y| \, e^{-\nu|x-z|\,|\xi|^{-1}} \, |f(z, \xi)|^2 \, ds_z \; . \tag{7}$$

Now we set

$$\xi = \xi u \;, \quad \xi = |\xi| \tag{8}$$

and we get, from (7)

$$\{\rho^p \mu_e Uf\}^2(x) \leq \int_0^\infty \rho^{2p} \omega \, d\xi \int_{|u|=1} \mu_e^2 \, du \int_y^x |x-y| e^{-\nu|x-z||\xi|^{-1}} \, |f(z, \xi)|^2 \, ds_z \tag{9}$$

then, from

$$du \; ds_z = \frac{dv_z}{|x - z|^2} \tag{10}$$

we obtain

$$(11)\quad \{\rho^p \mu_e Uf\}^2(x) \leq \int_\Omega \frac{|x-y|}{|x-z|^2}\, \mu_e^2\left(\frac{x-z}{|x-z|}\right) dv_z \int_0^\infty \rho^{2p}\, \omega\, |f(z,\xi)|^2\, d\xi$$

where we have majorized the exponential factor by one. We observe that (11) holds whether the domain Ω is convex or not. We observe also the point y is defined as the point of $\partial\Omega$ closest to x on the ray $x - su$, $s > 0$, $u = (x-z)/|x-z|$, $\xi = \xi u$.

3.2.2.2. We substitute Af to f in 3.2.2.1.(11) and we use 3.1.7.2.(1) so that setting temporarily $\omega^{1/2} f = \hat{f}$ we obtain

$$(1)\quad \{\rho^p \mu_e UAf\}^2(x) \leq \int_\Omega \frac{|x-y|}{|x-z|^2}\, \mu_e^2\left(\frac{x-z}{|x-z|}\right) dv_z \cdot \int_0^\infty \rho^{2p}\, d\xi \int_{R^3} |B(\xi,\zeta_1)|\, |\hat{f}(z,\zeta_1)|\, d\zeta_1 \int_{R^3} |B(\xi,\zeta_2)|\, |\hat{f}(z,\zeta_2)|\, d\zeta_2$$

then using the same argument as with 3.2.2.1.(5) and 3.1.7.2.(2) we have

$$(2)\quad \{\rho^p \mu_e UAf\}^2(x) \leq C_1\, \sigma^2 \int_\Omega \frac{|x-y|}{|x-z|^2}\, \mu_e^2(u)\, dv_z \cdot \int_{R^3} |f(z,\zeta)|^2\, d\zeta \int_0^\infty \rho^{2p}\, |B(\xi u,\zeta)|\, d\xi .$$

From (2) and 3.1.7.2.(4), and with ρ as in 3.1.6.3.1.(1) we have

$$(3)\quad \{\rho^p \mu_e UAf\}^2(x) \leq C_1\, C_2(p)\, \sigma^4 \int_\Omega \frac{|x-y|}{|x-z|^2} \left(1 + \mathrm{Log}\left|\sin\frac{\varphi}{2}\right|^{-1}\right) dv_z \cdot$$

$$\int_{R^3} \omega(|\zeta|)\,[\rho(|\zeta|)]^{2(p+1)}(1 + \mathrm{Log}|\sin\frac{\theta}{2}|^{-1})\,|\,f(z,\zeta)\,|^2\ d\zeta\ .$$

with the notation of the figure and in particular

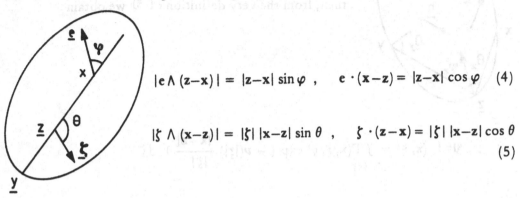

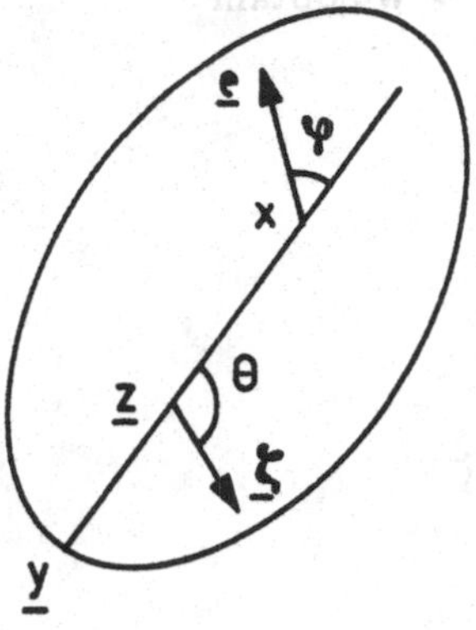

$$|e \wedge (z-x)| = |z-x| \sin\varphi\ , \qquad e\cdot(x-z) = |z-x|\cos\varphi \qquad (4)$$

$$|\zeta \wedge (x-z)| = |\zeta|\,|x-z| \sin\theta\ , \qquad \zeta\cdot(z-x) = |\zeta|\,|x-z|\cos\theta \qquad (5)$$

3.2.3. Continuity of T *in* $B_r(\partial\Omega)$

3.2.3.1. By inspection from 2.1.2.1.(6)(7) we conclude that

$$< J^- \, B\,E\phi^+ >_r \;\leqslant\; <\phi^+>_r\ , \qquad (1)$$

and that, for a continuous ϕ^+, $J^-\,BE\phi^+$ is continuous if Ω is convex. From 2.1.2.3.(3) and 3.1.6.3.1.$(v)_r$ $(vi)_s$ we have

$$< \mathcal{V}\phi^+ >_r \;\leqslant\; M_{4,r}\ M_{5,s}\ <\phi^+>_s \qquad (2)$$

for any $r \geqslant 0, s \geqslant 0$ such that 3.1.6.3.1. $(v)_r$ $(vi)_s$ hold. We shall use this estimate in conjunction with 3.1.5.3.(5).

3.2.3.2. We assume a continuous f^+ and try to estimate $\mathcal{V}^2 f^+$, pointwise. We refer to the figure for the notations and observe that

$$\zeta \wedge (x-z) = 0\ , \qquad \zeta\cdot(x-z) > 0\ , \qquad (1)$$

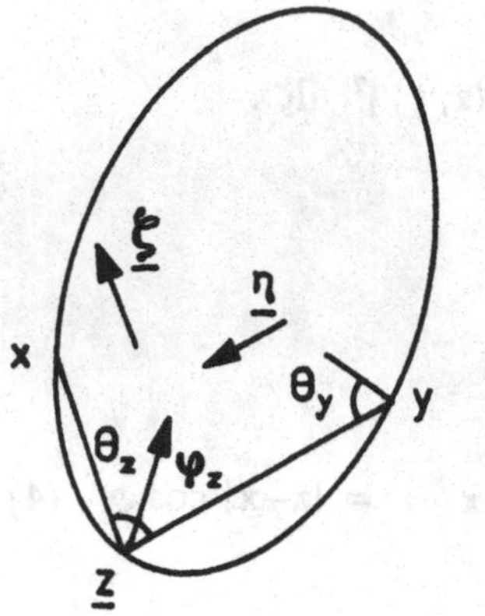

(2) $\eta \wedge (z-y) = 0 \;, \quad \eta \cdot (z-y) > 0 \;.$

then, from the very definition of $\mathcal{V}$ we obtain

(3)
$$\mathcal{V}^2 f^+(x,\xi) = \int_{R^3(x)} \Gamma(x,\xi,\zeta) \exp\{-\nu(|\zeta|)\frac{|x-z|}{|\zeta|}\}\, d\zeta \,.$$
$$\int_{R^3(z)} \Gamma(z,\zeta,\eta) \exp\{-\nu(|\eta|)\frac{|z-y|}{|\eta|}\} f^+(y,\eta)\, d\eta \,.$$

and, from 3.1.6.3.(2)

(4) $$|\mathcal{V}^2 f^+(x,\xi)| \leqslant \Gamma_1(x,\xi) \int_{R^3_-(x)} \Gamma_2(x,\zeta)\, d\zeta \int_{R^3_-(z)} \Gamma_1(z,\zeta)\, \Gamma_2(z,\eta)\, |f^+(y,\eta)|\, d\eta \,.$$

Setting

(5) $$\zeta = \zeta\, u \quad , \quad \eta = \eta\, v$$

and observing that

(6) $$d\zeta = \zeta^2\; d\zeta\, du = \zeta^2\; d\zeta\, \frac{\cos\theta_z\, dS_z}{|x-z|^2}$$

$$d\eta = \eta^2 \; d\eta \; dv = \eta^2 \; d\eta \; \frac{\cos\theta_y \; dS_y}{|z-y|^2} \tag{7}$$

we find

$$|\nu^2 f^+(x,\xi)| \leqslant \Gamma_1(x,\xi) \int_0^\infty \zeta^2 \, d\zeta \int_{\Sigma(x)} \frac{\cos\theta_z}{|x-z|^2} \, dS_z \int_0^\infty \eta^2 \, \Gamma_1(z,\zeta)\,\Gamma_2(x,\zeta)\,\Gamma_2(z,\eta)\, d\eta \cdot$$
$$\int_{\Sigma(z)} \frac{\cos\theta_y}{|z-y|^2} \, |f^+(y,\eta)| \, dS_y \tag{8}$$

where $\Sigma(x)$ is that part of $\partial\Omega$ which is "seen" from x, that is the set of $z \in \partial\Omega$ such that for any $\rho \in [0,1]$ the point $\rho x + (1-\rho) z \in \Omega$. Interchanging the order of integration we obtain

$$|\nu^2 f^+(x,\xi)| \leqslant \Gamma_1(x,\xi) \int_0^\infty \eta^2 \, d\eta \int_{\partial\Omega} dS_y \int_0^\infty \zeta^2 \, d\zeta \cdot$$
$$\cdot \int_{\Sigma(x,y)} \frac{\cos\theta_z \; \cos\theta_y}{|x-z|^2 \; |z-y|^2} \, |f^+(y,\eta)| \, \Gamma_1(z,\zeta)\,\Gamma_2(x,\zeta)\,\Gamma_2(z,\eta)\, dS_z \tag{9}$$

and $\Sigma(x,y)$ is the set of points of $\partial\Omega$ which are "seen" from both x and y. Now

$$\eta^2 \, d\eta \, \frac{\cos\varphi_z \; dS_z}{|y-z|^2} = d\eta \tag{10}$$

and, when z runs through $\Sigma(x,y)$, $v = |\eta|^{-1}\eta$ runs over part of the set of unit vectors in three dimensional Euclidean space. As a consequence (9) leads to

$$
(11)\quad
\begin{aligned}
|\mathcal{V}^2 f^+(x,\xi)| \leqslant{}& \Gamma_1(x,\xi) \int_{\partial\Omega} dS_y \int_{R^3_+} |f^+(y,\zeta)| \frac{\cos\theta_z \cos\theta_y}{|x-z|^2 \cos\varphi_z} d\zeta \cdot \\
&\cdot \int_0^\infty \Gamma_1(z,\zeta)\,\Gamma_2(x,\zeta)\,\Gamma_2(z,\eta)\,\varsigma^2\,\varsigma S .
\end{aligned}
$$

In order to estimate the right hand side of (11) we use either the hypothesis of 3.1.6.2.1. and 3.1.6.3.1.A. or 3.1.6.3.1.B. In the first case we have

$$
(12)\qquad \frac{\cos\theta_y \cos\theta_z}{|x-z|^2 \cos\varphi_z} \Gamma_2(z,\eta) \leqslant P \frac{\cos\theta_z}{|x-z|^2} \Gamma_2(|\eta|) ,
$$

while, in the second case

$$
(13)\qquad \frac{\cos\theta_y \cos\theta_z}{|x-z|^2 \cos\varphi_z} \Gamma_2(z,\eta) \leqslant \frac{\cos\theta_z}{|x-z|^2} \Gamma_{2,B}(|\eta|) .
$$

Let us define

$$
(14)\qquad \Gamma_{2,1}(|\varsigma|) = \begin{cases} P\,\Gamma_2(|\varsigma|) & \text{if } 3.1.6.2.1 \text{ holds,} \\ \Gamma_{2,B}(|\varsigma|) & \text{if } 3.1.6.3.1 \text{ B) holds,} \end{cases}
$$

and observe that, from the Liapounov condition

$$
(15)\qquad \frac{\cos\theta_z}{|x-z|^2} \leqslant L(\partial\Omega) \quad |x-z|^{\alpha-2}
$$

as this results from 2.2.1.5.(3), then we have

$$|\mathcal{V}^2 f^+(x,\xi)| \leq L\,\Gamma_1(x,\xi) \int_{\partial\Omega} dS_y \int_{R^3_+} \frac{|f^+(y,\eta)|}{|x-z|^{2-\alpha}} d\eta \int_0^\infty \Gamma_{2,1}(|\eta|)\,\Gamma_1(\zeta|)\,\Gamma_2(|\zeta|)\,|\zeta|^2\, d\zeta \tag{16}$$

and, using 3.1.6.3.1.(ii) we, finally, get

$$|\mathcal{V}^2 f^+(x,\xi)| \leq \Gamma_1(x,\xi) \int_{\partial\Omega} dS_y \int_{R^3_+} M_2(x,y,\eta)\,|f^+(y,\eta)|\,d\eta\,, \tag{17}$$

with

$$M_2 = L(\partial\Omega)\,M_1\,\frac{\Gamma_{2,1}(|\eta|)}{|x-z|^{2-\alpha}}\,. \tag{18}$$

3.2.3.3. We iterate on 3.2.3.2.(17) and this leads to

$$|\mathcal{V}^4 f^+(x,\xi)| \leq \Gamma_1(x,\xi) \int_{\partial\Omega} dS_y \int_{R^3_+} M_4'(x,y,\eta)\,|f^+(y,\eta)\,d\eta \tag{1}$$

with

$$M_4'(x,y,\eta) = \int_{\partial\Omega} dS_z \int_{R^3_+} M_2(x,z,\zeta)\,\Gamma_1(|\zeta|)\,M_2(z,y,\eta)\,d\zeta\,. \tag{2}$$

We use 3.2.3.2.(18) and refer to the figure of 3.2.3.2. and draw the accompanying figure from which we see that

$$M_2(x,z,\zeta) = \frac{L\,M_1\,\Gamma_{2,1}(|\zeta|)}{|x-t|^{2-\alpha}}\,, \tag{3}$$

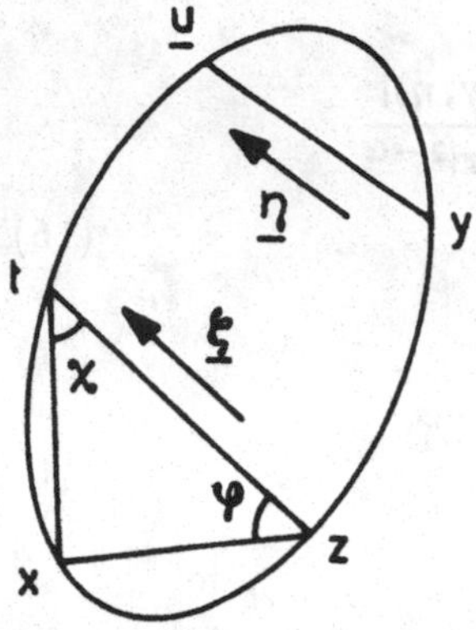

(4) $$M_2(z, y, \eta) = \frac{L\, M_1\ \Gamma_{2,1}(|\eta|)}{|z-u|^{2-\alpha}} \ ,$$

so that, setting

(5) $$\zeta = \zeta\ e \ ,$$

we find

$$M'_4(x, y, \eta) = L^2 M_1^2 \int_{\partial\Omega} dS_z \int_{\substack{|e|=1\\ e\cdot n_z \geqslant 0}} \frac{de}{(|x-t|\ |z-u|)^{2-\alpha}} \int_0^\infty \zeta^2\, \Gamma_{2,1}(\zeta)\, \Gamma_1(\zeta)\, \Gamma_{2,1}(\eta)\, d\zeta$$

(6)

$$\leqslant L^2 M_1^3 \sup(1, P)\ P_{2,1}(|\eta|) \int_{\partial\Omega} dS_z \int_{\substack{|e|=1\\ e\cdot n_z \geqslant 0}} \frac{de}{(|x-t|\ |z-u|)^{2-\alpha}}$$

where we have used 3.1.6.3.1.(ii) A) or B) with 3.2.3.2.(14). Referring to the figure

(7) $$\frac{1}{|x-t|} = \frac{\sin\chi}{\sin\varphi}\ \frac{1}{|x-z|} \leqslant \frac{1}{|x-z|\ \sin\varphi}$$

and

(8) $$M'_4 \leqslant L^2 M_1^3 \sup(1, P)\, \Gamma_{2,1}(|\eta|) \int_{\partial\Omega} \frac{dS_z}{|x-z|^{2-\alpha}\ |z-u|^{2-\alpha}} \int_{\substack{|e|=1\\ e\cdot n_z > 0}} \frac{de}{(\sin\varphi)^{2-\alpha}} \ .$$

Setting

$$C_4(\alpha) = \int_{|e|=1} \frac{de}{(\sin\varphi)^{2-\alpha}} = 2\pi \int_0^{\pi} \frac{\sin\varphi}{(\sin\varphi)^{2-\alpha}}\, d\varphi\,, \tag{9}$$

we obtain

$$M_4' \leqslant C_4(\alpha)\, L^2\, M_1^3 \sup(1,P)\, \Gamma_{2,1}(|\eta|) \int_{\partial\Omega} \frac{dS_z}{|x-z|^{2-\alpha}|z-u|^{2-\alpha}}\,. \tag{10}$$

Finally, using 3.1.8.3.(1), and observing that we may, at will, change α to $\alpha-\epsilon$, with $\epsilon > 0$ arbitrarily small such that $4-2\alpha \neq 2$, we obtain

$$|\mathcal{U}^4 f^+(x,\xi)| \leqslant \Gamma_1(x,\xi) \int_{\partial\Omega} dS_y \int_{R^3_+} m_4(x,y,\eta)\, |f^+(y,\eta)|\, d\eta \tag{11}$$

with

$$M_4(x,y,\zeta) = m_4(\alpha,\Omega,M_1,P)\, \frac{\Gamma_{2,1}(|\zeta|)}{|x-u|^{2-2\alpha}} \tag{12}$$

where u is as on the figure accompanying this paragraph, namely

$$u \in \partial\Omega\,,\ (y-u)\wedge\eta = 0\,,\quad \rho y + (1-\rho)u \in \Omega,\ \text{for } \rho\in[0,1]\,. \tag{13}$$

The constant m_4 is

$$m_4 = C_4(\alpha) L^2(\partial\Omega))^2 M_1^3 \sup(1,P)\ \ C'_{2-\alpha,2-\alpha}(\partial\Omega)\,. \tag{14}$$

3.2.3.4. We iterate the process of 3.2.3.3. Assume

(1) $$\mathcal{V}^{2^P} f^+(x,\xi) \leq \Gamma_1(x,\xi) \int_{\partial\Omega} dS_y \int_{R^3_+} M_{2^P}(x,y,\zeta) \mid f^+(y,\zeta) \mid d\zeta \;,$$

with

(2) $$M_{2^P}(x,y,\zeta) = m_{2^P}(\alpha,\Omega,M_1,P) \frac{\Gamma_{2,1}(|\zeta|)}{|x-u|^{2-2^{P-1}\alpha}},$$

with $2-2^{P-1}\alpha > 0$. From the computation of 3.2.3.3. we get

(3) $$\mathcal{V}^{2^P} f^+(x,\xi) \leq \Gamma_1(x,\xi) \int_{\partial\Omega} dS_y \int_{R^3_+} M'_{2^{P+1}}(x,y,\zeta) \mid f^+(y,\zeta) \mid d\zeta$$

with

(4) $$\begin{aligned} M'_{2^{P+1}} &\leq M_1 \sup(1,P)\, \Gamma_{2,1}(|\zeta|) \int_{\partial\Omega} dS_z \int_{\substack{|e|=1 \\ e\cdot n_z>0}} \frac{de}{|x-t|^{2-2^{P-1}\alpha}\, |z-u|^{2-2^{P-1}\alpha}} \\ &\leq M_1 \sup(1,P)\, C'_{2-2^{P-1}\alpha,\, 2-2^{P-1}\alpha}\, C_4(2^{P-1}\alpha)\, \Gamma_{2,1}(|\zeta|)\, |x-u|^{-2+2^P\alpha} \end{aligned}$$

provided $2-2^P\alpha > 0$. Changing α to $\alpha-\epsilon$, $\epsilon > 0$, arbitrarily small we may arrange that $2-2^P\alpha$ is never zero. If $2-2^P\alpha < 0$ we have

(5) $$M'_{2^{P+1}} \leq M_1 \sup(1,P)\, C'_{2-2^{P-1}\alpha,\, 2-2^{P-1}\alpha}\, C_4(2^{P-1}\alpha)\, \Gamma_{2,1}(|\zeta|)\;.$$

As a consequence we may find an N such that

(6) $$|\mathcal{V}^N f^+(x,\xi)| \leq C_5(\alpha,\Omega,M_1,P)\, \Gamma_1(x,\xi) \int_{\partial\Omega} dS_y \int_{R^3_+} \Gamma_{2,1}(|\zeta|) \mid f^+(y,\zeta) \mid d\zeta$$

then, from 3.1.6.3.1.(iv),

$$|\mathcal{V}^N f^+(x,\xi)| \leqslant C_5 \Gamma_1(x,\xi)\Big(\int_{\partial\Omega} dS_y \int_{R^3_+} \omega\, |f^+(y,\zeta)|^2\, d\zeta \int_{\partial\Omega} dS_y \int_{R^3_+} \omega^{-1}\Gamma_{2,1}\, d\zeta\Big)^{1/2}$$

$$\leqslant C_6\, \Gamma_1(x,\xi)\ \|f^+\|_{L^2(\partial\Omega,\,H)} \tag{7}$$

with

$$C_6 = (2\pi M_3\ |\partial\Omega|)^{1/2}\ C_5\ \sup(1, P) \tag{8}$$

3.2.3.5. Using 3.2.3.4.(27) and 3.1.6.3.1.$(v)_r$ we obtain from 3.1.5.3.(5) and 3.2.3.1.(2), choosing N suitably,

$$< T f^+ >_r \leqslant \{\sum_{n=0}^{N-1} (M_{4,r}\ M_{S,r})^n\} < f^+ >_r + C_6\, M_{4,r}\ \|T f^+\|_{L^2(\partial\Omega,\,H)} \tag{1}$$

From 3.1.5.3.(5) with N = 1 we have, referring to 3.1.7.1.(7), 2.1.2.3.(3) 2.1.3.8.(6) and 2.2.1.9.1.(2)

$$\|T f^+\|_{L^2(\partial\Omega,\,H)} \leqslant \|f^+\|_{L^2(\partial\Omega,\,H)} + M_7\ M_8\ \|T\|_{L^2(\partial\Omega,\,\bar H)} \|f^+\|_{L^2(\partial\Omega,\,\bar H)} \tag{2}$$

on the other hand,

$$\|f^+\|_{L^2(\partial\Omega,\,H)} \leqslant \Big(\int_{\partial\Omega} dS \int_{R^3_+} (1+|\xi|^2)^{-r}\, d\xi\Big)^{1/2} < f^+ >_r$$

$$\leqslant \Big(\frac{1}{2}\ |\partial\Omega|\ A(r)^{1/2} < f^+ >_r,\quad r > 3/2 \tag{3}$$

where A(r), is given by 3.2.1.3.(3). In the same way

$$\| f^+ \|_{L^2(\partial\Omega, \bar{H})} \leqslant \left(\int_{\partial\Omega} dS \int_{R^3_+} (1+|\xi|^2)^{-r} \, |\xi| \, d\xi \right)^{1/2} < f^+ >_r \tag{4}$$
$$\leqslant \left\{ \frac{1}{2} \, |\partial\Omega| \, A(r - \frac{1}{2}) \right\}^{1/2} < f^+ >_r ,$$

summing up we have

$$< T f^+ >_r \leqslant C_7^{(r)} < f^+ >_r \quad , \quad r > 2 , \tag{5}$$

with

$$C_7^{(r)} = \sum_{n=0}^{N-1} (M_{4,r} \, M_{5,r})^n + C_6 \, M_{4,r} \left(\frac{1}{2} |\partial\Omega|\right)^{1/2} [(A(r))^{1/2} \tag{6}$$
$$+ M_7 \, M_8 \, \| T \|_{L^2(\partial\Omega, \bar{H})} (A(r - \frac{1}{2}))^{1/2}] .$$

We observe that the estimate (5) does not prove that $T f^+ \in B_r(\partial\Omega)$ when $f^+ \in B_r(\partial\Omega)$. To obtain this stronger result we need to prove continuity of Tf^+. Assume $f^+ \in B_r(\partial\Omega)$, with $r > 2$; we know already that $\mathfrak{V}^n f^+ \in B_r(\partial\Omega)$ for any r (refer to 3.2.3.1.) and we need only prove, according to 3.1.5.3.(5) that $\mathfrak{V}^N T f^+ \in B_r(\partial\Omega)$. Now, we have already proven the estimate

$$< \mathfrak{V}^N \, T f^+ >_r \leqslant C_6 \, M_{4,r} \, \| T f^+ \|_{L^2(\partial\Omega, H)} . \tag{7}$$

Let $\{g_n^+\}$ be a sequence of functions of $B_r(\partial\Omega)$ which converges to Tf^+ in $L^2(\partial\Omega, H)$. It is easy to prove the existence of such a sequence. First from the theory of L^p spaces we know that we can find a convergent sequence $\{h_p^+\}$ of continuous functions. Truncating each $\{h_p^+\}$ for $|\xi| \geqslant q$ we obtain another sequence $\{h_{p,q}^+\}$ of functions of $B_r(\partial\Omega)$ which converges to h_p^+. By the diagonal process we can find $g_n^+ = h_{n,q(n)}^+$ such that g_n^+ converges to Tf^+ in $L^2(\Omega, H)$ and each $g_n^+ \in B_r(\partial\Omega)$. Obviously $\mathfrak{V}^N g_n^+ \in B_r(\partial\Omega)$ and, from (7), $\{\mathfrak{V}^N g_n^+\}$ is a Cauchy sequence in $B_r(\partial\Omega)$, which proves that $\mathfrak{V}^N T f^+ \in B_r(\partial\Omega)$.

3.2.3.5.1. We state the result that we have proven: for any $r > 2$, T is continuous from $B_r(\partial\Omega)$ into $B_r(\partial\Omega)$ and 3.7.3.5.(5) holds.

3.2.4. Estimation of $\{\rho^p \mu_e WAf\}$

3.2.4.1. Referring to 2.1.2.4.(5) we have

$$WA = \cup A + \Lambda \cup A \,, \tag{1}$$

$$\Lambda = E T G J^- B \,. \tag{2}$$

Our purpose is to obtain for WA an estimate analogous to the one obtained in 3.2.2.2.(3) for UA. We need worry only about ΛUA. To this end we set

$$J^- B \cup A f = \phi^- \,, \tag{3}$$

$$G\phi^- = \psi^+ \,, \tag{4}$$

$$\Lambda \cup A f = E T \psi^+ \,. \tag{5}$$

3.2.4.2. From 3.1.6.3.1. we find

$$|\psi^+(x,\xi)| \leqslant \Gamma_1(x,\xi) \{ \int_{R^3_-} \omega^{-1}(|\zeta|) \left(\frac{\zeta\cdot n}{|\zeta|}\right)^{-2a_0} P_2^2(x,\zeta)\, d\zeta\}^{1/2} \cdot$$
$$\cdot \{ \int_{R^3_-} \omega(|\zeta|)\, |\cos\varphi|^{2a_0} |\phi^-(x,\zeta)|^2\, d\zeta\}^{1/2} \,, \tag{1}$$

where

$$\cos\varphi = \frac{|\zeta\cdot n|}{|\zeta|} \,, \tag{2}$$

and where we use

$$(3)\qquad \begin{cases} a_0 = 0 & \text{if 3.1.6.3.1 A) holds,} \\ a_0 = 1 & \text{if 3.1.6.3.1 B) holds.} \end{cases}$$

From 3.1.6.3.1.(iii) either in the case of A or in the case B we deduce from (1) that the following holds, with the notation of 3.2.1.2.,

$$(4)\qquad |\psi^+(x,\xi)| \leq (2\pi M_3)^{1/2}\, \Gamma_1(x,\xi)\, \{\nu_n \phi^-\}(x) ,$$

where

$$(5)\qquad \nu_n = \left(\frac{\zeta}{|\zeta|} \cdot n^{a_0}\right).$$

we may, as a little reflection shows, use 3.2.2.1.(11) to estimate $\{\nu_n \phi^-\}(x)$. Referring to the figure for notations we

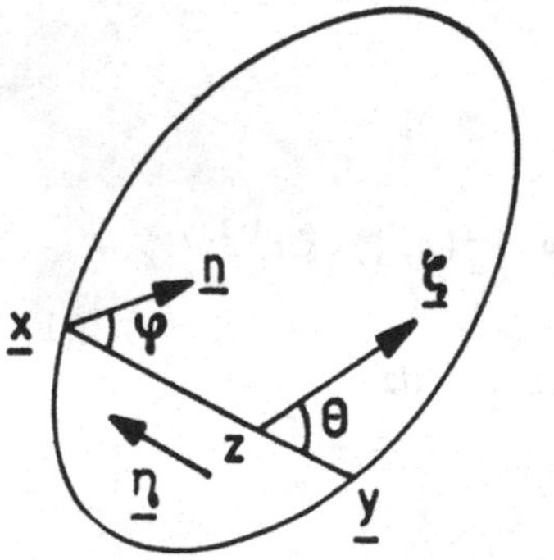

$$(6)\qquad \{\nu_n \phi^-\}^2(x) \leq \int_\Omega \frac{|x-y|\,|\cos\varphi|^{2a_0}}{|x-z|^2}\, dv_z \cdot \int_0^\infty \omega(|\eta|)\, |Af(z,\eta)|^2\, d|\eta|$$

where $(x-z)\wedge\eta = 0$, $(x-z)\cdot\eta > 0$. Remaking the computations of 3.2.2.2. we have

$$(7)\qquad \{\nu_n \phi^-\}^2(x) \leq C_1\, C_2(0)\, \sigma^4 \int_\Omega \frac{|x-y|\,|\cos\varphi|^{2a_0}}{|x-z|^2}\, dv_z .$$

$$\int_{R^3} \omega(|\zeta|)\ \rho^2(|\zeta|)\,(1 + \mathrm{Log}\,|\sin\frac{\theta}{2}|^{-1})\ |f(z,\zeta)|^2\ dS$$

with ρ as in 3.1.6.3.1.(1). Summing up (4) and (7) we get our starting estimate

$$|\psi^+(x,\xi)| \leq \{2\pi\ M_3\ C_1\ C_2(0)\}^{1/2}\ \sigma^2\ \chi(x)\ \Gamma_1(x,\xi) \tag{8}$$

with

$$\chi^2(x) = \int_\Omega \frac{|x-y|\,|\cos\varphi|^{2a_0}}{|x-z|^2}\,dv_z \int_{R^3} \omega(|\zeta|)\ \rho^2(|\zeta|)\,(1 + \mathrm{Log}\,|\sin\frac{\theta}{2}|^{-1})\ |f(z,\zeta)|^2\ d\zeta \tag{9}$$

and we recall that a_0 is given by (3).

3.2.4.3. We start from an f^+ such that

$$|f^+(x,\xi)| \leq \chi(x)\ \Gamma_1(x,\xi) \tag{1}$$

and we try to estimate Tf^+. To this end we use 3.1.5.3.(5) and, first, we estimate $\mathcal{V}f^+$. From $\mathcal{V} = G\,J^-\,BE$ we have

$$\mathcal{V}f^+(x,\xi) = \int_{R^3_-} \Gamma(x,\xi,\zeta)\ e^{-\hat{\nu}(|\zeta|)\,|x-y|\,|\zeta|^{-1}}\ f^+(y,\zeta)\ d\zeta \tag{2}$$

and referring to 3.1.6.3.(2) and 3.1.6.3.1.(i)
we get, using (10)

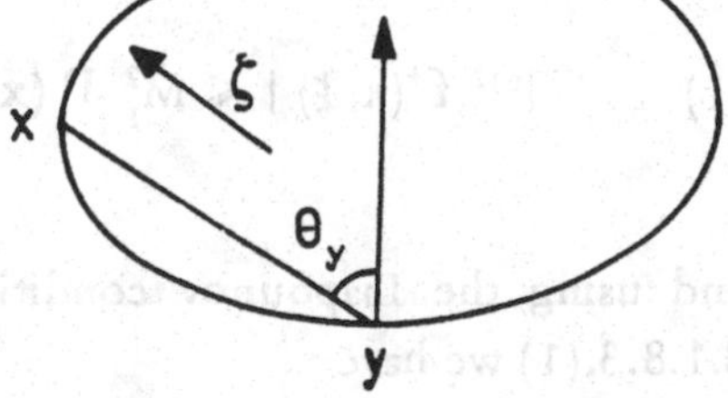

(3) $$|\mathcal{V} f^+(x,\xi)| \leq \Gamma_1(x,\xi) \int_{R^3_-} \Gamma_1(y,\zeta)\, \Gamma_2(x,\zeta)\, \chi(y)\, d\zeta \; .$$

Now from an argument already worked out in 3.2.3.2. we change the variables from ζ to $|\zeta|$, y and we have, according to

(4) $$d\zeta = |\zeta|^2 \, d|\zeta| \, \frac{\cos\theta_y \, dS_y}{|x-y|^2} \; ,$$

that

(5) $$|\mathcal{V} f^+(x,\xi)| \leq \Gamma_1(x,\xi) \int_{\partial\Omega} \frac{|\cos\theta_y|}{|x-y|^2} \chi(y)\, dS_y \int_0^\infty |\zeta|^2 \, P_1(|\zeta|)\, P_2(|\zeta|)\, d|\zeta| \; .$$

Using 3.1.6.3.1.(ii) we obtain the estimate we had in mind

(6) $$|\mathcal{V} f^+(x,\xi)| \leq M_1 \, \Gamma_1(x,\xi) \int_{\partial\Omega} \frac{|\cos\theta_y|}{|x-y|^2} \chi(y)\, dS_y \; .$$

3.2.4.4. We iterate on 3.2.4.3.(15)

(1) $$|\mathcal{V}^2 f^+(x,\xi)| \leq M_1^2 \, \Gamma_1(x,\xi) \int_{\partial\Omega} \chi(y)\, dS_y \int_{\partial\Omega} \frac{|\cos\theta_y|\,|\cos\theta_z|}{|x-z|^2 \, |z-y|^2} \, dS_z$$

and using the Liapounov condition, especially its consequence 3.2.3.2.(15), and 3.1.8.3.(1) we have

(2) $$|\mathcal{V}^2 f^+(x,\xi)| \leq M_1^2 \, C'_{2-\alpha,\,2-\alpha} \, \Gamma_1(x,\xi)\, L^2 \int_{\partial\Omega} \frac{\chi(y)}{|x-y|^{2-2\alpha}} \, dS_y \; .$$

For a bounded Ω with diameter D, we have

$$\frac{1}{|x-y|^{2-2\alpha}} \leqslant \frac{D^\alpha}{|x-y|^{2-\alpha}} \tag{3}$$

and using 3.2.3.2.(15) in (3) and 3.2.4.3.(15) we see that

$$|\mathcal{V}^n f^+(x,\xi)| \leqslant (M_1 L)^n (D^\alpha C'_{2-\alpha,2-\alpha})^{n-1} \Gamma_1(x,\xi) \int_{\partial\Omega} \frac{\chi(y)}{|x-y|^{2-\alpha}} dS_y \tag{4}$$

holds for $n = 1$ and 2. Now it is an easy matter to prove that, if (4) holds for n it holds for $n+1$ also.

3.2.4.5. Let f^+ be as in 3.2.4.3.(1), from 3.2.4.3. and 3.2.4.4. we have, using 3.1.5.3.(5) and 3.2.3.4.(7)

$$|T f^+(x,\xi)| \leqslant \Gamma_1(x,\xi) \{ \chi(x) + \sum_{n=1}^{N-1} (M_1 L)^n (D^\alpha C'_{2-\alpha,2-\alpha})^{n-1} \int_{\partial\Omega} \frac{\chi(y)}{|x-y|^{2-\alpha}} dS_y + \tag{1}$$

$$+ C_6 \| T f^+ \|_{L^2(\partial\Omega, H)} \}.$$

Let us substitute 3.2.3.5.(2) into the right hand side of (1), we obtain

$$|T f^+(x,\xi)| \leqslant \Gamma_1(x,\xi) \{\chi(x) + \sum_{n=1}^{N-1} (M_1 L)^n (D^\alpha C'_{2-\alpha,2-\alpha})^{n-1} \int_{\partial\Omega} \frac{\chi(y)}{|x-y|^{2-\alpha}} dS_y + \tag{2}$$

$$+ C_6 \|\Gamma_1 \chi\|_{L^2(\partial\Omega, H)} + C_6 M_7 M_8 \|\Gamma_1 \chi\|_{L^2(\partial\Omega, \tilde{H})} \}.$$

From 3.1.6.3.2.(i) (ii) we get

(3) $$\|\Gamma_1 \chi\|_{L^2(\partial\Omega, H)} \leqslant M_7 \|\chi\|_{L^2(\partial\Omega)}$$

(4) $$\|\Gamma_1 \chi\|_{\tilde{L}^2(\partial\Omega, \tilde{H})} \leqslant M_6 \|\chi\|_{L^2(\partial\Omega)}$$

and, consequently

(5) $$|T f^+(x, \xi)| \leqslant C_{11}\, \Gamma(x, \xi)\, \{\chi(x) + \int_{\partial\Omega} \frac{\chi(y)}{|x-y|^{2-\alpha}}\, dS_y + \|\chi\|_{L^2(\partial\Omega)}\},$$

with

(6) $$C_{11} = \sup\{1,\ C_{10},\ C_6\ M_7(1 + M_6\ M_8)\},$$

(7) $$C_{10} = \sum_{n=1}^{N-1} (M_1\ L)^n\ (D^\alpha\ C'_{2-\alpha,\ 2-\alpha})^{n-1}.$$

For further reference we set

(8) $$\lambda(x) = \chi(x) + \int_{\partial\Omega} \frac{\chi(y)}{|x-y|^{2-\alpha}}\, dS_y + \|\chi\|_{L^2(\partial\Omega)}.$$

3.2.4.6. In order to achieve the purpose of 3.2.4.1. we need only estimate $\{\rho^p\ \mu_e\ E f^+\}(x)$ when

(1) $$|f^+(x, \xi)| \leqslant \Gamma_1(\xi)\ \lambda(x).$$

Referring to the figure we see that

$$|E f^{+}(x,\xi)| \leqslant \lambda(y)\ P_1(y,\xi) \quad (2)$$

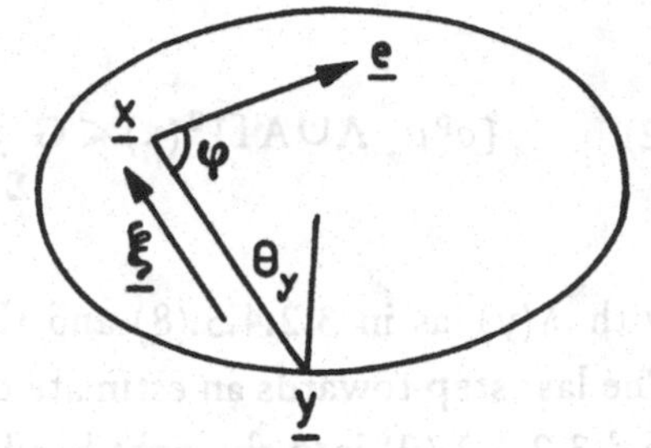

and, consequently

$$\{\rho^p \mu_e E f^{+}\}^2(x) \leqslant \int_{R^3} \rho^{2p}\ \omega(1 + \mathrm{Log}|\sin\frac{\varphi}{2}|^{-1})\ \lambda^2(y,\xi)\ \Gamma_1^2(y,\xi)\ d\xi\ . \qquad (3)$$

From an argument already used, namely in 3.2.3.2., we change the variables of integration from ξ to $|\xi|$ and y and note that

$$d\xi = |\xi|^2\ d|\xi|\ \frac{\cos\theta_y\ dS_y}{|x-y|^2}\ , \quad \cos\theta_y = \frac{x-y}{|x-y|}\cdot n_y \geqslant 0 \qquad (4)$$

which leads to

$$\{\rho^p \mu_e E f^{+}\}^2(x) \leqslant \int_0^\infty |\xi|^2 \rho^{2p}\ \omega\ d|\xi| \int_{\Sigma(x)} (1 + \mathrm{Log}|\sin\frac{\varphi}{2}|^{-1})\ \frac{\cos\theta_y}{|x-y|^2}\ \lambda^2(y)\ P_1^2(y,\xi)\ dS_y \qquad (5)$$

where $\Sigma(x)$ is that part of $\partial\Omega$ which is seen from x, that is, the set of $y \in \partial\Omega$ such that $(x-y)\cdot n_y \geqslant 0$. Making use of 3.1.6.3.1(iii) we obtain

$$\{\rho^p \mu_e\ E f^{+}\}^2(x) \leqslant M_2(p) \int_{\Sigma(x)} (1 + \mathrm{Log}|\sin\frac{\varphi}{2}|^{-1})\ \frac{|\cos\theta_y|}{|x-y|^2}\ \lambda^2(y)\ dS_y\ . \qquad (6)$$

3.2.4.7. Setting

$$G = 2\pi\ M_3\ M_2(p)\ C_{11}^2\ C_1\ C_2(0)\ \sigma^4 \qquad (1)$$

we have

$$(2) \qquad \{\rho^p \mu_e \Lambda \cup A f\}^2(x) \leqslant G \int_{\Sigma(x)} (1 + \mathrm{Log}|\sin\frac{\varphi}{2}|^{-1}) \frac{|\cos\theta_y|}{|x-y|^2} \lambda^2(y)\, dS_y$$

with $\lambda(y)$ as in 3.2.4.5.(8) and the notations of the figure accompanying 3.2.4.6. The last step towards an estimate of $\{\rho^p \mu_e \Lambda U A f\}(x)$ is to substitute 3.2.4.5.(8) and 3.2.4.2.(9) into the right hand side of (2). First we have

$$(3) \qquad \lambda^2 \leqslant 3\,\{\chi^2 + (\int_{\partial\Omega} \frac{\chi(y)}{|x-y|^{2-\alpha}} dS_y)^2 + \|\chi\|^2_{L^2(\partial\Omega)}\} ,$$

and then

$$(4) \qquad \{\rho^p \mu_e \Lambda \cup A f\}^2(x) \leqslant 3\,G\,(I + J + K\,\|\chi\|^2_{L^2(\partial\Omega)}) ,$$

with

$$(5) \qquad I = \int_{\Sigma(x)} (1 + \mathrm{Log}|\sin\frac{\varphi}{2}|^{-1}) \frac{|\cos\theta_y|}{|x-y|^2} \chi^2(y)\, dS_y ,$$

$$(6) \qquad J = \int_{\Sigma(x)} (1 + \mathrm{Log}|\sin\varphi/2|^{-1}) \frac{|\cos\theta_y|}{|x-y|^2} dS_y \int_{\partial\Omega} \frac{\chi(z_1)}{|y-z_1|^{2-\alpha}} dS_{z_1} \int_{\partial\Omega} \frac{\chi(z_2)}{|y-z_2|^{2-\alpha}} dS_{z_2} ,$$

$$(7) \qquad K = \int_{\Sigma(x)} (1 + \mathrm{Log}|\sin\frac{\varphi}{2}|^{-1}) \frac{|\cos\theta_y|}{|x-y|^2} dS_y .$$

3.2.4.7.1. We rearange J as follows using

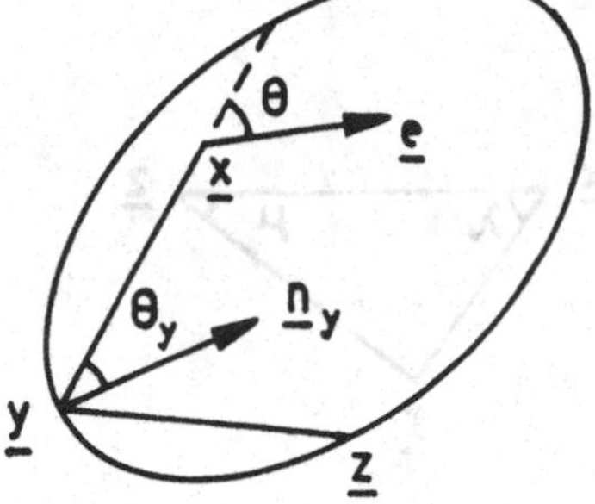

$$\chi(z_1)\ \chi(z_2) \leqslant \frac{1}{2}\,[\chi^2(z_1) + \chi^2(z_2)] \qquad (1)$$

$$\int_{\partial\Omega} \frac{dS_z}{|y-z|^{2-\alpha}} \leqslant D_1(\partial\Omega, \alpha) \qquad (2)$$

and we find

$$J \leqslant D_1 \int_{\partial\Omega} \chi^2(z)\ dS_z \int_{\Sigma(x)} (1 + \text{Log}|\sin\frac{\varphi}{2}|^{-1}) \frac{\cos\theta_y}{|x-y|^2\ |y-z|^{2-\alpha}}\ dS_y\ . \qquad (3)$$

Let us estimate

$$H = \int_{\Sigma(x)} (1 + \text{Log}|\sin\frac{\varphi}{2}|^{-1})\ \frac{\cos\theta_y}{|x-y|^2\ |y-z|^{2-\alpha}}\ dS_y \qquad (4)$$

We divide the range of integration Σ as

$$\Sigma(x) = \Sigma_1 \cup \Sigma_2 \qquad (5)$$

with

$$\Sigma_1 : \ \{ y \mid y \in \Sigma(x)\ ,\ |y-z| \geqslant |x-y| \} \qquad (6)$$

$$\Sigma_2 : \ \{ y \mid y \in \Sigma(x)\ ,\ |x-y| \geqslant |y-z| \} \qquad (7)$$

on Σ_1 we have $|y-z|^{-1} \leqslant 2\ |x-z|^{-1}$ and, consequently,

$$H \leqslant \left(\frac{2}{|x-z|}\right)^{2-\alpha} K + \int_{\Sigma_2(x)} (1 + \text{Log}|\sin\frac{\varphi}{2}|^{-1})\ \frac{\cos\theta_y}{|x-y|^2\ |y-z|^{2-\alpha}}\ dS_y\ . \qquad (8)$$

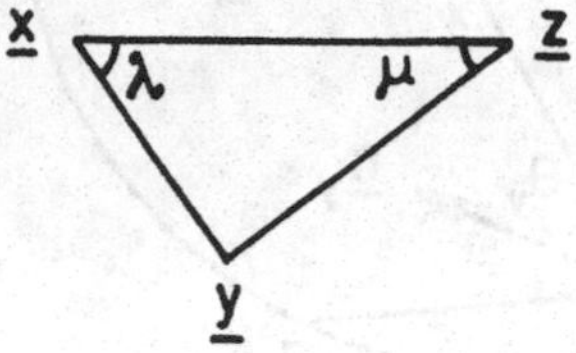

Referring to the figure we have

(9) $$y \in \Sigma_2 \Rightarrow \frac{1}{|y-z|} \leqslant \frac{2}{|x-z|}\, \frac{\sin \mu}{\sin \lambda}$$

and, consequently

(10) $$H \leqslant \left(\frac{2}{|x-z|}\right)^{2-\alpha} \{ K + M \}$$

(11) $$M = \int_{\Sigma(x)} (1 + \operatorname{Log}|\sin \frac{\varphi}{2}|^{-1}) \frac{\cos \theta_y}{|x-y|^2} (\sin \lambda)^{-(2-\alpha)}\, dS_y \; .$$

Now, setting

(12) $$x - y = |x-y|\, u$$

we change the variable of integration in K and M from y to u, using

(13) $$\frac{\cos \theta_y\, dS_y}{|x-y|^2} = du \; ,$$

and we find

(14) $$K = \int_{|u|=1} (1 + \operatorname{Log}|\sin \frac{\varphi}{2}|^{-1})\, du = 2\pi \int_0^{\pi} \sin \varphi\, (1 + \operatorname{Log}|\sin \frac{\varphi}{2}|^{-1})\, d\varphi \; ,$$

(15) $$M = \int_{|u|=1} (1 + \operatorname{Log}|\sin \frac{\varphi}{2}|^{-1}) (\sin \lambda)^{-2+\alpha}\, du \; .$$

We compute K as follows .

$$K = 8\pi \int_0^1 x(1 + \mathrm{Log}\,\frac{1}{x})\,dx = 8\pi\,\{\frac{1}{2} + \int_0^\infty y\,e^{-2y}\,dy\} = 6\pi\,, \tag{16}$$

and, concerning M we divide the range of integration in two parts $S_1 \cup S_2$ with

$$S_1 : \{u \mid |u| = 1, \quad \sin\frac{\varphi}{2} \leqslant |\sin\lambda|\}, \tag{17}$$

$$S_2 : \{u \mid |u| = 1, \quad \sin\varphi/2 > |\sin\lambda|\,, \tag{18}$$

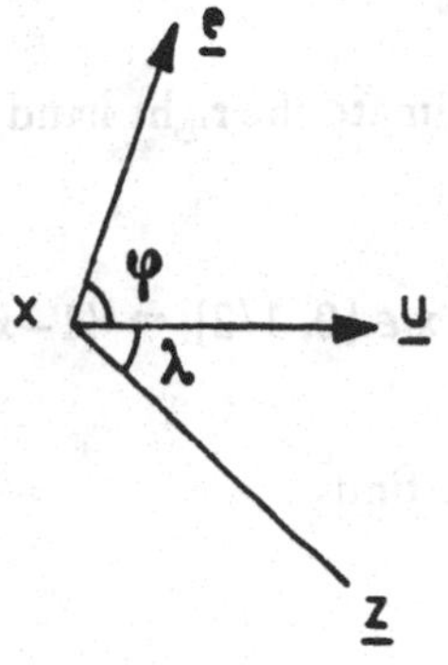

and we find

$$M \leqslant M_1 + M_2\,, \tag{19}$$

where

$$M_1 = \int_{|u|=1} (1 + \mathrm{Log}|\sin\frac{\varphi}{2}|^{-1})\,|\sin\frac{\varphi}{2}|^{-(2-\alpha)}\,du\,, \tag{20}$$

$$M_2 = \int_{|u|=1} (1 + \mathrm{Log}\,|\sin\lambda|^{-1})\,|\sin\lambda|^{-(2-\alpha)}\,du\,. \tag{21}$$

We compute

$$\begin{aligned} M_1 &= 8\pi \int_0^1 x^{-1+\alpha}(1 + \mathrm{Log}\ x^{-1})\ dx = 8\pi\{\alpha^{-1} + \int_0^\infty y e^{-\alpha y}\,dy\} \\ &= 8\pi(\alpha^{-1} + \alpha^{-2}) \end{aligned} \tag{22}$$

$$M_2 = 4\pi \int_0^{\pi/2} (1 + \mathrm{Log}|\sin\lambda|^{-1})(\sin\lambda)^{-1+\alpha}\, d\lambda,$$

(23)

$$= 4\pi \int_0^1 x^{-1+\alpha}(1-x^2)^{-1/2}(1 + \mathrm{Log}\, x^{-1})\, dx\,.$$

We estimate the right hand side of (23) from

(24) $x \in [0, 1/2] \Rightarrow (1-x^2)^{-1/2} \leqslant \frac{2}{\sqrt{3}}\,, \quad x \in [\frac{1}{2}, 1] \Rightarrow \mathrm{Log}\, x^{-1} < \mathrm{Log}\, 2$

and we find

$$M_2 \leqslant \frac{8\pi}{\sqrt{3}}\frac{1}{\alpha}\left(\frac{1}{2}\right)^{\alpha} + \frac{8\pi}{\sqrt{3}}\frac{1}{\alpha}\left\{ \left(x^{\alpha}\,\mathrm{Log}\, x^{-1}\right)_{x=0}^{x=\frac{1}{2}} + \int_0^{1/2} x^{\alpha-1}\, dx \right\}$$

(25)

$$+ \; 4\pi\,(1 + \mathrm{Log}\, 2) \int_{1/2}^{1} \frac{x^{\alpha-1}}{\sqrt{1-x^2}}\, dx$$

$$\leqslant \frac{8\pi}{\sqrt{3}}\frac{1}{\alpha}\left(\frac{1}{2}\right)^{\alpha}(2 + \mathrm{Log}\, 2) + \frac{4\pi}{\sqrt{3}}(1 + \mathrm{Log}\, 2)$$

so that, summing up

(26) $$M \leqslant \alpha^{-2}\, C(\alpha)$$

3.2.4.7.2. Going back to 3.2.4.7.(4) and using the results of 3.2.4.7.1. we have

$$\{\rho^p \mu_e \Lambda U A f\}^2 (x) \leq 18\pi G \| \chi \|^2_{L^2(\partial\Omega)} +$$

$$+ 3G \int_\Sigma (1 + \mathrm{Log} | \sin \frac{\varphi}{2} |^{-1}) \frac{\cos \theta_y}{|x-y|^2} \chi^2(y)\, dS_y \tag{1}$$

$$+ 3G D_1(\partial\Omega, \alpha)\, N_1(\alpha) \int_{\partial\Omega} \frac{\chi^2(y)}{|x-y|^{2-\alpha}} dS_y$$

with

$$N_1 = 2^{2-\alpha} \{6\pi + \alpha^{-2} C(\alpha)\} \tag{2}$$

3.2.4.7.3. We end the process which is going on by substituting 3.2.4.2.(9) into the right hand side of 3.2.4.7.2.(1) and estimating the result. We consider first $\|\chi\|_{L^2(\partial\Omega)}$

$$\| \chi \|^2_{L^2(\partial\Omega)} = \int_{\partial\Omega} dS_x \int_\Omega \frac{|x-y| (\cos\varphi)^{a_0}}{|x-z|^2} dv_z \,.$$

$$\cdot \int_{R^3} \omega(|\zeta|) (\rho(|\zeta|))^2 (1 + \mathrm{Log} | \sin \frac{\theta}{2} |^{-1}) \,| f(z,\zeta) |^2\, d\zeta \tag{1}$$

$$= \int_\Omega dv_z \int_{R^3} \rho^2 \omega \,| f(z,\zeta) |^2\, d\zeta \int_{\Sigma(z)} \frac{|x-y|}{|x-z|^2} (\cos\varphi)^{a_0} (1 + \mathrm{Log} | \sin \frac{\theta}{2} |^{-1})\, dS_x$$

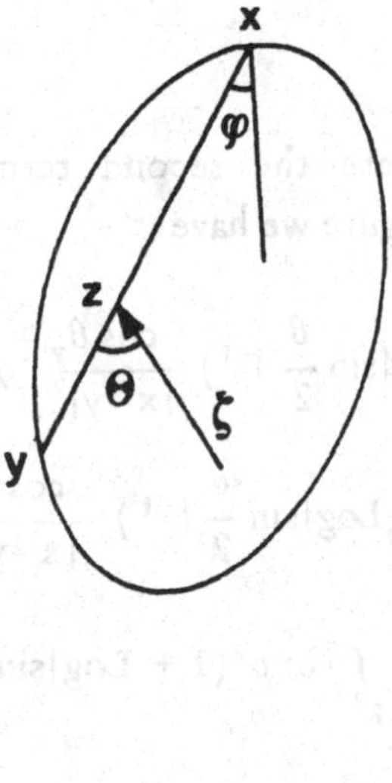

If $a_0 = 0$, that is if only 3.1.6.3.1.(A) and not 3.1.6.3.1. B) holds we require that $\partial\Omega$ satisfy 3.1.6.2.2. We use this by writing, with $D = \text{diam } \Omega$,

(2) $$|x-y| = |x-y|^{a_0} \, |x-y|^{1-a_0} \leqslant D^{a_0} (R \cos\varphi)^{1-a_0} ,$$

and

(3) $$\frac{|x-y| \, (\cos\varphi)^{a_0}}{|x-z|^2} \leqslant D^{a_0} \, R^{1-a_0} \, \frac{\cos\varphi}{|x-z|^2} ,$$

so that

(4) $$\| \chi \|^2_{L^2(\partial\Omega)} \leqslant D^{a_0} \, R^{1-a_0} \int_\Omega dv_z \int_{R^3} \rho^2 \omega \, | f(z,\zeta) |^2 \, d\zeta \, . \int_{\Sigma(z)} (1 + \text{Log} | \sin \frac{\theta}{2} |^{-1}) \frac{\cos\varphi}{|x-z|^2} \, dS_x$$

$$= D^{a_0} \, R^{1-a_0} \, K \, ||| \rho f |||^2 = 6\pi \, D^{a_0} \, R^{1-a_0} \, ||| \rho f |||^2 \, .$$

3.2.4.7.4. Now we estimate the second term in the right hand side of 3.2.4.7.2.(1). Referring to the figure we have

(1) $$\int_{\Sigma(x)} (1 + \text{Log} | \sin \frac{\theta}{2} |^{-1}) \frac{\cos\theta_y}{|x-y|^2} \, \chi^2(y) \, dS_y = \int_{\Sigma(x)} (1 + \text{Log} | \sin \frac{\theta}{2} |^{-1}) \frac{\cos\theta_y}{|x-y|^2} \, dS_y \, .$$

$$\cdot \int_\Omega \frac{|y-t| \, (\cos\varphi)^{a_0}}{|y-z|^2} \, dv_z \int_{R^3} \omega \rho^2 (1 + \text{Log} | \sin \frac{\psi}{2} |^{-1}) \, | f(z,\zeta) |^2 \, dS$$

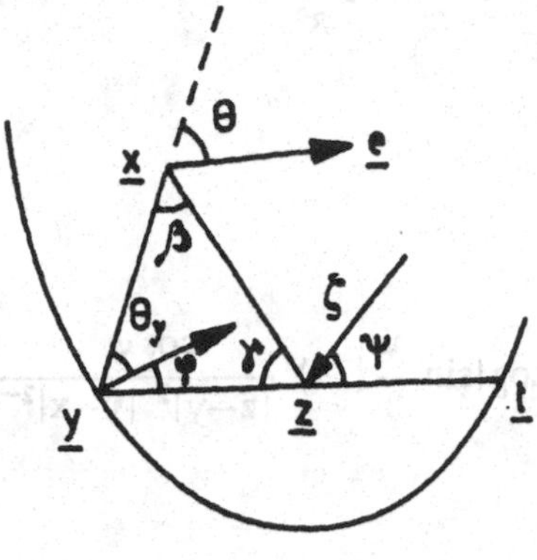

and, with the analogue of 3.2.4.7.3.(3) after interchange of the integration,

$$\int_{\Sigma(x)} (1 + \mathrm{Log}\,|\sin\frac{\theta}{2}|^{-1}) \frac{\cos\theta_y}{|x-y|^2}\, \chi^2(y)\, dS_y = \tag{2}$$

$$D^{a_0} R^{1-a_0} \int_\Omega dv_z \int_{R^3} \omega(|\zeta|)\, \rho^2(|\zeta|)\, J_1(x, z, e, \zeta)\, |f(z, \zeta)|^2\, d\zeta$$

where

$$J_1 = \int_{\Sigma(x)\, \cap\, \Sigma(z)} (1 + \mathrm{Log}|\sin\frac{\theta}{2}|^{-1})\, (1 + \mathrm{Log}\sin|\frac{\psi}{2}|^{-1}) \frac{\cos\varphi \quad \cos\theta_y}{|z-y|^2\ |y-x|^2}\, dS_y\ . \tag{3}$$

We recall that $\Sigma(x) \cap \Sigma(z)$ is the set of points on $\partial\Omega$ which are seen from both x and z.

3.2.4.7.5. Through a computation analogous to the one of 3.2.4.7.4. we obtain

$$\int_{\partial\Omega} \frac{\chi^2(y)}{|x-y|^{2-\alpha}}\, dS_y = D^{a_o} R^{1-a_o} \int_\Omega dv_z \int_{R^3} \omega(|\zeta|)\, \rho^2(|\zeta|)\, J_2(x, z, e, \zeta)\, |\, f(z, \zeta)\, |^2\, d\zeta$$

(1)

where

(2)
$$J_2 = \int_{\Sigma(z)} (1 + \mathrm{Log}|\sin \frac{\psi}{2}|^{-1})\, \frac{\cos\varphi}{|z-y|^2\, |y-x|^{2-\alpha}}\, dS_y \ ,$$

and, referring to 3.2.4.7.2(1)

(3)
$$\{\rho^p \mu_e\, \Lambda \cup A f\}^2(x) \leqslant 2^2 3^3 \pi^2\, G\, D^{a_o} R^{1-a_o} |||\rho\, f|||^2 + $$
$$+ 3\, G\, D^{a_o} R^{1-a_o} \int_\Omega dv_z \int_{R^3} \omega\rho^2\, J(x, z, e, \zeta)\, |\, f(z, \zeta)\, |^2\, d\zeta \ ,$$

with

(4)
$$J = \int_{\partial\Omega} (1 + \mathrm{Log}\ |\sin \frac{\theta}{2}|^{-1})\, (1 + \mathrm{Log}\, |\sin \frac{\psi}{2}\, |^{-1})\, (\frac{\cos\theta_y}{|x-y|^2} + $$
$$+ \frac{N_1 D_1}{|x-y|^{2-\alpha}})\, \frac{\cos\varphi}{|z-y|^2}\, dS_y$$

3.2.4.7.6. Summing up 3.2.2.2.(3) and 3.2.4.7.5.(3) and setting

(1) $C_{12} = \sup(2^2 3^3 \pi^2\, G\, D^{a_o} R^{1-a_o}\ ,\ 3GD^{a_o} R^{1-a_o}\ ,\ C_1\, C_2(p) D\, \sigma^4$

we obtain

(2) $\{\rho^p\, \mu_e\, W A f\}^2(x) \leqslant C_{12}\, \{\, |||\rho\, f|||^2 + \int_\Omega dv_z \int_{R^3} \omega\, \rho^2\, J(x, z, e, \zeta)\, |f(z, \zeta)\, |^2\, d\zeta$

$$+ \int_{\Omega} \frac{1 + \mathrm{Log}\,|\sin \varphi/2|^{-1}}{|x-z|^2}\, dv_z \int_{R^3} \omega(|\zeta|)(\rho(|\zeta|))^{2(p+1)}(1 + \mathrm{Log}\,|\sin \frac{\theta}{2}|^{-1})\, |f(z,\zeta)|^2\, d\zeta$$

with the notations of the figure accompanying 3.2.2.2. We observe that the estimate (33) for J on page 224 of reference is valid only when both x and z belong both to $\partial\Omega$ and further computations are needed to derive the estimate (34) in theorem 8 page 225 of that reference with a different value of the constant C_{12}. We prefer to go on with (2) as it stands.

We observe that, with a continuous f, such that the right hand side of (2) is finite, and a convex domain Ω, UAf and J^-BUAf are continuous. From 3.2.4.2.(8) and (9) we see that GJ^-BUAf is continuous and belongs to $B_r(\partial\Omega)$ according to 3.1.6.3.1.(v). Then, from 3.2.3.5.1., TGJBUAf is continuous and the same is true of WAf.

3.3. Regularisation of the Solution of the Linear Problem. Second Step

3.3.1. Estimate of $(WA)^2 f$

3.3.1.1. We substitute WAf to f in 3.2.4.7.6.(2) and we make $\mu_e = 1$, then we get for a convex domain

$$\{\rho^p (WA)^2 f\}^2(x) \leqslant C_{12} \{ |||\rho WAf|||^2 + \int_\Omega dv_z \cdot$$

(1) $$\cdot \int_{\partial\Omega} \left(\frac{\cos\theta_y}{|x-y|^2} + \frac{N_1 D_1}{|x-y|^{2-\alpha}} \right) \frac{\cos\varphi}{|y-z|^2} \{\mu_u \rho WAf\}^2(z)\, dS_y$$

$$+ \int_\Omega \frac{\{\mu_w\, \rho^{p+1}\, WAf\}^2(z)}{|x-z|^2}\, dv_z \}$$

with the notations of the figure, and for the unit vectors u and w

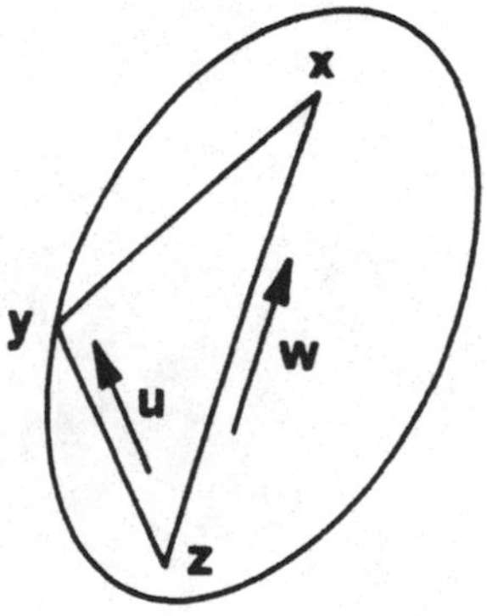

(2) $u \wedge (y-z) = 0\,, \qquad u \cdot (y-z) = |y-z|\,,$

(3) $w \wedge (x-z) = 0\,, \qquad w \cdot (x-z) = |x-z|\,.$

3.3.1.2. We start by estimating $|||\rho WAf|||^2$ from 3.2.4.7.6.(2) and

(1) $$|||\rho WAf|||^2 = \int_\Omega \{\rho\, WAf\}^2(x)\, dv_x$$

So that

$$|||\rho W A f|||^2 \leqslant C_{12} \{ |\Omega| \; |||\rho f|||^2 + \int_\Omega dv_z \int_{R^3} \omega \rho^2 |f(z,\zeta)|^2 d\zeta \cdot$$

$$\int_\Omega dv_x \int_{\partial\Omega} \frac{\cos\varphi}{|z-y|^2} \left(\frac{\cos\theta_y}{|x-y|^2} + \frac{N_1 D_1}{|x-y|^{2-\alpha}} \right) (1 + \mathrm{Log}|\sin\frac{\psi}{2}|^{-1}) \, dS_y \tag{2}$$

$$+ \int_\Omega dv_z \int_{R^3} \omega(|\zeta|) \, \rho^4(|\zeta|) \, |f(z,s)|^2 d\zeta \int_\Omega \frac{1 + \mathrm{Log}|\sin\theta/2|^{-1}}{|z-x|^2} dv_x \}$$

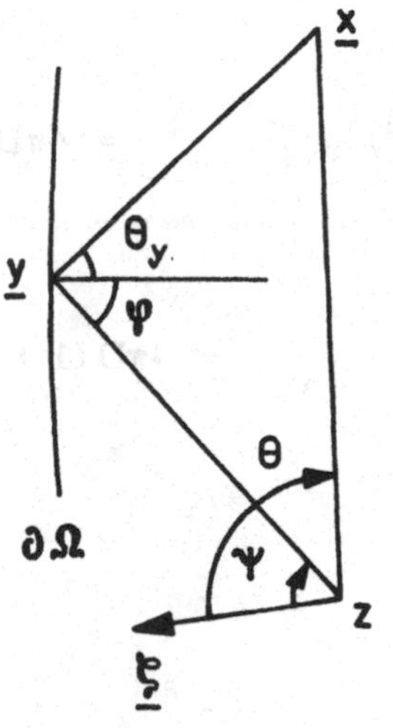

The notation for the angles $\psi, \theta, \varphi, \theta_y$ is indicated on the figure. Remembering that Ω is contained in a sphere of radius D centered at every point in Ω we have

$$\int_\Omega \frac{1 + \mathrm{Log}|\sin\theta/2|^{-1}}{|x-z|^2} dv_x \leqslant 2\pi D \int_0^\pi (1 + \mathrm{Log}|\sin\theta/2|^{-1}) \sin\theta \, d\theta = 6\pi D. \tag{3}$$

On the other hand

(4) $$\int \frac{dv_x}{|x-y|^{2-\alpha}} \leqslant \frac{4\pi}{1+\alpha} D^{\alpha+1} ,$$

and

$$\int_\Omega dv_x \int_{\partial\Omega} \frac{\cos\varphi}{|z-y|^2} \left(\frac{\cos\theta_y}{|x-y|^2} + \frac{N_1 D_1}{|x-y|^{2-\alpha}} \right) (1 + \operatorname{Log}|\sin\frac{\psi}{2}|^{-1}) \, dS_y$$

$$\leqslant 4\pi D \left(1 + \frac{N_1 D_1}{1+\alpha} D^\alpha\right) \int_{\partial\Omega} \frac{\cos\varphi}{|y-z|^2} (1 + \operatorname{Log}|\sin\frac{\psi}{2}|^{-1}) \, dS_y$$

(5) $$= 4\pi D\left(1 + \frac{N_1 D_1}{1+\alpha} D^\alpha\right) \int_{|u|=1} (1 + \operatorname{Log}|\sin\frac{\psi}{2}|^{-1}) \, du$$

$$= 4\pi D \left(1 + \frac{N_1 D_1}{1+\alpha} D^\alpha\right) 2\pi \int_0^\pi (1 + \operatorname{Log}|\sin\frac{\psi}{2}|^{-1}) \sin\psi \, d\psi$$

$$= 4\pi D \left(1 + \frac{N_1 D_1}{1+\alpha} D^\alpha\right) 6\pi .$$

Summing up and taking into account $\rho \leqslant \rho^2$, we obtain

(6) $$|||\rho\, W A f|||^2 \leqslant C_{12} \left(|\Omega| + 4\pi D \left[1 + \frac{N_1 D_1 D^\alpha}{1+\alpha}\right] 6\pi\right) |||\rho^2 f|||^2$$

3.3.1.3. From 3.2.4.7.6.(2) we estimate

$$\int_\Omega dv_z \int_{\partial\Omega} \left(\frac{\cos\theta_y}{|x-y|^2} + \frac{N_1 D_1}{|x-y|^{2-\alpha}}\right) \{\mu_u \, \rho \, WAf\}^2(z) \frac{\cos\varphi}{|y-z|^2} \, dS_y \tag{1}$$

$$\leqslant C_{12} (J_1 + J_2 + J_3)$$

with

$$J_1 = |||\rho f|||^2 \int_\Omega dv_z \int_{\partial\Omega} \left(\frac{\cos\theta_y}{|x-y|^2} + \frac{N_1 D_1}{|x-y|^{2-\alpha}}\right) \frac{\cos\varphi}{|y-z|^2} \, dS_y \,, \tag{2}$$

$$J_2 = \int_\Omega dv_z \int_{\partial\Omega} \left(\frac{\cos\theta_y}{|x-y|^2} + \frac{N_1 D_1}{|x-y|^{2-\alpha}}\right) \frac{\cos\varphi}{|y-z|^2} \, dS_y \,. \tag{3}$$

$$\cdot \int_\Omega dv_{z1} \int_{R^3} \omega(|\zeta|) \, \rho^2(|\zeta|) \, J(z, z_1, u, \zeta) \, | f(z_1, \zeta) |^2 \, d\zeta \,,$$

where we refer to 3.3.1.1.(2) for the meaning of u,

$$J_3 = \int_\Omega dv_z \int_{\partial\Omega} \left(\frac{\cos\theta_y}{|x-y|^2} + \frac{N_1 D_1}{|x-y|^{2-\alpha}}\right) \frac{\cos\varphi}{|y-z|^2} \, dS_y \, \cdot \tag{4}$$

$$\int_\Omega \frac{(1 + \operatorname{Log}|\sin\frac{\chi}{2}|^{-1})}{|z-z_1|^2} \, dv_{z1} \int_{R^3} \omega(|\zeta|) \, \rho^4(|\zeta|) \, (1 + \operatorname{Log}|\sin\frac{\theta}{2}|^{-1}) \, | f(z_1, \zeta)|^2 \, d\zeta$$

With the notation of the figure. Now we are going to estimate separately J_1, J_2 and J_3

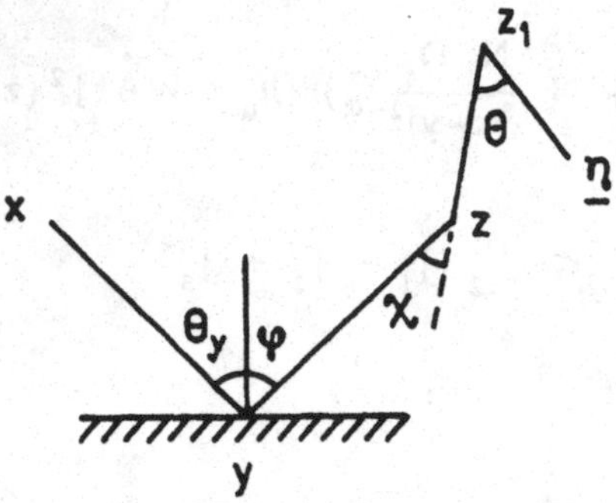

3.3.1.3.1. We begin by estimating J_1

(1)
$$J_1 \leqslant |||\rho f|||^2 \int_{\partial\Omega} \left(\frac{\cos\theta_y}{|x-y|^2} + \frac{N_1 D_1}{|x-y|^{2-\alpha}} \right) dS_y \int_{\Omega} \frac{dv_z}{|y-z|^2}$$

$$\leqslant 2\pi D\ |||\rho f|||^2 \int_{\partial\Omega} \frac{\cos\theta_y}{|x-y|^2}\, dS_y + 2\pi\, N_1 D_1\ |||\rho f|||^2 \int_{\partial\Omega} \frac{dS_y}{|x-y|^{2-\alpha}}$$

Let us consider

(2)
$$F(x) = \int_{\partial\Omega} \frac{dS_y}{|x-y|^{2-\alpha}} \quad ,$$

it is obviously bounded by

(3)
$$F(x) \leqslant |\partial\Omega|\, |d(x)|^{-2+\alpha}$$

where $d(x)$ is the distance from x to $\partial\Omega$. Let us assume that x is close to some point $x_0 \in \partial\Omega$. We choose x_0 as the origin of coordinates according to 2.2.1.5 and assume that x has coordinates (0, 0, h) with this system. We divide $\partial\Omega$ as $Br_1 \cup B'r_1$ with $Br_1 \cap B_{r1} = \phi$ and Br_1 the intersection of $\partial\Omega$ with the ball of radius r_1 centered at x. We assume $r_1 < r_0$ with r_0 as in 2.2.1.5., and $h \leqslant r_1/2$. We have

$$\int_{B^1_{r_1}} \frac{dS_y}{|x-y|^{2-\alpha}} \leqslant \left(\frac{2}{r_1}\right)^{2-\alpha} |\partial\Omega| \tag{4}$$

and, according to 2.2.1.7.(7) (8)

$$\begin{aligned}\int_{B_{r_1}} \frac{dS_y}{|x-y|^{2-\alpha}} &\leqslant \iint_{x^2+y^2 \leqslant r_1^2} \frac{dx\,dy}{\{x^2 + y^2 + [h - f(x, y)]^2\}^{1-\alpha/2}} \\ &\leqslant \sqrt{2}\, 2\pi \int_0^{r_1} r^{\alpha-1}\, dr = \frac{2\pi\sqrt{2}}{\alpha}\, r_1^\alpha .\end{aligned} \tag{5}$$

Summing up, we have

$$F(x) \leqslant \left(\frac{2}{r_1}\right)^{2-\alpha} |\partial\Omega| + \frac{2\pi\sqrt{2}}{\alpha}\, \gamma_1^\alpha , \tag{6}$$

and we set for further reference

$$\sup_{x\in\Omega} \int_{\partial\Omega} \frac{dS_y}{|x-y|^{2-\alpha}} = D_2(\alpha, \partial\Omega) < +\infty . \tag{7}$$

Coming back to (1) we find

$$J_1 \leqslant 2\pi\,(4\pi D + N_1 D_1 D_2)\, |||\rho f|||^2 . \tag{8}$$

3.3.1.3.2. To estimate J_2 we change the order of integrations and change z to z_1, and z_1 to z, namely

(1)
$$J_2 = \int_{\Omega} dv_z \int_{R^3} \omega(|\zeta|)\, \rho^2(|\zeta|)\, |f(z,\zeta)|^2\, d\zeta \int_{\partial\Omega} \left(\frac{\cos\theta_y}{|x-y|^2} + \frac{N_1\, D_1}{|x-y|^{2-\alpha}} \right) dS_y \cdot$$
$$\cdot \int_{\Omega} J(z_1, z, u_1, \zeta)\, \frac{\cos\varphi}{|y-z|^2}\, dv_{z_1}$$

where

(2)
$$y - z_1 = |y - z_1|\, u_1$$

and, from 3.2.4.7.4.(3) and 3.2.4.7.5.(2), according to the notation of the figure, we have

(3)
$$J(z_1, z, u_1, \zeta) = \int_{\partial\Omega} \left(1 + \mathrm{Log}\left|\sin\frac{\chi_1}{2}\right|^{-1}\right)\left(1 + \mathrm{Log}\left|\sin\frac{\psi_1}{2}\right|^{-1}\right) \frac{\cos\varphi_1}{|z-y_1|^2} \cdot$$
$$\left(\frac{\cos\theta_y}{|z_1 - y_1|^2} + \frac{N_1\, D_1}{|z_1 - y_1|^{2-\alpha}} \right) dS_{y_1}$$

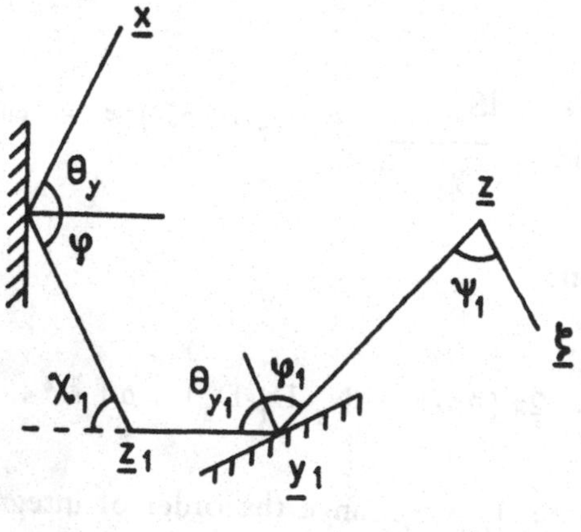

and, consequently

$$J_2 = \int_{\Omega} dv_z \int_{R^3} \omega \rho^2 \, |f(z, \zeta)|^2 \, K(x, z, \zeta) \, d\zeta \tag{4}$$

with

$$\begin{aligned} K(x, z, \zeta) &= \int_{\partial\Omega} \Big(\frac{\cos\theta_y}{|x-y|^2} + \frac{N_1 D_1}{|x-y|^{2-\alpha}}\Big) dS_y \,\cdot \\ &\cdot \int_{\partial\Omega} (1 + \mathrm{Log}|\sin\frac{\psi_1}{2}|^{-1}) \frac{\cos\varphi_1}{|z-y_1|^2} \, dS_{y_1} \,\cdot \\ &\cdot \int_{\Omega} (1 + \mathrm{Log}|\sin\frac{\chi_1}{2}|^{-1}) \Big(\frac{\cos\theta_{y_1}}{|y_1-z_1|^2} + \frac{N_1 D_1}{|y_1-z_1|^{2-\alpha}}\Big) \frac{\cos\varphi}{|y-z_1|^2} \, dv_{z_1} \end{aligned} \tag{5}$$

From 3.1.8.1.(1) we have

$$\begin{aligned} &\int_{\Omega} (1 + \mathrm{Log}|\sin\frac{\chi_1}{2}|^{-1}) \Big(\frac{\cos\theta_{y_1}}{|y_1-z_1|^2} + \frac{N_1 D_1}{|y_1-z_1|^{2-\alpha}}\Big) \frac{\cos\varphi}{|y-z_1|^2} \, dv_{z_1} \\ &\leqslant C_3(\Omega) (1 + N_1 D_1 D^{\alpha}) |y-y_1|^{-1} \end{aligned} \tag{6}$$

$$\begin{aligned} K(x, z, \zeta) &\leqslant C_3 (1 + N_1 D_1 D^{\alpha}) \int_{\partial\Omega} \Big(\frac{\cos\theta_x}{|x-y|^2} + \frac{N_1 D_1}{|x-y|^{2-\alpha}}\Big) dS_y \,\cdot \\ &\cdot \int_{\partial\Omega} (1 + \mathrm{Log}|\sin\frac{\psi_1}{2}|^{-1}) \frac{\cos\varphi_1}{|z-y_1|^2 \, |y_1-y|} \, dS_{y_1} \end{aligned} \tag{7}$$

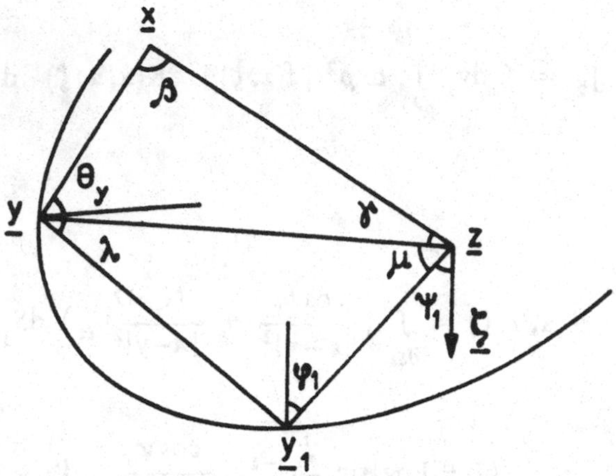

We evaluate the inner integral by writing

$$\partial\Omega = \Sigma_1 \cup \Sigma_2 \tag{8}$$

$$\begin{aligned} \Sigma_1 &: \quad \{y_1 \mid y_1 \in \partial\Omega\,,\ |y_1 - y| \geq |y_1 - z|\} \\ \Sigma_2 &: \quad \{y_1 \mid y_1 \in \partial\Omega\,,\ |y_1 - z| \geq |y_1 - y|\} \end{aligned} \tag{9}$$

$$y_1 - z = |y_1 - z|\ u \tag{10}$$

and observing that

$$y_1 \in \Sigma_1 \Rightarrow \frac{1}{|y_1 - y|} \leq \frac{2}{|z - y|} \tag{11}$$

$$y_1 \in \Sigma_2 \Rightarrow \frac{1}{|y_1 - y|} \leqslant \frac{2}{|z-y|} \, \frac{1}{\sin \mu} \tag{12}$$

So that

$$\begin{aligned} &\int_{\partial\Omega} (1 + \mathrm{Log}|\sin \frac{\psi_1}{2}|^{-1}) \frac{\cos \varphi_1}{|z-y_1|^2 \, |y_1 - y|} \, dS_{y_1} \leqslant \\ &\frac{2}{|z-y|} \int_{|u|=1} (1 + \mathrm{Log}|\sin \frac{\psi_1}{2}|^{-1}) (\sin \mu)^{-1} \, du \end{aligned} \tag{13}$$

Using Hölder inequality we find

$$\begin{aligned} &\int_{|u|=1} (1 + \mathrm{Log}\,|\sin \frac{\psi_1}{2}|^{-1}) (\sin \mu)^{-1} \, du \leqslant \{ \int_{|u|=1} (1 + \mathrm{Log}|\sin \frac{\psi_1}{2}|^{-1})^p \, du \}^{1/p} \\ &\cdot \{ \int_{|u|=1} (\sin \mu)^{-q} \, du \}^{1/q} \\ &= 2\pi \{ 4 \int_0^1 x \, (1 + \mathrm{Log}\, x^{-1})^p \, dx \}^{1/p} \{ \int_0^\pi (\sin \mu)^{1-q} \, d\mu \}^{1/q} = \frac{N_2}{2} \end{aligned} \tag{14}$$

With some numerical constant for N_2, and, substituting (14) into (13) and (7) we obtain

$$K(x, z, \zeta) \leqslant C_3 (1 + N_1 D_1 D^\alpha) N_2 \int_{\partial\Omega} (\frac{\cos \theta_y}{|x-y|^2} + \frac{N_1 \, D_1}{|x-y|^{2-\alpha}}) \frac{1}{|y-z|} \, dS_y \; . \tag{15}$$

We estimate the integral by writing

$$\partial\Omega = \Sigma'_1 \cup \Sigma'_2 \tag{16}$$

$$
(17) \qquad \begin{aligned} \Sigma_1' &= \{ y \mid y \in \partial\Omega, \quad |y-z| \geqslant \delta |y-x| \} \\ \Sigma_2' &= \{ y \mid y \in \partial\Omega, \quad |y-z| \leqslant \delta |y-x| \} \end{aligned}
$$

with some $\delta < 1$ and we find, using 3.3.1.3.1.(7) ,

$$
(18) \qquad \begin{aligned} \int_{\partial\Omega} \left(\frac{\cos\theta_y}{|x-y|^2} + \frac{N_1 D_1}{|x-y|^{2-\alpha}} \right) \frac{1}{|y-z|} \, dS_y &\leqslant \frac{(4\pi D + N_1 D_1 D_2)(1+\delta)}{\delta |x-z|} + \\ &+ \int_{\Sigma_2'} \left(\frac{\cos\theta_y}{|x-y|^2} + \frac{N_1 D_1}{|x-y|^{2-\alpha}} \right) \frac{1}{|y-z|} \, dS_y \, . \end{aligned}
$$

On the other hand

$$
(19) \qquad y \in \Sigma_2 \Rightarrow \frac{\cos\theta_y}{|x-y|^2} + \frac{N_1 D_1}{|x-y|^{2-\alpha}} \leqslant \frac{(1 + N_1 D_1 D^{\alpha})(1+\delta)^2}{|x-z|^2}
$$

and

$$
(20) \qquad \begin{aligned} \int_{\partial\Omega} \left(\frac{\cos\theta_y}{|x-y|^2} + \frac{N_1 D_1}{|x-y|^{2-\alpha}} \right) \frac{1}{|y-z|} \, dS_y &\leqslant \frac{(4\pi + N_1 D_1 D_2)(1+\delta)}{\delta |x-z|} \\ &+ \frac{(1 + N_1 D_1 D^{\alpha})(1+\delta)^2}{|x-z|^2} \int_{\Sigma_2'} \frac{dS_y}{|z-y|} \end{aligned}
$$

where

$$\Sigma_2' : \{ y \mid y \in \partial\Omega , |z-y| \leqslant \frac{\delta}{1+\delta} | x-z| \} . \tag{21}$$

We choose δ sufficiently small for Σ_2' being included in a Liapounov sphere or void. When Σ_2' is not void let y_0 be the point of $\partial\Omega$ nearest to z and choose y_0 as the origin of a coordinate system like the one in 2.2.1.5. Set $(0, 0, h)$ for the coordinates of z and assume δ sufficiently small for the following inequality hold

$$\frac{\delta D}{1+\delta} \leqslant r_1 \leqslant r_0 , \tag{22}$$

with the same r_0 as in 2.2.1.7.(7), we have

$$\int \frac{dS_y}{|z-y|} \leqslant \sqrt{2} \iint_{\Delta} \frac{dx\ dy}{\{ x^2 + y^2 + [h-f(x, y)]^2 \}^{1/2}} . \tag{23}$$

with

$$\Delta : \{ (x, y) \mid \{x^2 + y^2 + [h-f(x, y)]^2 \}^{1/2} \leqslant \frac{\delta}{1+\delta} |x-z| \} \tag{24}$$

and, consequently

$$\int_{\Sigma_2'} \frac{dS_y}{|z-y|} \leqslant 2\pi \sqrt{2} \frac{\delta}{1+\delta} |x-z| . \tag{25}$$

Summing up (20) and (21) we find

$$\int_{\partial\Omega} \left(\frac{\cos\theta_y}{|x-y|^2} + \frac{N_1\ D_1}{|x-y|^{2-\alpha}} \right) \frac{1}{|y-z|} dS_y \leqslant \tag{26}$$

$$\left\{\frac{(4\pi D + N_1 D_1 D_2)(1+\delta)}{\delta} + 2\pi\sqrt{2}(1 + N_1 D_1 D^{\alpha})\,\delta(1+\delta)\right\}\frac{1}{|x-z|}$$

and, as a consequence, from (4) (5) (26),

$$J_2 \leqslant C_3(1 + N_1 D_1 D^{\alpha})\, N_2[(4\pi D + N_1 D_1 D_2) + 2\pi\delta^2\sqrt{2}(1 + N_1 D_1 D^{\alpha})]\,\frac{1+\delta}{\delta} \tag{27}$$

$$\cdot \int_{\Omega} \frac{\{\rho f\}^2(z)}{|x-z|}\, dv_z$$

3.3.1.3.3. Now we estimate J_3 as follows

$$J_3 = \int_{\Omega} dv_z \int_{R^3} \omega(|\zeta|)(\rho(|\zeta|))^4\, |f(z,\zeta)|^2\, d\zeta \int_{\partial\Omega} \left(\frac{\cos\theta_y}{|x-y|^2} + \frac{N_1 D_1}{|x-y|^{2-\alpha}}\right) dS_y$$

(1)

$$\int_{\Omega} (1 + \mathrm{Log}|\sin\frac{\chi}{2}|^{-1})(1 + \mathrm{Log}|\sin\frac{\theta}{2}|^{-1})\,\frac{\cos\varphi}{|y-z_1|\,|z_1-z|^2}\, dv_{z_1}$$

with the notations of the figure accompanying 3.3.1.3. when z and z_1 have been interchanged. The inner integral is estimated, according to 3.1.8.1., to be less than $C_3(\Omega)\,|y-z|^{-1}$ so that

$$J_3 \leqslant C_3 \int_{\Omega} dv_z \int_{R^3} \omega\rho^4\, |f(z,\zeta)|^2\, d\zeta\,. \tag{2}$$

$$\int_{\Omega} \left(\frac{\cos\theta_y}{|x-y|^2} + \frac{N_1 D_1}{|x-y|^{2-\alpha}} \right) \frac{1}{|y-z|} dS_y$$

and, from 3.3.1.3.2.(26) we have

$$J_3 \leqslant C_3[(4\pi D + N_1 D_1 D_2) + 2\pi\delta^2 \sqrt{2}(1 + N_1 D_1 D^{\alpha})] \frac{1+\delta}{\delta} \int_{\Omega} \frac{\{\rho^2 f\}^2(z)}{|x-z|} dv_z \tag{3}$$

3.3.1.3.4. Summing up 3.3.1.3.1.(8) 3.3.1.3.2.(27) 3.3.1.3.3.(3) and 3.3.1.3.(1) we obtain

$$\int_{\Omega} dv_z \int_{\partial\Omega} \left(\frac{\cos\theta_y}{|x-y|^2} + \frac{N_1 D_1}{|x-y|^{2-\alpha}} \right) \{\mu_u u \rho W A f\}^2(z) \frac{\cos\varphi}{|y-z|^2} dS_y$$

$$\leqslant 2\pi(4\pi D + N_1 D_1 D_2) C_{12} |||\rho f|||^2 + \tag{1}$$

$$2C_3 C_{12}[(4\pi D + N_1 D_1 D_2) + 2\pi\delta \sqrt{2}(1 + N_1 D_1 D^{\alpha})] \frac{1+\delta}{\delta} \int_{\Omega} \frac{\{\rho^2 f\}^2(z)}{|x-z|} dv_z$$

where we have taken into account $\rho \geqslant 1$ and δ is a constant which may be chosen according to 3.3.1.3.2.(22).

3.3.1.4. Now we estimate the last term in 3.3.1.1.(1). From 3.2.3.7.6. we have

$$\int_{\Omega} \frac{\{\mu_w \rho^{p+1} W A f\}^2(z)}{|x-z|^2} dv_z \leqslant C_{12}(H_1 + H_2 + H_3) \tag{1}$$

with

(2) $$H_1 = |||\rho f|||^2 \int_\Omega \frac{dv_z}{|x-z|^2} \leqslant 4\pi D\ |||\rho f|||^2$$

(3) $$H_2 = \int_\Omega \frac{dv_z}{|x-z|^2} \int_\Omega dv_{z_1} \int_{R^3} \omega \rho^2 J(z, z_1, w, \zeta)\ |f(z_1, \zeta)|^2\ d\zeta$$

(4) $$H_3 = \int_\Omega \frac{dv_z}{|x-z|^2} \int_\Omega \frac{(1 + \mathrm{Log}|\sin\frac{\varphi}{2}|^{-1})}{|z-z_1|^2} dv_{z_1} \int_{R^3} \omega\rho^{2(p+2)}(1+\mathrm{Log}|\sin\frac{\theta}{2}|^{-1})\,|f(z_1,\zeta)|^2\ d\zeta$$

where, in H_2 ,

(5) $$x - z = |x - z|\ w$$

while, for H_3 , the notations are indicated on the figure.

3.3.1.4.1. Let us estimate H_2 after having interchanged z and z_1, we have

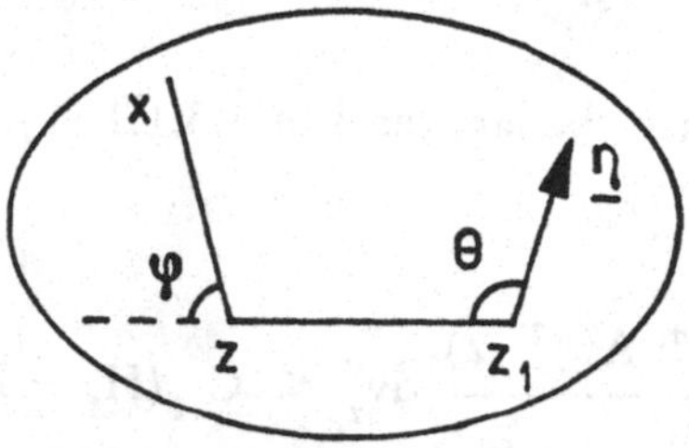

$$J = \int_{\partial\Omega} (1 + \mathrm{Log}|\sin \psi_1/2|^{-1})(1 + \mathrm{Log}|\sin \chi_1/2|^{-1}) \cdot$$

$$\cdot \frac{\cos \varphi_1}{|z-y_1|^2}\left(\frac{\cos \theta_{y1}}{|z-y_1|^2} + \frac{N_1 D_1}{|z_1-y_1|^{2-\alpha}}\right) dS_{y_1} \qquad (1)$$

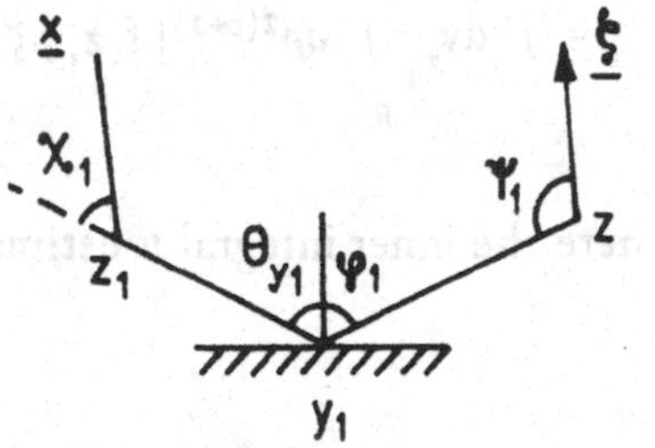

and

$$H_2 = \int_{\Omega} dv_z \int_{R^3} \omega(|\zeta|)\,\rho^2(|\zeta|)\,|f(z,\zeta)|^2\, d\zeta \int_{\partial\Omega} (1 + \mathrm{Log}|\sin \frac{\psi_1}{2}|^{-1}) \frac{\cos \varphi_1}{|z-y_1|^2}\, dS_{y_1} \cdot$$

$$\cdot \int_{\Omega} \left(\frac{\cos \theta_{y_1}}{|y_1-z_1|^2} + \frac{N_1 D_1}{|y_1-z_1|^{2-\alpha}}\right) \frac{1 + \mathrm{Log}|\sin \chi_1/2|^{-1}}{|x-z_1|^2} dv_{z_1} \qquad (2)$$

From 3.1.8.1.(1) the inner integral is less than $C_3(1 + N_1 D_1 D^\alpha)$ and we have

$$H_2 \leqslant C_3(1 + N_1 D_1 D^\alpha) \int_{\Omega} dv_z \int_{R^3} \omega\rho^2 |f(z,\zeta)|^2 d\zeta \int_{\partial\Omega} \frac{1 + \mathrm{Log}|\sin \psi_1/2|^{-1}}{|x-y_1|} \frac{\cos \varphi_1}{|z-y_1|^2} dS_{y_1} \qquad (3)$$

From 3.3.1.3.2.(13) we know that

$$\int_{\partial\Omega} \frac{1 + \mathrm{Log}|\sin \frac{\psi_1}{2}|^{-1}}{|x-y_1|} \frac{\cos \varphi_1}{|z-y_1|^2} dS_{y_1} \leqslant \frac{N_2}{|x-z|} \qquad (4)$$

and, consequently

$$H_2 \leqslant C_3(1 + N_1 D_1 D^\alpha) N_2 \int_{\Omega} \frac{\{\rho f\}^2(z)}{|x-z|} dv_z \qquad (5)$$

3.3.1.4.2. We estimate H_3 according to

(1)
$$H_3 = \int_\Omega dv_{z_1} \int_{R^3} \omega\rho^{2(p+2)} | f(z_1, \zeta) |^2 d\zeta \int \frac{(1 + \mathrm{Log}|\sin \varphi/2|^{-1})(1 + \mathrm{Log}|\sin \theta/2|^{-1})}{|x-z|^2 \; |z-z_1|^2} dv_z$$

where the inner integral is estimated from 3.1.8.1.(1) so that

(2)
$$H_3 \leqslant C_3 \int_\Omega \frac{\{\rho^{p+2} f\}^2 (z)}{|x-z|} dv_z$$

3.3.1.4.3. Summing up 3.3.1.4.(1) (2) (3) (4) and 3.3.1.4.1.(5) 3.3.1.4.2.(2) we obtain

(1)
$$\int \frac{\{\mu_w \rho^{p+1} \mathcal{W} A f\}^2 (z)}{|x-z|^2} d\vartheta_z \leqslant C_{12} \, 4\pi D \, |||\rho f|||^2 +$$
$$+ C_{12} C_3 [1 + N_2 (1 + N_1 D_1 D^\alpha)] \int_\Omega \frac{\{\rho^{p+2} f\}^2 (z)}{|x-z|}$$

3.3.1.5. Now we go back to 3.3.1.1.(1) and sum up with 3.3.1.2.(6) 3.3.1.3.4.(1) 3.3.1.4.3.(1) and this leads to

(1)
$$\{\rho^p (\mathcal{W} A)^2 f^2\} (x) \leqslant C_{13} \{ |||\rho^2 f|||^2 +$$
$$+ \int_\Omega \frac{\{\rho^{p+2} f\}^2 (z)}{|x-z|} dv_z \}$$

with

$$C_{13} = C_{12}^2 \{4\pi D + 2\pi(4\pi + N_1 D_1 D_2) + 6\pi [|\Omega| + 4\pi D(1 + \frac{N_1 D_1 D^\alpha}{1+\alpha})] +$$

$$+ C_3 [1 + N_2(1 + N_1 D_1 D^\alpha)] + 2 C_3 [4\pi + N_1 D_1 D_2 + \tag{2}$$

$$+ 2\pi\delta \sqrt{2}(1 + N_1 D_1 D^\alpha)] \frac{1+\delta}{\delta}\}$$

where, once more, we have taken into account the fact that $\rho \leqslant \rho^{p+1}$ and δ is as in 3.3.1.3.2.(22).

3.3.2. Estimation of $(WA)^4$

3.3.2.1. From 3.3.1.5.(1) we get

$$\{\rho^p (WA)^4 f\}^2(x) \leqslant C_{13} \{ |||\rho^2 (WA)^2 f|||^2 + \tag{1}$$

$$+ \int_\Omega \frac{\{\rho^{p+2} (WA)^2 f\}^2(z)}{|x-z|} dv_z \}$$

and, from 3.3.1.5.(1) again,

$$|||\rho^2 (WA)^2 f|||^2 \leqslant C_{13} \{ |\Omega| \, |||\rho^2 f|||^2 + \int_\Omega \{\rho^4 f\}^2(z) \, dv_z \int_\Omega \frac{dv_x}{|x-z|} \}$$

$$\leqslant C_{13} \{ |\Omega| \, |||\rho^2 f|||^2 + 2\pi D^2 \, |||\rho^4 f|||^2 \} \tag{2}$$

$$\leqslant C_{13} (|\Omega| + 2\pi D^2) \, |||\rho^4 f|||^2$$

while

$$\int_\Omega \frac{\{\rho^{p+2}(WA)^2 f\}^2(z)}{|x-z|}\, dv_z \leqslant C_{13}\ \{|||\rho^2 f|||^2 \int_\Omega \frac{dv_z}{|x-z|} +$$

(3)
$$+ \int_\Omega \{\rho^{p+4} f\}^2(z)\, dv_z \int_\Omega \frac{dv_{z1}}{|x-z_1|\ |z_1-z|}\ \}$$

$$\leqslant C_{13}\ (2\pi D^2\ |||\rho^2 f|||^2 + C_{1,1} \int_\Omega (1 + |\mathrm{Log}|x-z||)\ \{\rho^{p+4} f\}^2(z)\, d\nu_z)$$

taking 3.1.8.2. into account. Substituting (2) and (3) into (1) we find

(4) $$\{\rho^p (WA)^4 f\}^2(x) \leqslant C_{14}\ \{|||\rho^4 f|||^2 + \int_\Omega (1 + |\mathrm{Log}|x-z||)\ \{\rho^{p+4} f\}^2(z)\, d\nu_z\}$$

with

(5) $$C_{14} = C_{13}^2\ \sup\ \{|\Omega| + 2\pi D^2,\ C_{1,1}\}$$

3.3.2.2. From 3.3.2.1.(4) we have*

(1)
$$\{(WA)^5 f\}^2(x) \leqslant C_{14}\{|||\rho^4 WAf|||^2 +$$
$$+ \int_\Omega (1 + |\mathrm{Log}\,|x-z|\,|)\ \{\rho^4\ WAf\}^2(z)\, d\vartheta_z\}.$$

We use $W = U + \Lambda U$ and $\rho^p U = U\rho^p$ and we conclude that

(2)
$$\{(WA)^5 f\}^2(x) \leqslant C_{14}\ \{|||U\rho^4 Af|||^2 + \int_\Omega (1 + |\mathrm{Log}|x-z|\,|)\ \{U\rho^4 Af\}^2(z)\, d\vartheta_z +$$

*) The calculations up to 3.3.2.4. may be simplified by noting that the logarithmic term in 3.3.2.1.(3) is superfluous.

$$+ \ |||\rho^4 \Lambda \cup Af|||^2 + \int_{\Omega} (1 + |\mathrm{Log}|x-z||) \{\rho^4 \Lambda \cup Af\}^2(z)\, d\nu_z \}$$

3.3.2.2.1. From the definition 3.1.7.2.(1) we have

$$\|\rho^p Af\|_H^2 = \int_{R^3} \rho^{2p}\, d\xi \int_{R^3} |B(\xi,\zeta)\,|\omega|\, f(\zeta)|^2\, d\zeta \int_{R^3} |B(\xi,\zeta_1)|\, d\zeta_1$$

$$\leqslant \sigma^2 C_1 \int_{R^3} \rho^{2p}\, d\xi \int_{R^3} |B(\xi,\zeta)\,|\,\omega\,|\, f(\zeta)\,|^2\, d\zeta \tag{1}$$

$$\leqslant \frac{1}{2}\, \sigma^2 C_1\, C_1'(2,p)\ \ \|f\|_H^2$$

according to 3.1.7.2.(2) and (3), and from 2.1.3.3.(9) and $\nu \geqslant \nu_0$ we find

$$|||\cup \rho^p Af|||^2 \leqslant \frac{1}{2}\, \nu_0^{-2}\, \sigma^2 C_1\, C_1'(2,p)\ \ |||f|||^2 . \tag{2}$$

On the other hand, with the notations of the figure, namely

$$y \in \partial\Omega , \quad \xi \Lambda (z-y) = 0 , \quad \xi \cdot (z-y) > 0 , \tag{3}$$

and using the same computation as in 2.1.3.5.(3), we find

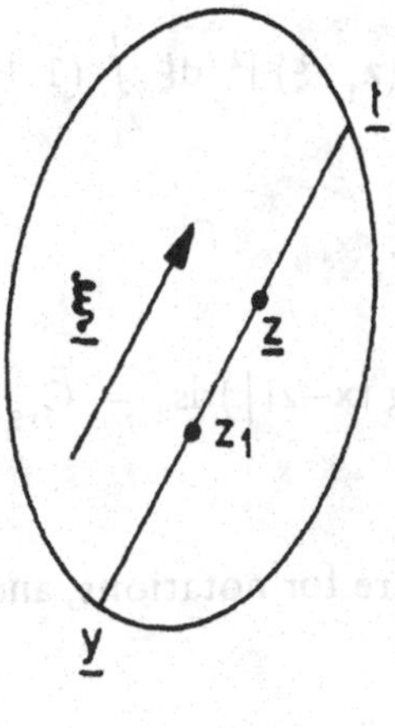

$$(4)\qquad \int_\Omega (1+|\mathrm{Log}|x-z||)\,\{U\rho^p Af\}^2(z)\,dv_z \leqslant \int_\Omega (1+|\mathrm{Log}|x-z||)\,dv_z \int_{R^3} \frac{\omega}{2|\xi|\nu}\,d\xi \int_y^z |g(z_1,\xi)|^2\,ds_{z_1},$$

where $g = \rho^p Af$ and $s_{z1} = |z_1 - y|$. Setting $\xi = \xi u$ and using

$$(5)\qquad ds_{z_1}\,d\xi = \xi^2\,d\xi\,du\,ds_{z_1} = \xi^2\,d\xi\,\frac{dv_{z_1}}{|z-z_1|^2}$$

we obtain

$$\int_\Omega (1+|\mathrm{Log}|x-z||)\,\{Ug\}^2(z)\,dv_z \leqslant \int_0^\infty \frac{\omega\,\xi}{2\nu}\,d\xi\,\cdot$$

$$(6)\qquad \int_\Omega (1+|\mathrm{Log}|x-z||)\,dv_z \int_\Omega |g(z_1,\xi)|^2 \frac{dv_{z_1}}{|z-z_1|^2}$$

$$= \int_\Omega dv_{z_1} \int_0^\infty \frac{\omega\,\xi}{2\nu}\,d\xi \int_\Omega (1+|\mathrm{Log}|x-z||)\,|g(z_1,\xi)|^2 \frac{dv_z}{|z-z_1|^2}$$

$$= \int_\Omega dv_{z_1} \int_{R^3} (2|\xi|\nu)^{-1}\omega\,|g(z_1,\xi)|^2\,d\xi \int_{z_1}^t (1+|\mathrm{Log}|x-z||)\,ds_z\,.$$

We observe that, in an obvious way,

$$(7)\qquad \sup_{\substack{x\in\Omega,\ y\in\partial\Omega\\ z_1\in\Omega}} \int_{z_1}^t (1+|\mathrm{Log}|x-z||)\,ds_z = C_{15}(\Omega) < +\infty\,,$$

where reference is made to the figure for notations, and, consequently

$$\int_{\Omega} (1+|\mathrm{Log}\,|x-z||\;\{\cup \rho^{p} A f\}^2(z)\; dv_z \leqslant$$
$$C_{15} \int_{\Omega} \{(2\,|\xi|\,\nu)^{-1/2}\; \rho^{p} A f\}^2(z)\; dv_z \,. \tag{8}$$

Using $\nu \geqslant \nu_0$ and remaking the computation of (1) we get

$$\int_{\Omega} (1+|\mathrm{Log}|x-z||)\,\{\cup\rho^{p} A f\}^2(z)\, dv_z \leqslant \frac{1}{4\nu_0}\,\sigma^2\, C_{15}\, C_1\, C_1'(1,p)\; |||f|||^2 \tag{9}$$

3.3.2.3. Let us now estimate the last two terms in 3.3.2.2.(2). To this end we estimate

$$I = \int_{\Omega} F(z)\;\{\rho^{p}\; EG\phi^{-}\}^2(z)\; d\vartheta_z\,,\quad F(z) \geqslant 0\,. \tag{1}$$

From the very definitions of E and G we have

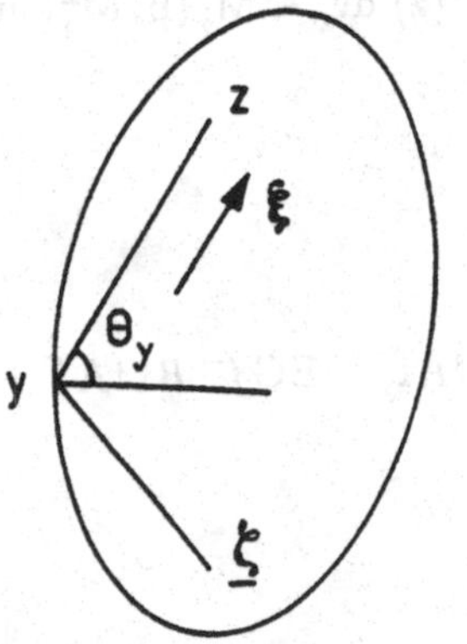

$$I \leqslant \int_{\Omega} F(z)\; dv_z \int_{R^3} \omega\rho^{2p}\; d\xi\;\{\int_{R^3} \Gamma(y,\xi,\zeta)\;\; \phi^{-}(y,\zeta)\; d\zeta\;\}^2 \tag{2}$$

$$\leq \int_{\Omega} F(z)\, dv_z \int_{R^3} \omega\rho^{2p}\, d\xi \int_{R^3} \omega\, |\zeta\cdot n|\, |\phi^-(y,\zeta)|^2\, d\zeta \cdot$$

$$\cdot \int_{R^3} \omega_1^{-1} |\zeta\cdot n|^{-1}\, \Gamma^2(y,\zeta,\zeta_1)\, d\zeta_1$$

and using 3.1.6.3.1. and 3.1.6.3.2. we find

$$\text{(3)}\quad \begin{aligned} I &\leq M_9^2 \int_{\Omega} F(z)\, dv_z \int_{R^3} \omega\, \rho^{2p}\, \Gamma_1^2\, d\xi \int_{R^3} \omega|\zeta\cdot n|\, |\phi^-(y,\zeta)|^2\, d\zeta \\ &= M_2(p)\, M_9^2 \int_{\Omega} F(z)\, dv_z \int_{\partial\Omega} \frac{\cos\theta_y}{|z-y|^2}\, dS_y \int_{R^3} \omega|\zeta\cdot n|\, |\phi^-(y,\zeta)|^2\, d\zeta\ , \end{aligned}$$

where we have set $\xi = \xi u$ and $d\xi = \xi^2 d\xi \cos\theta_y / |z-y|^2\, dS_y$. Interchanging the integrations we obtain

$$\text{(4)}\quad \int_{\Omega} F(z)\, \{\rho^p\, E\, G\, \phi^-\}^2(z)\, dv_z \leq M_2(p)\, M_9^2 (\sup \int_{\Omega} \frac{F(z)}{|y-z|^2}\, dv_z)\, \|\phi^-\|^2_{L^2(\Omega,\tilde{H})}$$

Now, we observe that

$$\text{(5)}\quad \Lambda U A f = ETGJ^- BUAf = EGJ^- BUAf + EGJ^- BETGJ^- BUAf$$

and, as a consequence

$$\text{(6)}\quad \begin{aligned} &\int_{\Omega} F(z)\, \{\rho^p \Lambda U A f\}^2(z)\, dv_z \leq \\ &2\, M_2(p)\, M_9^2 (\sup_{y\in\partial\Omega} \int_{\Omega} \frac{F(z)}{|y-z|^2}\, dv_z\,) \cdot \end{aligned}$$

$$\cdot \{ \| \mathcal{J}^- B \cup A f \|^2_{L^2(\partial\Omega, \bar{H})} + \| \mathcal{J}^- BETG \mathcal{J}^- B \cup A f \|^2_{L^2(\partial\Omega, \bar{H})} \} . \tag{6}$$

Referring to 2.1.3.5.(7) and 2.1.3.8.(5) we see that

$$\int_\Omega F(z) \{\rho^p \Lambda \cup A f\}^2 (z) \, dv_z \leq 2 M_2(p) M_9^2 .$$

$$\cdot \left(\sup_{y \in \partial\Omega} \int_\Omega \frac{F(z)}{|y-z|^2} dv_z\right) \{ (2\nu_0)^{-1} |\!|\!| A f |\!|\!|^2 + \tag{7}$$

$$+ (2\nu_0)^{-1} \| TG \mathcal{J}^- B \cup A f \|^2_{L^2(\partial\Omega, \bar{H})} \}$$

and using 2.3.1.2. and 3.1.7.1.(7) we get

$$\int_\Omega F(z) \{\rho^p \Lambda \cup A f\}^2 (z) \, dv_z \leq \nu_0^{-1} M_2(p) M_9^2 .$$

$$\cdot \left(1 + \frac{1}{2} |T|^2 M_6^2 M_8^2\right) \left(\sup_{y \in \partial\Omega} \int_\Omega \frac{F(z)}{|y-z|^2} dv_z \right) |\!|\!| A f |\!|\!|^2 \ . \tag{8}$$

Using

$$|\!|\!| \rho^4 \Lambda \cup A f |\!|\!|^c \leq \nu_0^{-1} M_2(4) M_g^2 \left(1 + \frac{1}{2} |T|^2 M_6^2 M_8^2\right) k_0'^2 \ |\!|\!| f |\!|\!|^2 , \tag{9}$$

we find

$$\int_\Omega (1 + |\mathrm{Log}|x-z||) \{\rho^4 \Lambda \cup A f\}^2 (z) dv_z \leq \nu_0^{-1} M_2(4) M_g^2 \left(1 + \frac{1}{2} |T|^2 M_6^2 M_8^2\right) k_0'^2 C_{16} |\!|\!| f |\!|\!|^2 , \tag{10}$$

where

$$C_{16}(\Omega) = \sup_{x \in \Omega} \int_{\Omega} (1 + |\text{Log}\,|x - z||)\,|x - z|^{-2}\, dv_z \,. \tag{11}$$

3.3.2.4. We state the result which has been proven. The hypothesis are as in 3.1.6. Then, for any f which is continuous and belongs to $L^2(\Omega, H)$, $(WA)^5 f$ is continuous and

$$\{ (WA)^5 f \}(x) \leqslant C_{17} \; |||f||| \,, \tag{1}$$

with

$$\begin{aligned} C_{17} = C_{14}^{1/2} \{ & \sigma^2 C_1 [\, 2^{-1} \nu_0^{-2} C_1'(2,4) + 4^{-1} \nu_0^{-1} C_1'(1,4)] \; + \\ & + \nu_0^{-1} M_2(4) M_9^2 (1 + \frac{1}{2} |T|^2 M_6^2 M_8^2)\, k_0'^2 (1 + C_{16}) \}^{1/2} \,. \end{aligned} \tag{2}$$

The continuity statement follows readily from the one in 3.2.4.7.6.

3.4. Regularisation of the Linear Solution. Last Step

3.4.1. Estimation of W in $B_r(\Omega)$

3.4.1.1. From the definition

$$E\,\phi^+(x,\xi) = e^{-\nu(|\xi|)\,|x-y|\,|\xi|^{-1}}\;\phi^+(y,\xi)\,. \qquad (1)$$

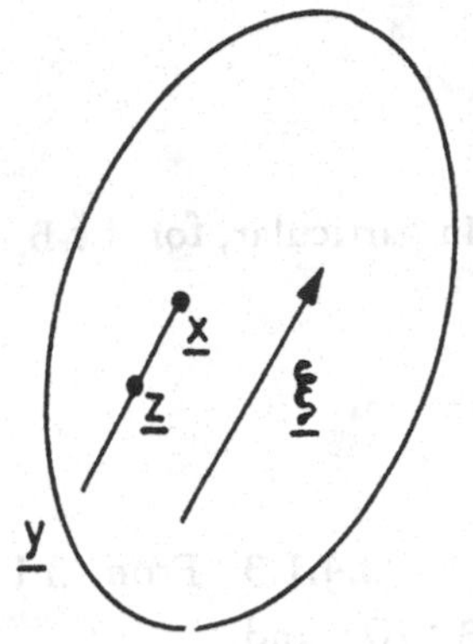

For a convex Ω, y depends continuously on x and ξ so that, if ϕ^+ is continuous $E\,\phi^+$ is continuous also. We observe that a continuous ϕ^+ is uniformly bounded in the vicinity of $|\xi| = 0$ and that $E\,\phi^+$ tends to zero when $|\xi|$ tends to zero. On the other hand, for $\phi^+ \in B_r(\partial\Omega)$ we have

$$(1+|\xi|^2)^{r/2}\,\omega^{1/2}(|\xi|)\,|\,E\,\phi^+(x,\xi)\,| \leqslant (1+|\xi|^2)^{r/2}\,\omega^{1/2}(|\xi|)\,|\,\phi^+(y,\xi)\,| \qquad (2)$$

so that

$$<E\,\phi^+>_r \;\leqslant\; <\phi^+>_r \qquad (3)$$

3.4.1.2. From the definition of U, we have referring to the figure of 3.4.1.1.

$$U f(x,\xi) = \int_y^x |\xi|^{-1}\,e^{-\nu(|\xi|)|x-z||\xi|^{-1}}\;f(z,\xi)\;ds_z \qquad (1)$$

with $s_z = |z-y|$ and

$$|\omega^{1/2}(|\xi|)\,(1+|\xi|^2)^{r/2}\nu U f(x,\xi)\,| \leqslant \int_y^x \nu|\xi|^{-1}\;e^{-\nu|\xi|^{-1}|z-x|}\;\omega^{1/2}(1+|\xi|^2)^{r/2}\,|f(z,\xi)|\,ds_z$$

$$\leqslant\; <f>_r \int_0^{|x-y|} |\xi|^{-1}\,\nu\;e^{-|\xi|^{-1}\nu s}\,ds\,. \qquad (2)$$

Now, for a convex Ω, Uf is continuous when f is and we have, for $f \in B_r(\Omega)$

(3) $$<\nu Uf>_r \leqslant <f>_r .$$

In particular, for $f \in B_r(\Omega)$, $\nu J^- BUf \in B_r(\partial\Omega)$

(4) $$<\nu J^- BUf>_r \leqslant <f>_r .$$

3.4.1.3. From 3.1.6.3. we know that G is continuous from $B_S(\partial\Omega)$ into $B_r(\partial\Omega)$ and

(1) $$<G\phi^- >_r \leqslant M_{4,r} \; M_{5,s} <\phi^- >_r .$$

For $r > 2$, according to 3.2.3.1., T is continuous in $B_r(\partial\Omega)$ and, consequently, taking into account 3.1.6.4.(1), we see from 3.4.1.2.(4) that $f \in B_r(\Omega)$, $r > 1$, implies that $ETGJ^- BUf \in B_{r+1}(\Omega)$ with

(2) $$<ETGJ^- BUf>_{r+1} \leqslant a^{-1} M_{4,r+1} \; M_{5,r+1} \; C_7^{(r+1)} <f>_r , \quad r > 1 .$$

For $0 \leqslant r \leqslant 1$ the result follows in a slightly different way and

(3) $$<ETGJ^- BUf>_{r+1} \leqslant a^{-1} M_{4,r+2} \; M_{5,r+1} \; C_7^{(r+2)} <f>_r , \quad 0 \leqslant r \leqslant 1 .$$

As a consequence, for $r \geqslant 0$, W is continuous from $B_r(\Omega)$ into $B_{r+1}(\Omega)$ and

(4) $$<Wf>_{r+1} \leqslant C_8^{(r)} <f>_r , \qquad r \geqslant 0 ,$$

with

$$C_8^{(r)} = \begin{cases} a^{-1}(1 + M_{4,r+1} M_{5,r+1} C_7^{(r+1)}) , & r > 1 , \\ a^{-1}(1 + M_{4,r+2} M_{5,r+1} C_7^{r+2}) , & 0 \leqslant \gamma \leqslant 1 . \end{cases} \tag{5}$$

or

$$< \nu W f >_r \leqslant C_9^{(r)} < f >_r \qquad r \geqslant 0 \tag{6}$$

with

$$C_9^{(r)} = b\, C_8^{(r)} . \tag{7}$$

3.4.2. Estimation of $\mathcal{E}$ *in* B_r

From 2.1.2.4.(10), 3.2.3.5. and 3.4.1.1. we see that $\mathcal{E}$ is cortinuous from $B_r(\partial\Omega)$ into $B_r(\Omega)$ provided $r > 2$, furthermore

$$< \mathcal{E} \phi^+ >_r \geqslant C_7^{(r)} < \phi^+ >_r , \quad r > 2 . \tag{1}$$

3.4.3. Estimation of $(WA)^{6+p}$ *in* B_r

3.4.3.1. Let us consider a continuous f belonging to $L^2(\Omega, H)$, from 3.3.2.4. we know that $(WA)^5 f$ belongs to $C_0(\Omega, H)$ and we see that $A(WA)^5$ belongs to $B_1(\Omega)$, namely

$$< A (W A)^5 f >_1 \leqslant k_0\, C_{17} \, |||f||| . \tag{1}$$

Applying 3.4.1.3.(4) we get $(WA)^6 f \in B_2(\Omega)$ and

(2) $$< (WA)^6 f >_2 \leqslant k_0 C_{17} C_8^{(1)} |||f|||$$

3.4.3.2. Now, from 1.5.3.6. and 3.4.1.3.(4) we see that WA is continuous from $B_r(\Omega)$ into B_{r+2} with

(1) $$< WA f >_{r+2} \leqslant k_r C_8^{(r+1)} < f >_r \qquad r \geqslant 0$$

and, iterating on 3.4.3.1.(2) we conclude that, whenever f is continuous and belongs to $L^2(\Omega, H)$ $(WA)^{6+p} f$ belongs to $B_{2p+1}(\Omega)$ and

(2) $$< (WA)^{5+p} f >_{2p} \leqslant C_{18}^{(p)} |||f|||$$

with

(3) $$C_{18}^{(p)} = k_0 C_{17} C_8^{(1)} C_8^{(3)} \ldots C_8^{(2p-1)} k_2 k_4 \ldots k_{2p-2}$$

3.5. Solution of the Non-linear Problem

3.5.1. First Step

From 3.1.5.2.(2), 3.4.3.2.(1)(2), and 3.4.1.3.(4), we have

$$< \mathscr{C} f >_r \leqslant \sum_{n=0}^{N-1} a^{-1} C_8^{(r)} (k_r C_8^{(r+1)})^n < f >_r + C_{18}^{(p)} |\mathscr{C}| \, |||f||| . \tag{1}$$

with $p = [r/2] + 1$ and $N = 5 + p$, while $|\mathscr{C}|$ is the norm of $\mathscr{C}$ in $L^2(\Omega, H)$. This holds for $f \in B_r(\Omega)$, $r > 3/2$ in order that $f \in L^2(\Omega, H)$. We observe that $|\mathscr{C}|$ is the constant $C(\Omega, L, G)$ of 2.3.5.7.(3). From 3.2.1.3.(3) we conclude that

$$< \mathscr{C} f >_r \leqslant C_{19}^{(r)} < f >_r , \tag{2}$$

$$C_{19}^{(r)} = \sum_{n=0}^{N-1} a^{-1} C_8^{(r)} (k_r C_8^{(r+1)})^n + C_{18}^{(p)} C(\Omega, L, G) \, A(r) , \tag{3}$$

and we may state that $f \in B_r(\Omega)$, $r > 3/2 \Rightarrow \mathscr{C} f \in B_r(\Omega)$. This result may be improved using

$$\mathscr{C} f = W f + W A \mathscr{C} f , \tag{4}$$

and 3.4.1.3.(4) 3.4.3.2.(1) as

$$f \in B_r(\Omega) , \quad r > 3/2 \Rightarrow \mathscr{C} f \in B_{r+1}(\Omega) \tag{5}$$

with

$$< \mathscr{C} f >_{r+1} \leqslant (C_8^{(r)} + k_r C_8^{(r+1)} C_{19}^{(r)}) < f >_r , \quad r > 3/2 , \tag{6}$$

and

(7) $$< \nu \mathscr{C} f >_r \leqslant a^{-1} (C_8^{(r)} + k_r C_8^{(r+1)} C_{19}^{(r)}) <f>_r , \quad r > 3/2 .$$

From 3.1.5.1.(1) we conclude that, for $r > 3/2$, $\mathscr{C} Q(f, g) \in B_r(\Omega)$ when f and g both belong to $B_r(\Omega)$ and that 3.1.4.1.(1) holds with

(8) $$\gamma = \alpha_2 a(C_8^{(r-1)} + k_{r-1} C_8^{(r)} C_{19}^{(r-1)})$$

From 3.4.2.(1), and (6) we see that $\mathscr{E}$ and $\mathscr{C} A \mathscr{E}$ and continuous from $B_r(\partial\Omega)$ into $B_r(\Omega)$ and that 3.1.4.1.(2) holds with $r > 1/2$ and

(9) $$\beta = C_7^{(r)} (1 + K_{r-1} C_{19}^{(r)})$$

Now we may apply the reasoning of 3.1.4. and conclude that 3.1.2.3.(1) has a unique solution in the ball $S_{r,R}$ as stated 3.1.4.2.

3.5.2. Second Step

Let f be the unique solution of 3.1.2.3.(1), the existence of which has been proven. We set

(1) $$f = \mathscr{C} g + E \phi^+ ,$$

with

(2) $$g = Q(f, f) + A \mathscr{E} G_1 (1 + f)^- \in B_{r-1} (\Omega) ,$$

(3) $$\phi^+ = T G_1 (1 + f^-) \in B_r(\partial\Omega) .$$

We have

$$\mathscr{C}g = Wg + WACg = Uh + E\,\psi^+ \tag{4}$$

with

$$h = g + ACg \in B_r(\bar{\Omega}), \quad \psi^+ = TGJ^- BUg + TGJ^- BUA\mathscr{C}g \in B_r(\partial\Omega). \tag{5}$$

This argument proves that C_g is continuously differentiable along rays and that

$$(\xi \cdot \nabla + \nu)\,\mathscr{C}g = g + A\mathscr{C}g, \tag{6}$$

so that, in $B_{r\cdot 1}(\Omega)$,

$$(\xi \cdot \nabla + L)\ \mathscr{C}g = g\,. \tag{7}$$

In the same way, again in $B_{r-1}(\Omega)$

$$(\xi \cdot \nabla + \nu)\ E\,\phi^+ = 0 \tag{8}$$

and, consequently

$$(\xi \cdot \nabla + L)\,f = Q(f,f) + A\mathscr{E}G_1(1+f)^- - AE\phi^+ = Q(f,f)\,. \tag{9}$$

Then in solution f solves the non linear Beltzmann equation, in a classical sense in $B_{r-1}(\Omega)$. We cannot state that f is derivable but $\xi \cdot \nabla f$ exists and belongs to $B_{r-1}(\Omega)$.

Now, from (1) and (4) we have

$$J^+ B f = \psi^+ + \phi^+$$

(10)

$$= T G_1 (1 + f)^- + T G J^- B \cup g + T G J^- B \cup A \mathscr{C} g$$

while

$$J^- B f = J^- B \cup g + J^- B \cup A C G + J^- B E T G_1 (1 + f)^-$$

(11)

$$+ J^- B E T G J^- B \cup g + J^- B E T G J^- B \cup A \mathscr{C} g$$

and

(12) $$G J^- B f = (1 + \vartheta T) [G J^- B \cup g + G J^- B \cup A T] + \vartheta T G_1 (1 + f)^-$$

so that

$$J^+ B f - G J^- B f = (1 - \vartheta) T G_1 (1 + f)^- + (T - (1 + \vartheta T)) [G J^- B \cup g +$$

(13)

$$+ G J^- B \cup A T].$$

But $T - (1 + \vartheta T) = 0$ and

$$J^+ B f - G J^- B f = G_1 (1 + f)^- \qquad (14)$$

holds in $B_r(\partial\Omega)$.

As a consequence our unique solution of 3.1.2.3.(1) is a classical solution of 3.1.1.(8).

3.5.3. Uniqueness

Consider a classical solution to 3.1.1.(8) and require that, for this solution, the following relation holds

$$((1, f)) = ((1, \& G_1 (1 + f)^-)) = 0 , \qquad (1)$$

we see that f solves 3.1.2.3.(1) and such a solution is unique with the requirement

$$<f>_r \leqslant R , \qquad (2)$$

with $R \in [R_1, R_*[$ as in 3.1.4.2.

Let $F = \omega(1 + f)$ be the corresponding solution of 3.1.1.(6) (7), from 3.1.2.1.(4) we see that the relation (1) is equivalent to

$$\int_\Omega d\nu_x \int_{R^3} F \; d\xi = |\Omega| + ((1, \& G_1 (\omega^{-1} F)^-)) . \qquad (3)$$

Observing that F is real and referring to 2.3.4.1.(12) we have

$$\int_\Omega d\vartheta_x \int_{R^3} F \; d\xi = |\Omega| + \{G_1 (\omega^{-1} F)^-, J^+ B W^* 1\} \qquad (4)$$

3.5.4. Statement of the Main Result

The collision operator is the one for rigid spheres. The reflection operator K splits as follows

$$(\omega^+)^{-1} \; K \; \omega^- = G + G_1 \qquad (1)$$

where G satisfies a number of requirements to be given below while G_1 satisfies 3.1.4.2.(4) and, of course, conveys a continuous function to a continuous one.

As for G and its adjoint G* they must satisfy 1.4.4.5.(1) (2) (3) and 1.4.5.4.(1), they must be bounded from $L^2(\partial\Omega, X)$ to $L^2(\partial\Omega, Y)$ where X and Y are any of H and H, they must meet the requirement of 2.3.5.6.(10) and the corresponding one with G* in place of G, furthermore the set of hypothesis in 3.1.6.3., 3.1.6.3.1. and 3.1.6.3.2. must be satisfied.

Concerning Ω it must be bounded, convex, with a smooth boundary in Liapounov sense and furthermore 3.1.6.2.1. and 3.1.6.2.2. hold.

Then the boundary value problem

$$\text{(2)} \qquad \begin{aligned} \xi \cdot \nabla F &= J(F, F) \\ F^{+} &= K F^{-} \end{aligned}$$

admits a classical solution (in the sense that F is derivable along rays) which is unique in

$$\text{(3)} \qquad < \omega^{-1} F - 1 >_r \leqslant R$$

with $R \in [R_1, R_*[$ as in 3.1.4.2., provided we add the requirement that 3.5.3.(4) holds.

REFERENCES

[1] H. GRAD (1958), *Principles of the Kinetic Theory of Gases*, In Handbuch der Physik, Vol. XII, Thermodynamick der Gaze, p. 205-294.

[2] C. CERCIGNANI (1972), *On the Boltzmann Equation for Rigid Spheres.* Transport Theory and Statistical Physics, 2, 3, p. 211-225, 1972.

[3] T. CARLEMAN (1957), *Problémes mathématiques dans la théorie unitique des gaz.* Publications mathématiques de l'Institut Mittag-Leffler.

[4] C. CERCIGNANI (1969), *Mathematical Methods in Kinetic Theory.* Plemum Press, 1969.

[5] H. GRAD (1963), *Asymptotic Theory of the Boltzmann Equation.* In LAURMANN Rarefied Gas Dynamics 1962, Vol. 1, p. 26-59.

[6] C. CERCIGNANI (1967), *On the Boltzmann Equation with Cut-off Potentials,* Phys. Fluids, 10, 10, p. 2097-2104, 1967.

[7] H. DRANGE (1972), *The linearized Boltzmann Collision Operator for Cut-off Potentials.* Dept. Appl. Math. Univ. Berger Rept. 37.

[8] M.N. KOGAN (1969) *Rarefied Gas Dynamics.* Plemum Press 1969.

[9] R.G. BARANTSEV (1972), *Some Problems of Gas-solid Surface Interaction. In Kucheman Progress in aerospace Suonce Volume 13, p. 1-80. Pergamon 1972.*

[10] I. KUSCER (1971), *Reciprocity in Scattering of Gas Molecules by Surfaces.* Surface Science, 25, p. 225-237, 1971.

[11] C. CERCIGNANI, M. LAMPIS (1971), *Kinetic Models for Gas Surface Interactions.* Transport Theory and Statistical Physics, 1, 2, p. 101-114, 1971.

[12] C. CERCIGNANI, *Scattering Kernels for Gas-surface Interactions.* Transport Theory and Statistical Physics, 2, 1, p. 27-53, (1972).

[13] J.S. DARROZES, J.P. GUIRAUD (1966), *Generalisation formelle du theorème H en prèsence de parois.* Comptes Rendus 262, A, p.1368-1371, 1966.

[14] K.M. CASE (1972), *Uniqueness Theorems for the Linearized Boltzmann Equation.* Physics Fluids, 15, 3, p. 377-379, 1972.

[15] J.P. GUIRAUD (1974), *An H Theorem for a Gas of Rigid Spheres in a Bounded Domain.* Communication to "Colloque international de CNRS n. 236 Théories cinétiques classiques et relativistes", Paris, Juin 1974 p. 29-58.

[16] J.H. CHONG, L. SIROVICH (1970), *Structure of Three Dimensional Supersonic and Hypersonic Flow.* Phys. Fluids 13, 8, p. 1990-1999, 1970.

[17] N. NARASIMHA (1968), *Asymptotic Solutions for the Distribution Function in Non-equilibrium Flow. Part I, The Weak Shock.* Journal Fluid Mechanics, 34, 1, p. 1-24, 1968.

[18] J.P. GUIRAUD (1970), *Problème aux Limites Intérieur pour l'équation de Boltzmann linéaire.* Journal de Mécanique, 9, 3, p. 443-490, 1970.

[19] J.M. SCHECHTER (1971), *Principles of Functional Analysis.* Academic Press, 1971.

[20] F. RIESZ, B.Sz NAGY (1953), *Leçon d'analyse fonctionelle.* Académie des Sciences de Hongrie, 1953.

[21] T. KATO (1966), *Perturbation Theory for Linear Operators.* Springer 1966.

[22] N. DUNFORD, J.T. SCHWARZ, *Linear Operators, Part 1, Part 2.* Interscience 1958, 1963.

[23] ZAANEN (19), *Linear Analysis*

[24] J. L. LIONS, E. MAGENES (1968), *Problèmes aux limites non homogènes et applications.* Vol. 1, Dunod 1968.

[25] C. CERCIGNANI (1968), *Existence, Uniqueness and Convergence of the Solutions of Models in Kinetic Theory.* Journal of Mathematical Physics, 9, 4, p. 633-639, 1968.

[26] C. CERCIGNANI (1967), *Existence and Uniqueness in the Large for Boundary Value Problems in Kinetic Theory.* Journal of Mathematical Physics, 8, 8, p. 1653-1656, 1974.

[27] C. RIGOLOT-TURBAT (1971), *Problème de Kramers pour l'équation de Boltzmann en théorie cinétique des gaz.* Comptes Rendus Ac. Sci. Paris, 273, A, p.58-61, 1971.

[28] J.P. GUIRAUD (1972), *Problème aux limites intèrieur pour l'équation de Boltzmann en régime stationaire, faiblement non linéaire.* Journal de Mécanique, 11, 2, p.183-231, 1972.

[29] H. GRAD (1965), *Asymptotic Equivalence of the Navier-Stokes and Non-linear Boltzmann Equations.* Proceedings of Symposia in Applied Mathematics, Vol. XVII. Applications of Partial Differential Equations in Mathematical Physics. p. 154-183.

[30] Y.P. PAO, *Boundary Value Problems for the Linearized and Weakly Non-linear Boltzman Equation.* Journal of Math. Physics, 8, 9, 0. 1893-1898, 1967.

[31] C. TURBAT (1968), *Equation de Boltzmann sur la droite,* Comptes Rendus Ac.Sci., Paris T266, p.258-260.

[32] C. TURBAT (1969), *Equation de Boltzmann sur la droite,* Comptes Rendus Ac.Sci., Paris T269, p.481-484.

[33] C. RIGOLOT-TURBAT (1971), *Equation de Boltzmann linéaire sur la droite,* Comptes Rendus Ac.Sci., Paris T272, p. 697-700.

[34] C. RIGOLOT-TURBAT (1971), *Equation de Boltzmann non linéaire sur la droite,* Comptes Rendus Ac.Sci., Paris T272, p.763-766.

[35] C. RIGOLOT-TURBAT (1976), *Equation de Boltzmann linéaire, non statiounairee, sur la demi-droite, avec conditions de reflexion indépendantes du temps.* Comptes Rendus T283, p.683-686.

SURFACE INTERACTION AND APPLICATIONS

Silvio Nocilla
Politecnico di Torino

1. Satellites in the Upper Atmosphere, Absorption, Molecular Beams

Let us consider a *missile moving in the upper atmosphere* at an altitude between 150 and 300 km, where the atmosphere is very rarefied, with density of the order of 10^{-9}, 10^{-10}times the normal one. It is known from the kinetic theory of gases that the mean free-path λ of the air molecules is given by the formula:

$$\lambda = \frac{1}{\sqrt{2}\pi} \frac{m}{\sigma_e^2} \frac{1}{\rho(1 + K_s/T)} \tag{1.1}$$

where:

m = mass of a molecule = $M \cdot 1.68\ 10^{-24}$ gr (M = molecular weight)
ρ = density
T = absolute temperature
K_S = Sutherland constant = 110°K
σ_e = cross section for the molecular collisions (such that a shock arises when the distance between the centers of the two molecules is less than σ_e), independent on ρ and T. It is assumed $\sigma_e^2 = 10^{-15}$ cm².

On the basis of Eq. (1.1) we obtain for the standard upper atmosphere the values of the mean free-path indicated in the *table.* The *density* ρ , the *temperature* T and the mean molecular weight M have been measured [1]. The pressure p is calculated through the formula $p = \rho RT$. As regards the number flux N see later, Eq. (1.14).

For satellites having dimension d of the order of the meter, the Knudsen number:

$$K_n = \lambda/d \tag{1.2}$$

has the same numerical values indicated in the column of the mean free paths, that is variables between 10 and 1.000 about for altitudes between 150 and 300 km. For large values of the Knudsen number:

Z Km	T day °K	T night °K	mean density gr/cm^3	M	pressure μ(Hg)	mean free-path m	unit number flux $number/cm^2 sec$
0	288		$1.22 \cdot 10^{-3}$	29.3	$760 \; 10^3$	6.10^{-8}	$2.70 \cdot 10^{23}$
50	214		$1.13 \cdot 10^{-6}$	29.0	725	$5.9 \cdot 10^{-5}$	$2.22 \cdot 10^{20}$
100	205	210	$4.97 \cdot 10^{-10}$	28.2	$2.36 \cdot 10^{-1}$	0.14	$0.97 \cdot 10^{17}$
150	800	560	$2.30 \cdot 10^{-12}$	25.0	$5.20 \cdot 10^{-3}$	50	$0.98 \cdot 10^{15}$
200	1100	900	$3.50 \cdot 10^{-13}$	22.9	$1.30 \cdot 10^{-3}$	$3.3 \cdot 10^2$	$2.65 \cdot 10^{14}$
250	1176	1010	$1.25 \cdot 10^{-13}$	20.8	$0.58 \cdot 10^{-3}$	$9.3 \cdot 10^2$	$0.95 \cdot 10^{14}$
300	1222	1021	$0.40 \cdot 10^{-13}$	19.2	$1.96 \cdot 10^{-4}$	$2.7 \cdot 10^3$	$3.30 \cdot 10^{13}$
350	1250	1022	$0.22 \cdot 10^{-13}$	18.0	$1.15 \cdot 10^{-4}$	$8 \cdot 10^3$	$2.00 \cdot 10^{13}$
400	1280	1022	$0.65 \cdot 10^{-14}$	17.1	$0.37 \cdot 10^{-4}$	$2.2 \cdot 10^4$	$0.62 \cdot 10^{13}$

Z = altitude; T = temperature; M = mean molecular weight

Table for the upper atmosphere

$$K_n > 10 \text{ about}$$

we can neglect the shocks between gas molecules. We take into account only the shocks between gas molecules and surface (free-molecule flow regime).

What are the *physical properties of these shocks?* What are the mathematical instruments able to describe them and to carry out the *necessary statistics?* What are the *macroscopic magnitudes* having practical interest and which can be measured? These are the basic problems in the surface interaction. In order to give an appropriate answer we first observe that in virtue of the physical and chemical conditions of the surface, and essentially of the surface force-field there is a probability that the impinging molecules be "captured" by the solid surface, and stay on it. During this staying period or *sitting time* other gas molecules impinge on the surface, undergo vicissitudes analogous to the previous one, and are staying together, on the same solid surface. Now it is clear that this capture of molecules from the surface cannot continue indefinitely, because otherwise the gas impinging on the satellite would condensate continually on it. We must admit, in agreement with the experience, that the molecules, after a period of staying on the surface, are re-emitted into the surrounding gas, and that in stationary conditions as we suppose, a *statistical equilibrium* arises *between impinging molecules, staying molecules, and re-emitted* (or re-evaporated) *molecules.*

There are thus three basic phases in the interaction phenomena:

– incidence of the molecules on the surface
– sitting on the surface
– re-emission from the surface

Each of them determines in turn the conditions for the development of the subsequent phases. The incidence phases is usually known, whereas the other two phases constitute the fundamental unknown question of the problem.

The phenomenon we have now qualitatively described does not differ substantially from the well known phenomenon of the *"absorption"*, or more precisely *"physisorption"*. As regards the absorption I want to remind another aspect, which is more specifically related to the *chemical absorption,* as a suitable element characterising the chemical and physical state of the surface of the missile. Precisely, the outside surfaces of the missiles have a chemical and physical structure which is very complicated and perhaps not well known at the molecular range, with chemically and physically absorbed substances. Moreover, this chemical and physical structure may change when the missile passes from the altitude zero, at the ordinary pressure, to those ones of the highly rarefied atmosphere. Now it is well known that the presence of mono- or plurimolecular layers chemically or physically absorbed greatly affects the surface force fields.

The same process of incidence and sitting of the molecules on the surface, together with their subsequent re-emission, occurs when a *molecular beam* impinges on a rigid surface and is subsequently re-emitted.

We have so pointed out a *standard basic phenomenon* which originates from the behaviour of the gas molecules interacting with the atoms of a rigid surface and which, on account of this, must be studied in the light of this interaction process. Evidently, the same basic phenomenon takes very different aspects, especially from the experimental point of view, whether *absorption, or molecular beams, or aerospace applications* are concerned. However it is equally evident that the unitary ground of these three and apparently so different fields on the one hand is an effective instrument for studying them, on the other hand it requires that the knowledge acquired about each one must be taken into account in studying the other ones.

According to the description we made of the basic interaction phenomenon, *three magnitudes* are basically to be considered in its statistical equilibrium:

N = number of molecules striking the unit surface in the unit time (*number flux)*

τ*= *sitting time* of the absorbed molecules on the surface

N*= *number of molecules absorbed* (or condensed) in the unit surface

In statistical equilibrium conditions the three above magnitudes are not dependent on each other, but they are related by the following fundamental condition:

$$N^* = N \cdot \tau^* \qquad (1.3)$$

As concerns the sitting time, if for the absorbed molecules we assume the Frenkel [2] model according to which they behave as armonic oscillators moving in the direction normal to the absorbing surface, the sitting time is given by the formula:

$$\tau^* = \tau^{**} \exp(Q/RT_w) \qquad (1.4)$$

where Q is the absorption heat and R the gas constant of the incident gas, T_w the absorbing solid temperature and τ^{**} a parameter having the dimension of a time, strictly related to the vibration period of the atoms which constitute the absorbing surface, and having values of the same order as this period, namely 10^{-12} or 10^{-14} seconds about. *The value of the sitting time* τ^* is very much variable, for any temperature, according to the absorbed gas. For instance (see De Boer [3] page 35) for room temperature we have the following values:

10^{-12} sec for *hydrogen* on various surfaces

10^{-10} sec for *argon* Ar, *oxygen* O_2, *nitrogen* N_2, *carbon monoxide* CO on various surfaces

and hence 1000 (thousand) times about the oscillation period. For gases formed by *heavier* molecules we can have very greater sitting times, till 10^{-2} sec.

As regards the other two magnitudes indicated in the formula (1.3) we report now some first informations.

The amount of *absorbed molecules* N^* depends, as it is well known, on the surface temperature T_w and on the surrounding rarefied pressure p. As an example we report one of the Langumir absorption isotherms (fig. 1): a^* is the ratio between

Fig. 1 – Absorption isotherm

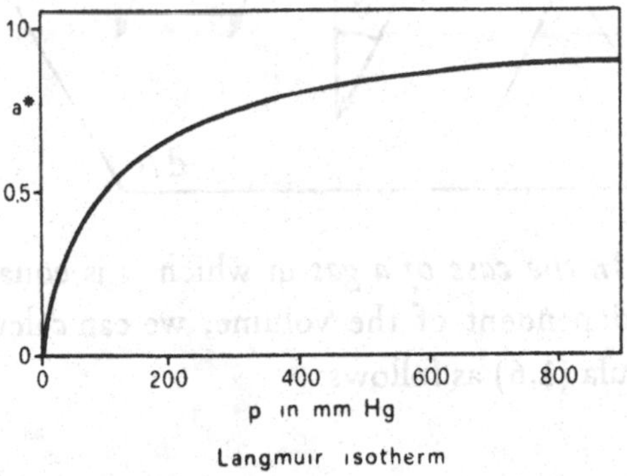

the number of the actually absorbed molecules per cm^2 surface and the number of molecules per cm^2 which would form a completely filled unimolecular layer. It is known too that the absorption heat Q is defined as the amount of heat which is lost by the gas when its molecules are absorbed or "condensed" over the surface, or has to be provided to the absorbed molecules for desorbing or "re-evaporating" them.

Finally, as regards the *number flux* N we observe first of all that in the surface interaction phenomena we have two distinct number fluxes: the first one for the impinging molecules, we call N_i and the second one for the re-emitted ones, we call N_r. But in statistical equilibrium as we suppose it results always:

$$N_i = N_r = N \tag{1.5}$$

In the case of a uniform and monoenergetic stream impinging on a surface dA (see fig. 2) with velocity U and with the angle ϑ with the normal n we can define the *number density* ν (number of molecules per unit volume) independently of the dimensions and of the form of this volume. The number flux N is obtained by:

$$N = U \cos\vartheta \cdot \nu \qquad (\nu = \rho/m) \tag{1.6}$$

and in particular for normal incidence:

$$N = U \cdot \nu \tag{1.7}$$

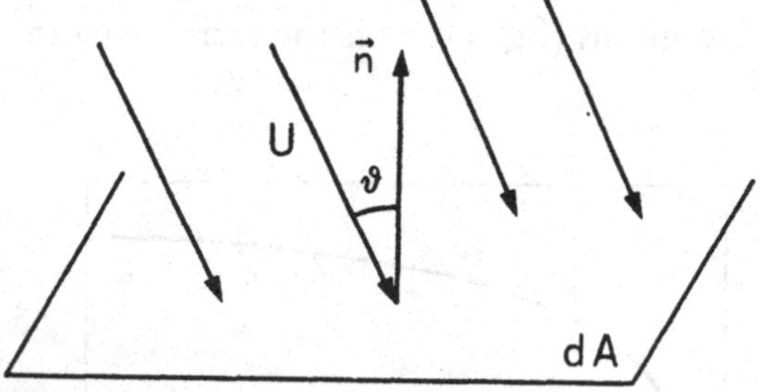

Fig. 2 – Uniform and monoenergetic stream

In the case of a gas in which it is equally possible to define a number density ν independent of the volume, we can calculate the number flux N by generalizing formula (1.6) as follows:

$$dN = -\ w\, d\nu^{(u,v,w)} \qquad (W \leqslant 0) \tag{1.8}$$

where $d\nu^{(u,v,w)}$ indicates the fraction of the ν molecules contained in the volume:

$$dS^{(\vec{x})} = dx\ dy\ dz$$

of the ordinary space having velocity in the volume

$$ds^{(V)} \equiv dS^{(u,v,w)} = du\ dv\ dw$$

of the velocity space.

For a gas in *Maxwellian equilibrium* we have, as well known:

$$d\nu^{(u,v,w)} = \frac{\nu}{\pi^{3/2}} f(V)\ du\ dv\ dw / c^3 \tag{1.9}$$

with:

$$f(V) = \exp(-V^2 / c^2) \tag{1.10}$$

where c is the most probable velocity. The total incident number flux is given by:

$$N = \int_{w \leq 0} (-w)\ d\nu^{(u,v,w)} \tag{1.11}$$

Taking into account Eqs. (1.9) and (1.10) and carrying out the integration in *cartesian coordinates* we have:

$$N = \int_{-\infty}^{+\infty} du \int_{-\infty}^{+\infty} dv \int_{-\infty}^{0} f(V) \cdot (-w)\ dw \tag{1.12}$$

and finally:

$$N = \nu \sqrt{\frac{RT}{2\pi}} = \frac{\nu \cdot c}{2 \cdot \sqrt{\pi}} \tag{1.13}$$

By comparing this formula with Eq. (1.7) we see that if the number density ν is the same in the two examples, in order to obtain also the same number flux N it must be:

$$U = \frac{c}{2\sqrt{\pi}} = 0.281 \cdot c$$

As an *example* we report here some numerical values of the number flux N. *In normal conditions* (temperature 20° C and pressure 760 mm Hg) we have (see for instance De Boer [3] page 7):

hydrogen (H_2) $\begin{cases} N = 11.0 \ 10^{23} \text{ molecules/cm}^2 \text{ sec} \\ N/N_{Av} = 1.82 \text{ gram molecules/cm}^2 \text{ sec} \end{cases}$

nitrogen (N_2) $\begin{cases} N = 2.94 \ 10^{23} \text{ molecules/cm}^2 \text{ sec} \\ N/N_{Av} = 0.487 \text{ gram molecules/cm}^2 \text{ sec} \end{cases}$

oxygen (0_2) $\begin{cases} N = 2.75 \ 10^{23} \text{ molecules/cm}^2 \text{ sec} \\ N/N_{Av} = 0.456 \text{ gram molecules/cm}^2 \text{ sec} \end{cases}$

At the *altitude of 250 Km,* assuming for the temperature T the mean value of 1093° K (see table of page 2) and for the density ρ a value of $1.02\ 10^{-10}$ times smaller than the normal one, we obtain that the above values must be multiplied by the factor:

$$1.02 \cdot 10^{-10} \sqrt{\frac{1093}{288}} \cong 2 \cdot 10^{-10}$$

Then we have the following values for the molar fractions:

hydrogen (H_2) $N/N_{Av} = 3.64 \ 10^{-10}$ gram molec/cm^2 sec

nitrogen (N_2) $N/N_{Av} = 0.974 \ 10^{-10}$ gram molec/cm^2 sec

oxygen (O_2) $N/N_{Av} = 0.912 \ 10^{-10}$ gram molec/cm^2 sec

2. Density Statistics and Flux Statistics

The start point for the construction of a *density statistics* is the consideration of the molecules contained in the element $dS^{(\vec{X})}$ of the physical space at time t.

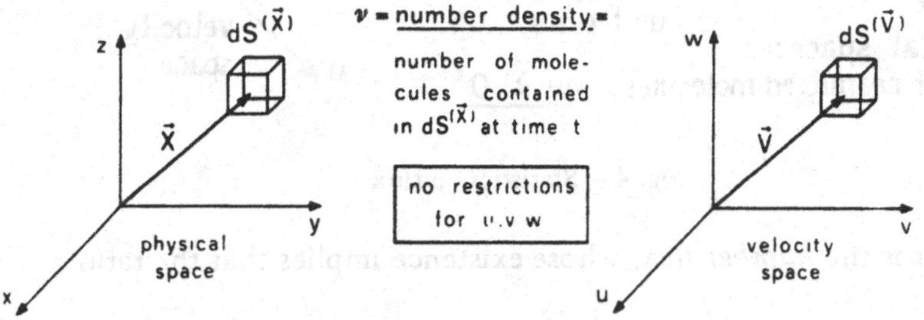

Fig. 3 – Statistics on density

This number is the ***number density* ν** whose existence implies that the ratio:

$$\frac{\text{number of molecules contained in a volume}}{\text{volume}}$$

be independent of the form of this volume (cube, parallelepiped, sphere, cylinder and so on). In the first lesson we considered two elementary examples in which this condition was satisfied. There are naturally infinite other cases in which the same property is true, as for instance the gases usually studied by the Boltzmann Equation. In these cases we can construct a statistics by means of the formula, strictly analogous to the Eq. (1.9) ($\vec{X}$ is the generical vector in the ordinary space):

$$\frac{d\nu^{(\vec{V})}}{\nu} = f(\vec{X}, \vec{V}, t) \frac{dS^{(\vec{V})}}{c^3} \tag{2.1}$$

where c is a suitable normalization constant. The distribution function f is defined, for any $\vec{X}$ and any time t, *in the whole velocity space.*

On the contrary the start point for the construction of a *flux statistics* is the consideration of the molecules leaving (or impinging) *in a time range* dt the elementary surface dA.

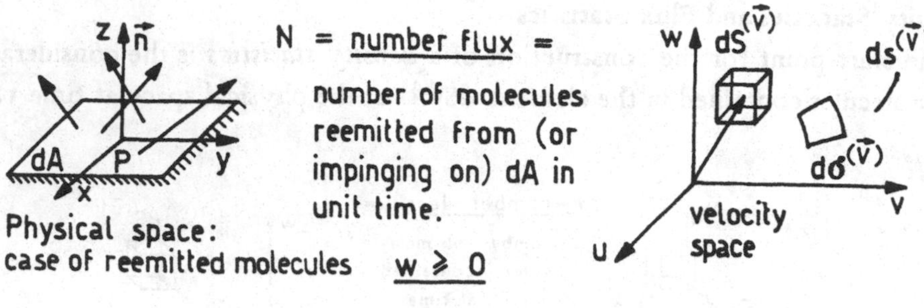

Fig. 4 – Statistics on flux

This number is the *number flux,* whose existence implies that the ratio:

$$\frac{\text{number of molecules crossing a surface}}{\text{surface}}$$

be independent of the form of this surface (square, rectangle, circle and so on). It is fundamental to point out that, although the number flux is usually related to the number density, *the consideration of the number flux N does not require the existence of a number density* ν in a volume element adjacent to the surface element dA. The construction of the flux statistics is carried out through the formula:

$$\frac{dN^{(\vec{V})}}{N} = g(P, \vec{V}, t)\,\frac{dS^{(\vec{V})}}{\cdot b^3} \tag{2.2}$$

where b is a suitable normalization velocity. The function $g(P,\vec{V},t)$ differently from $f(\vec{X},\vec{V},t)$, is defined (see fig. 4), for every point P on the surface and every time t, *in the half-space* $w \geq 0$ *of the velocity space,* corresponding to the external normal to the surface element dA relative to the point P when we consider the molecules *leaving* the surface; in the half-space $w \leq 0$, corresponding to the internal normal, when we consider the *impinging* molecules.

Let us now ask *which statistics must be employed in the interaction phenomena?* It appears obvious from the considerations developed in the first lesson that the *flux statistics naturally occurrs* in the surface interaction phenomena, since the number flux and not the number density naturally occurs in these problems. But I think that *this choice* cannot only be *suitable*, but also *necessary*. In fact it is evident that the study of the interaction with the surface element dA, must involve *only* molecules impinging, and subsequently re-emitted *from the element* dA. Now, when the statistics is performed on the density this condition does not appear to be satisfied, because among the molecules contained at time t in the element $dS^{(\vec{X})}$ adjacent to the surface element dA (see fig. 5) there is also the white one which, because of its velocity $\vec{V}_i$, impinges on a point P_i outside the surface element dA hence extraneous to the interaction phenomenon taking place on dA ; and the black one coming from point P_r hence equally extraneous to the interaction taking place on dA. In order to overcome to this difficulty it appears necessary to make suitable choices on the form of the volume element adjacent to dA.

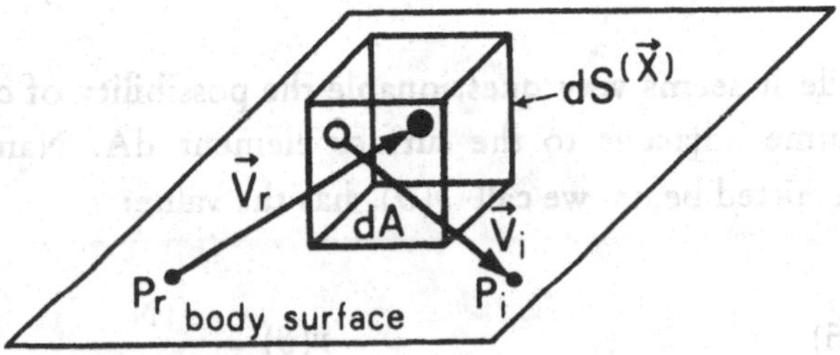

Fig. 5. – White and black molecules, though contained in $dS^{(\vec{X})}$, are extraneous to the interaction phenomena taking place in dA

A further argument in this question can be done by the following example. Let us consider a monoenergetic uniform beam incident on a surface element dA and subject, in the passage through dA to a deviation ϑ , whereas the velocity V keeps unchanged (see fig. 6). For the impinging molecules we can define near dA both the number flux N and the number density ν such that:

$$N = \nu \cdot V \tag{2.3}$$

For the deviated, or re-emitted molecules, we can consider the number flux:

$$N_r = N \tag{2.4}$$

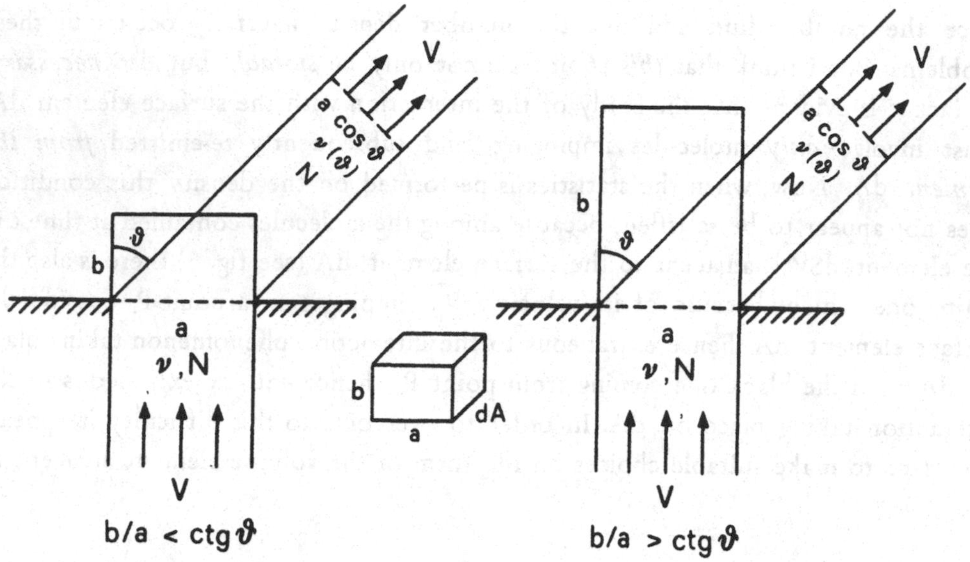

Fig. 6 – Uniform beam deflected by a surface element dA

while it seems very questionable the possibility of considering a number density in a volume adjacent to the surface element dA. Namely the number density of the re-emitted beam, we call $\nu(\vartheta)$, has the value:

$$\nu(\vartheta) = \frac{N}{V \cos \vartheta} \tag{2.5}$$

But this number density, as it appears from Eq. (2.5), fundamentally depends on the deviation angle ϑ and cannot be considered as a number density *adjacent to the surface element dA.* Hence we consider for instance a volume element having the form of a parallelepiped, with basis dA and a height b. This element has the volume:

$$dS = dA \cdot b$$

Now a simple calculation shows that the *effective density* in the volume element dS defined as follows:

$$\nu_r = \frac{\text{number of re-emitted molecules contained in dS}}{dS}$$

strictly depends on the height b in the following way:

$$\frac{\nu_r}{\nu} = \frac{1}{\cos\vartheta}\left(1 - \frac{1}{2}\,\frac{b}{a}\,\mathrm{tg}\,\vartheta\right) \qquad \text{for } b/a < \mathrm{ctg}\,\vartheta$$

$$\frac{\nu_r}{\nu} = \frac{a}{b}\;\frac{1}{2\sin\vartheta} \qquad \text{for } b/a > \mathrm{ctg}\,\vartheta \tag{2.6}$$

The curve of the ratio ν_r/ν as a function of the ratio b/a for some values of ϑ is plotted in fig. 7.

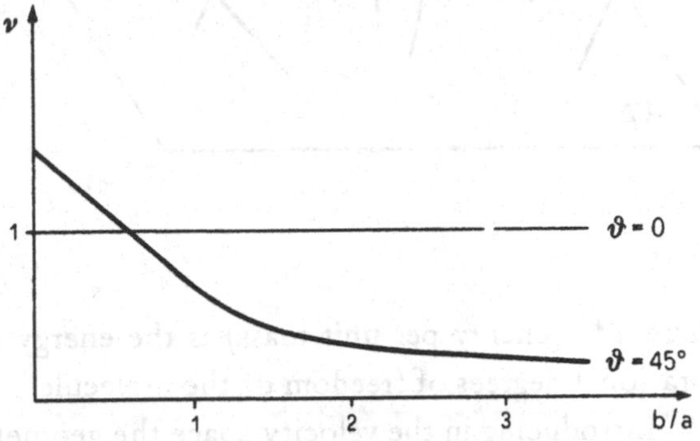

Fig. 7 – The ratio ν_r/ν as depending on the shape-parameter b/a of the volume element adjacent to the surface element dA

From the above considerations it appears that whereas from a density statistics in a volume dS it is possible to construct a flux statistics through a surface in this volume, from a flux statistics through a surface dA it is not possible, in general, to construct a density statistics in a volume adjacent to this surface.

The consideration of the flux statistics allows to introduce the concepts of *free-molecule sources and sinks* [4], characterized by the three following physical basic element (fig. 8):

1) its *location in the ordinary space* and the *orientation* of its *normal* $\vec{n}$
2) the *velocity distribution function* $g_r(\vec{V})$, for the source, or $g_i(\vec{V})$, for the sink
3) its temperature T_w

For each of these sources, or sinks, it is possible to introduce *in unitary way* also the elementary ***momentum flux*** and ***energy flux*** related to the elementary number flux $dN(\vec{V})$, given by Eq. (2.2), by the formulae:

(2.7) $$dQ^{(\vec{V})} = m\vec{V}\ dN^{(\vec{V})}$$

(2.8) $$dE^{(\vec{V})} = m(\frac{1}{2}V^2 + e^*)\ dN^{(\vec{V})}$$

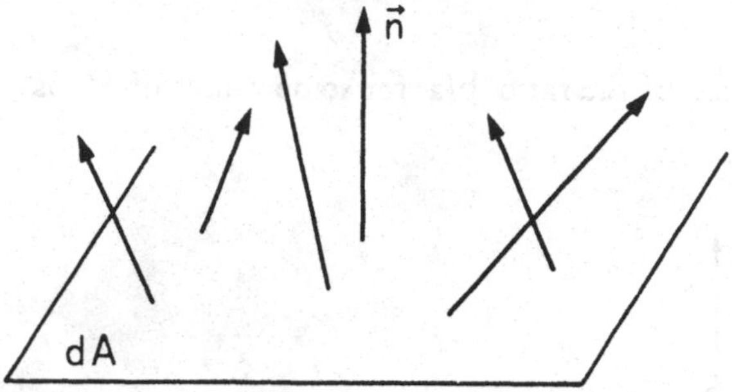

Fig. 8 – Free-molecule surface source

where e^* (energy per unit mass) is the energy corresponding to the rotational and vibrational degrees of freedom of the molecules.

Introducing in the velocity space the geometry in polar coordinates indicated in fig. 9 we have:

(2.9) $$u = V \sin\Theta \cos\phi \qquad\qquad 0 \leqslant \Theta \leqslant \pi$$
$$v = V \sin\Theta \sin\phi \qquad \text{with} \qquad 0 \leqslant \phi < 2\pi$$
$$w = V \cos\Theta$$

from which:

(2.10) $$dS^{(V)} = V^2 \sin\Theta\ dV\ d\Theta\ d\phi = V^2\ dV\ d\Omega$$

where:

(2.11) $$d\Omega = \sin\Theta\ d\Theta\ d\phi = \text{elementary solid angle}$$

and:

$$\vec{V} = V\,\vec{\omega} \tag{2.12}$$

with:

$$\vec{\omega} = \vec{\omega}(\Theta, \phi) = (\sin\Theta\cos\phi,\ \sin\Theta\sin\phi,\ \cos\Theta) \tag{2.13}$$

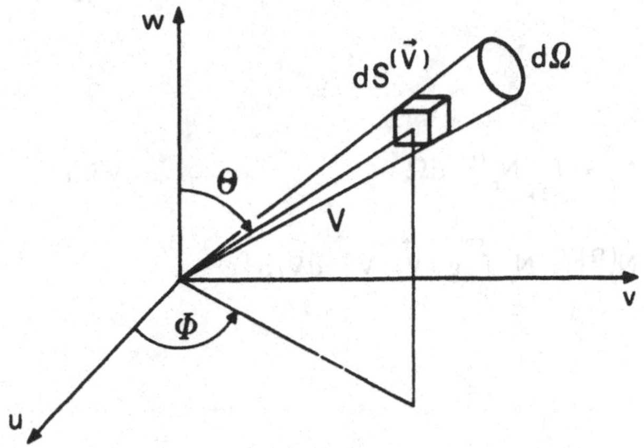

Fig. 9 – Geometry in polar coordinates in the velocity space

unitary vector in the direction of the velocity $\vec{V}$. Taking into account Eq. (2.10) we obtain from Eq. (2.2) the velocity distribution function along each direction

$$\frac{1}{d\Omega}\ \frac{dN^{(\vec{V})}}{N} = g(\vec{V})\, V^2\, dV$$

According to the above geometry each of the fluxes relative for instance to the re-emitted molecules, usually calculated in cartesian co-ordinates as follows:

(2.14) $$\int_{w \geqslant 0} (\ldots)\, dS^{(\vec{V})} = \int_0^{\infty} dw \int_{-\infty}^{+\infty} dv \int_{-\infty}^{+\infty} (\ldots)\, du$$

can be calculated in polar coordinates as follows:

(2.15) $$\int_{w \geqslant 0} (\ldots)\, dS^{(\vec{V})} = \int_{\Omega +} \left[\int_0^{\infty} (\ldots)\, V^2\, dV \right] d\Omega$$

Applying this procedure to the *number, momentum* and *energy flux* we have:

(2.16) $$\begin{cases} N_r = \int_{\Omega +} N_r^{(\Omega)}\, d\Omega \\ N_r^{(\Omega)} = N_r \int_0^{\infty} g_r(\vec{V})\, V^2\, dV/b^3 \end{cases}$$ with:

(2.17) $$\begin{cases} \vec{Q}_r = \int_{\Omega +} \vec{Q}_r^{(\Omega)}\, d\Omega \\ \vec{Q}_r^{(\Omega)} = mN_r \int_0^{\infty} g_r(\vec{V})\, V^3\, dV/b^3 \,.\, \vec{\omega} \end{cases}$$ with:

(2.18) $$\begin{cases} E_r = \int_{\Omega +} E_r^{(\Omega)}\, d\Omega \\ E_r^{(\Omega)} = m\,N_r \int_0^{\infty} (V^2/2 + e_r^*)\, g_r(\vec{V})\, V^2\, dV/b^3 \end{cases}$$ with:

3. Application of the Drift-Maxwellian Velocity Distribution F ınction to the Gas-surface Interaction Phenomena

Let us consider a surface element dA with a normal $\vec{n}$ moving in a rarefied gas with a velocity $\vec{U}$. In the motion relative to dA the velocity distribution function is drift-Maxwellian, characterized as follows:

$$\frac{d\nu_i^{(\vec{V})}}{\nu} = f_i(\vec{V})\ \frac{dS^{(\vec{V})}}{c^3} \tag{3.1}$$

with:

$$f_i(\vec{V}) = \frac{1}{\pi^{3/2}} \exp\left(-\left|\vec{V}-\vec{U}_i\right|^2/c^2\right) ; \qquad \vec{U}_i = -\vec{U} \tag{3.2}$$

The index i refers to the fact that we consider now *incident* molecules. We can go from the above density statistics to the corresponding flux statistics for the molecules impinging on the surface element dA through the basic formula:

$$dN_i^{(\vec{V})} = -w\ d\nu_i^{(\vec{V})}, \qquad -w = V\cos\Theta \geqslant 0 \tag{3.3}$$

Taking into account Eqs. (3.1) and (3.2) we have:

$$dN_i^{(\vec{V})} = -w\ \frac{1}{\pi^{3/2}}\ \nu\ \exp\left(-|\vec{V}-\vec{U}_i|^2/c^2\right) dS^{(\vec{V})}/c^3 \tag{3.4}$$

Introducing the total flux through the unit surface in unit time:

$$N_i = \int_{w\leqslant 0} dN_i^{(\vec{V})} \tag{3.5}$$

we obtain finally:

$$\frac{dN_i^{(\vec{V})}}{N_i} = g_i(\vec{V})\ dS^{(\vec{V})}/c^3 \tag{3.6}$$

with:

$$g_i(\vec{V}) = \frac{2}{\pi} \frac{-w \exp(-|\vec{V}-\vec{U}_i|^2/c^2)}{\bar{c}\,\chi} \qquad (w \leqslant 0) \tag{3.7}$$

where $\pi/2\,\chi$ indicates the integral:

$$\frac{\pi}{2}\chi = \int_{w \leqslant 0} \frac{w}{c} \exp(-|\vec{V}-\vec{U}_i|^2/c^2)\, dS^{(\vec{V})}/c^3 \tag{3.8}$$

Carrying out the integration in cartesian coordinates (as already made, Eq. (1.12)) and setting:

$$\begin{aligned} s_i &= U_i/c \\ \sigma_i &= s_i \cos \vartheta_i \end{aligned} \tag{3.9}$$

we obtain:

$$\chi = \chi(\sigma) = \exp(-\sigma^2) + \sqrt{\pi}.\sigma.(1 + \operatorname{erf} \sigma) \tag{3.10}$$

The total flux N_i given by Eq. (3.5) is obtained through Eqs. (3.4) and (3.8) and has the value:

$$N_i = N\,\chi(\sigma) \qquad (N = \frac{\nu . c}{2 . \sqrt{\pi}}) \tag{3.12}$$

where N is the total flux corresponding to $U_i = 0$, already considered and given by Eq. (1.13).

Fig. 10 shows the function $\chi(\sigma)$ given by Eq. (3.10).

Fig. 10 – Function $\chi(\sigma)$, formula (3.10)

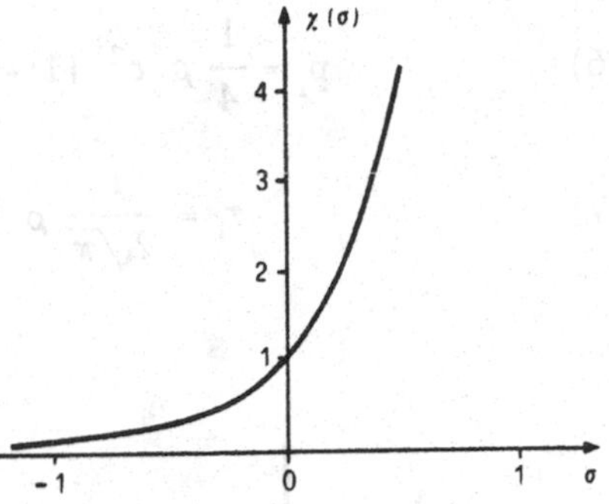

The basic formulae (3.6) and (3.7) allow to calculate the following two fundamental magnitudes: the ***momentum flux*** (m is the mass of a molecule):

$$\vec{Q}_i = \int_{w \leqslant 0} m \, \vec{V} \, dN_i^{(\vec{V})} = m \, N_i \int_{w \leqslant 0} \vec{V} \, g_i(\vec{V}) \, dS^{(\vec{V})} / c^3 \qquad (3.13)$$

the *energy flux:*

$$E_i = \int_{w \leqslant 0} \left(\frac{1}{2} m \, V^2 + e_i^*\right) dN_i^{(\vec{V})} =$$

$$= m \, N_i \int_{w \leqslant 0} (V^2/2 + e_i^*) \, g_i(\vec{V}) \, dS^{(\vec{V})} / c^3 \qquad (3.14)$$

where e_i^* (energy per unit mass) is the energy corresponding to the rotational and vibrational degrees of freedom, which we cannot discuss here in detail.

The above fluxes are *global fluxes,* that are obtained through the contribution of the molecules impinging on the surface from *any direction.* They are calculated carrying out the integrations in cartesian coordinates (see Eq. (1.12)). The integration gives the following results:

$$\vec{Q}_i = p_i \, \vec{n} + \tau_i \, \vec{t} \qquad (3.15)$$

where $\vec{n}$ and $\vec{t}$ are the unitary vectors normal and tangent to the surface indicated in fig. 11, and where the normal and tangential stresses p_i and τ_i have the values:

(3.16) $$p_i = \frac{1}{4}\rho\, c^2\, [1 + \operatorname{erf}\sigma_i + 2\frac{\sigma_i}{\sqrt{\pi}}\chi(\sigma_i)]$$

(3.17) $$\tau_i = \frac{1}{2\sqrt{\pi}}\rho\, c^2\, \chi(\sigma_i)\, s_i \sin\vartheta_i$$

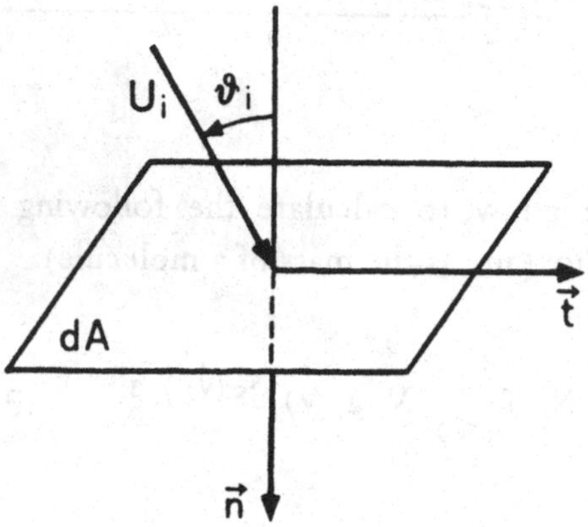

Fig. 11

where ρ indicates the density defined by:

(3.18) $$\rho = m\,\nu$$

It is interesting to remark that *for grazing incidence* (ϑ_i = 90°) the normal stress p_i has the value:

(3.19) $$p_i = \frac{1}{4}\rho\, c^2 = \sqrt{\pi}\, m\, c\, N$$

independent on the drift velocity U_i, and hence the same value as for $U_i = 0$; the tangential stress τ_i on the contrary depends on the drift velocity U_i as follows:

(3.20) $$\tau_i = \frac{1}{2\sqrt{\pi}}\rho\, c\, U_i = m\, N\, U_i$$

For the energy flux the analogous integration in cartesian coordinates gives the following result:

$$E_i = N\ k\ T\ f(s_i, \vartheta_i) \qquad \begin{matrix}(k = \text{Boltzman constant}) \\ (kT = 1/2\ mc^2)\end{matrix} \tag{3.21}$$

with:

$$f(s_i, \vartheta_i) = (2 + \delta + s_i^2)\ \chi\ (\sigma_i) + \frac{\sqrt{\pi}}{2}\ \sigma_i(1 + \operatorname{erf} \sigma_i) \tag{3.22}$$

For *grazing incidence* ($\vartheta_i = 90°$) the energy flux (3.21) reduces to:

$$\begin{aligned} E_i &= N\ k\ T\ (2 + \delta + s_i^2) \\ &= 1/2\ mc^2\ N(2 + \delta + s_i^2) \end{aligned} \tag{3.23}$$

As a second application of the drift-Maxwellian velocity distribution function we consider the re-emission model proposed by myself in 1962. The fundamental characteristic of the model is to describe *in unitary way* the properties of the *fluxes of re-emitted molecules,* of *their momenta* and of *their energies, in all directions starting from the surface element* *dA*. This requires that the calculations of these fluxes be carried out in *polar coordinates,* starting from a common velocity distribution function for the flux statistics.

The statistics employed in the model are characterized as follows:

$$dN_r^{(\vec{V})} = N_r\ g_r^{(\vec{V})}\ dS^{(\vec{V})}/b^3 \qquad (N_r = N_i) \tag{3.24}$$

$$g_r^{(\vec{V})} = \frac{2}{\pi}\ \frac{w}{b}\ \frac{\exp(-|\vec{V} - \vec{U}_r|^2/b^2)}{\chi(\sigma_r)} \tag{3.25}$$

$$\begin{cases} s_r = U_r/b \\ \sigma_r = s_r \cos\vartheta_r \end{cases} \tag{3.26}$$

where $\vec{U}_r$ and b are arbitrary parameters having the dimension of a velocity; ϑ_r is another arbitrary parameter, which represents the angle between the drift velocity $\vec{U}_r$ and the outward normal to the surface. The function $\chi(\sigma)$ is defined by Eq. (3.10) and plotted in fig. 10.

The elementary *momentum flux* and *energy flux* are related to the elementary number flux (3.24) by the formulae:

$$d\vec{Q}_r^{(\vec{V})} = m\vec{V}\, dN_r^{(\vec{V})} \tag{3.27}$$

$$dE_r^{(\vec{V})} = \frac{1}{2}(mV^2 + e_r^*)\, dN_r^{(\vec{V})} \tag{3.28}$$

With the general technique indicated previously we obtain the following results:

$$N_r^{(\Omega)} = \frac{1}{\pi} N_r \frac{\exp(-s_r^2)}{\chi(\sigma_r)} \cos\Theta\, [1 + F(X_r)] \tag{3.29}$$

with:

$$F(X) = X^2 + \sqrt{\pi}\,(\frac{3}{2} X + X^3) \exp X^2\, (1 + \mathrm{erf}\, X)$$

$$\vec{Q}_r^{(\Omega)} = \frac{3}{4\sqrt{\pi}}\, m\, b\, N_r \frac{\exp(-s_r^2)}{\chi(\sigma_r)} \cos\Theta\, [1 + G(X_r)] \,.\, \vec{\omega} \tag{3.30}$$

with:

$$G(X) = (1 + 4X^2 + \frac{4}{3} X^4) \exp X^2\, (1 + \mathrm{erf}\, X) +$$

$$+ \frac{4}{3\sqrt{\pi}}(5X + 2X^3) - 1$$

$$E_r^{(\Omega)} = \frac{1}{\pi} m\, b^2\, N_r \frac{\exp(-s_r^2)}{\chi(\sigma_r)} \cos\Theta \;\; [1 + K(X_r)] + + \frac{e_r^*}{b^2} [1 + F(X_r)] \tag{3.31}$$

with:

$$K(X) = \frac{9}{4} X^2 + \frac{1}{2} X^4 + + \frac{\sqrt{\pi}}{2} (\frac{15}{4} X + 5\, X^3 + X^5) \exp\, X^2 \;\; (1 + \operatorname{erf} X)$$

and where X_r has the value:

$$X_r = s_r (\sin\vartheta_r \;\; \sin\Theta \;\; \cos\phi + \cos\vartheta_r \; \cos\Theta) \qquad (0 \leqslant \Theta \leqslant \pi/2) \tag{3.32}$$

and hence also depends on the generic re-emission direction $\vec{\omega}(\Theta, \phi)$.

In the incidence plane, that is the one containing the normal $\vec{n}$ and the drift velocity $\vec{U}_r$ (see fig. 12) it results $\phi = 0$ for $u > 0$ and $\phi = \pi$ for $u < 0$. Consequently (3.32) becomes:

$$X_r = s_r \cos(\Theta \mp \vartheta_r) \tag{3.32'}$$

for $u \gtrless 0$, that is $\phi = \genfrac{}{}{0pt}{}{0}{\pi}$

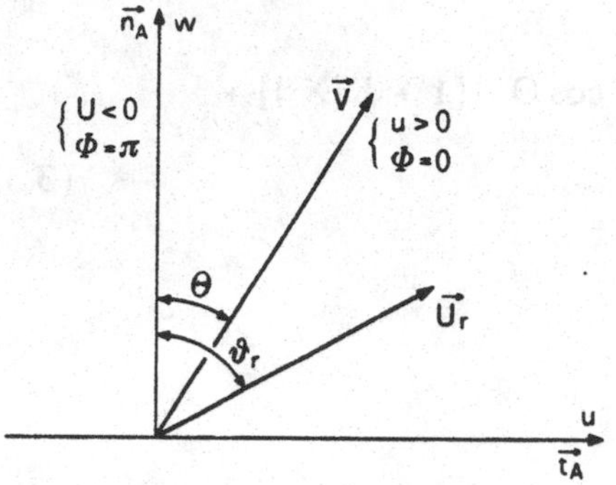

Fig. 12 – Geometry in the incidence plane

As regards the above re-emission model we remark:

first: for $U_r = 0$ we have $s_r = 0$, $X_r = 0$, and hence the classical *diffuse re-emission* considered in its most complete aspect, and not only as "cosinus law" as usual. Thus for instance it appears immediately from the formulae (3.30) and (3.31) that in this case not only the number flux, but also the momentum and the energy fluxes satisfy to the cosinus law. Furthermore the general obtained formulae give also the velocity distribution functions in any direction.

second: the model is a *spatial model* since it gives the various velocity distribution functions in each azimutal plane, characterized by the angle ϕ .

third: in particular, for each direction (Θ, ϕ) it is possible to calculate the most probable velocity, the mean velocity, and so on. For instance

the most probable velocity q_M, i.e. corresponding to the maximum point of the generical Gaussian, is given by:

(3.33)
$$q_M = q_M(\Theta, \phi, s_r, \vartheta_r) = \frac{1}{2}\left(X_r + \sqrt{X_r^2 + 6}\right)$$

where X_r is given by Eq. (3.32), and where q means the velocity ratio:

$$q = V/b$$

the mean velocity $\bar{q}$ defined by:

$$\bar{q} = \frac{\int_0^\infty q^3\, g_r(\vec{V})\, dq}{\int_0^\infty q^2\, g_r(\vec{V})\, dq} = \frac{\int_0^\infty q^4 \exp(-q^2 + 2\,q\,X_r)\, dq}{\int_0^\infty q^3 \exp(-q^2 + 2\,q\,X_r)\, dq} \tag{3.34}$$

The figures 13, 14 and 15 show some values calculated through the above formulae in the incidence plane, that is in the conditions indicated by Eq. (3.32') (pag. 27).

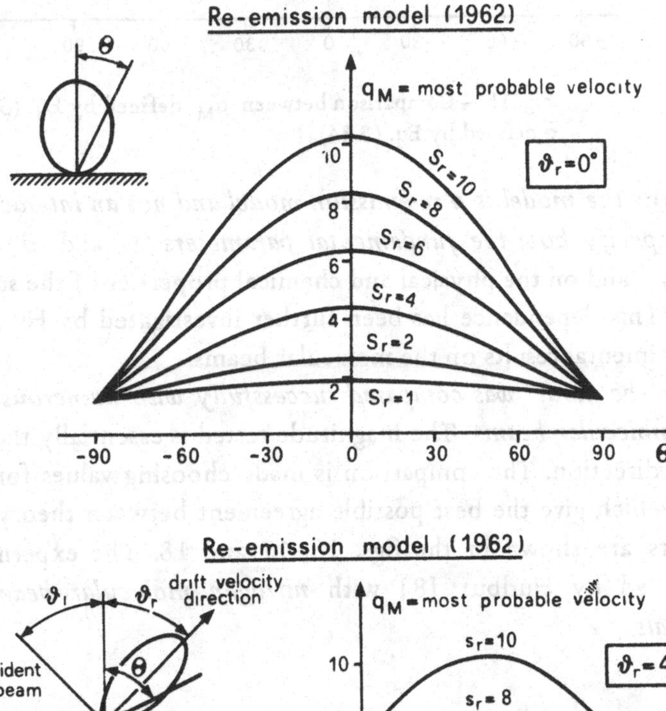

Fig. 13 – q_M versus Θ for $\vartheta_r = 0°$

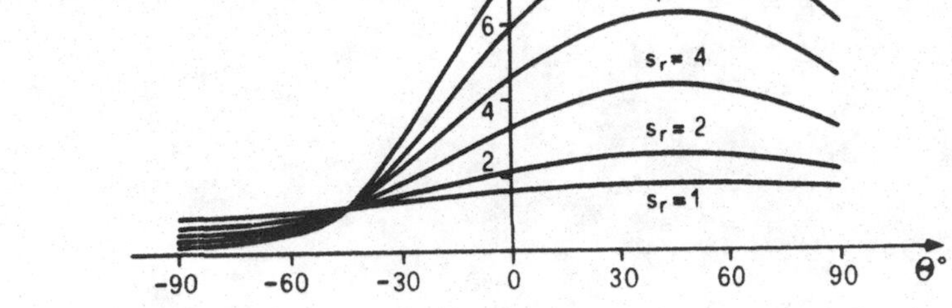

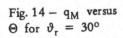

Fig. 14 – q_M versus Θ for $\vartheta_r = 30°$

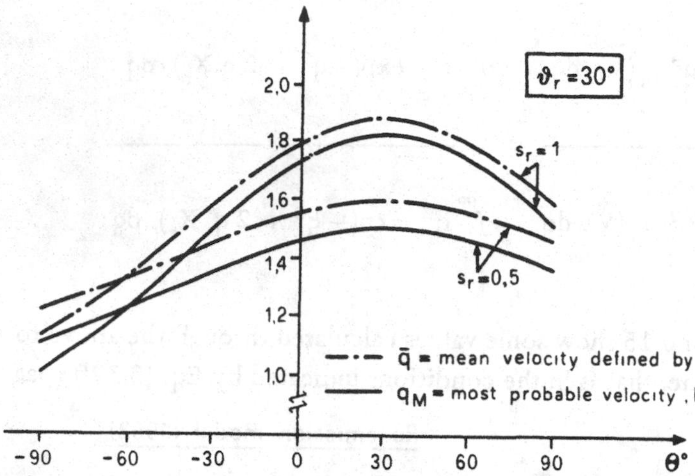

Fig. 15 – Comparison between q_M defined by Eq. (3.33) and $\bar{q}$ defined by Eq. (3.34)

fourth: *the model is a re-emission model and not an interaction model, since it does not specify how the fundamental parameters* s_r and ϑ_r depend on the function $g_i(\vec{V})$, and on the physical and chemical properties of the surface and of the gas.

This dependence has been further investigated by Hurlbut [7] on the basis of experimental results on the molecular beams.

fifth: *the model was compared successfully with numerous experimental results on the molecular beams.* The magnitude tested is essentially the number flux $N_r^{(\Omega)}$ in each direction. The comparison is made choosing values for the parameters s_r and ϑ_r which give the best possible agreement between theory and experiment. Some results are shown in the figs. 16, 17 and 18. The experimental values had been obtained by Hurlbut [8] with *nitrogen molecular beams on lithium fluoride crystals.*

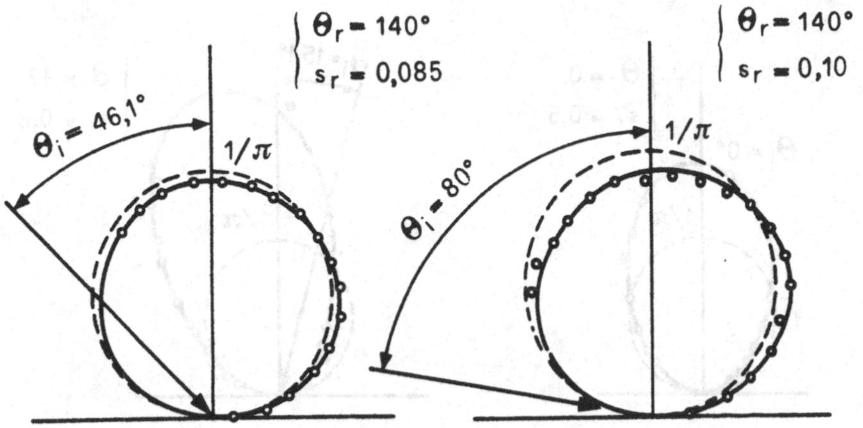

Fig. 16 – Comparison of the re-emission model with experiments of *polished lithium fluoride crystals*

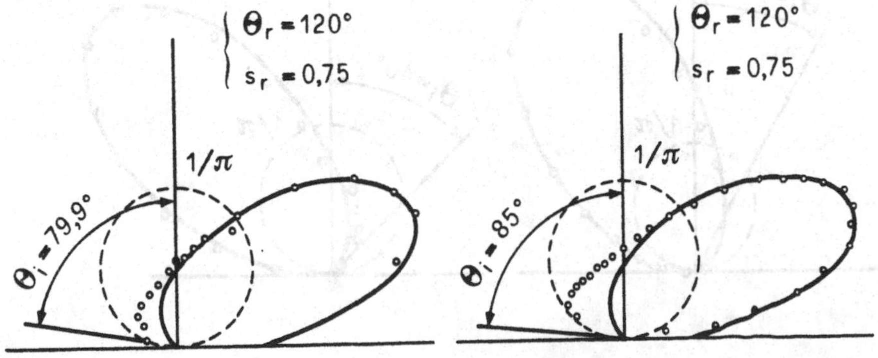

Fig. 17 – Comparison of the re-emission model with experiments for *cleaved lithium fluoride crystals*, for very great, *glancing incidence*

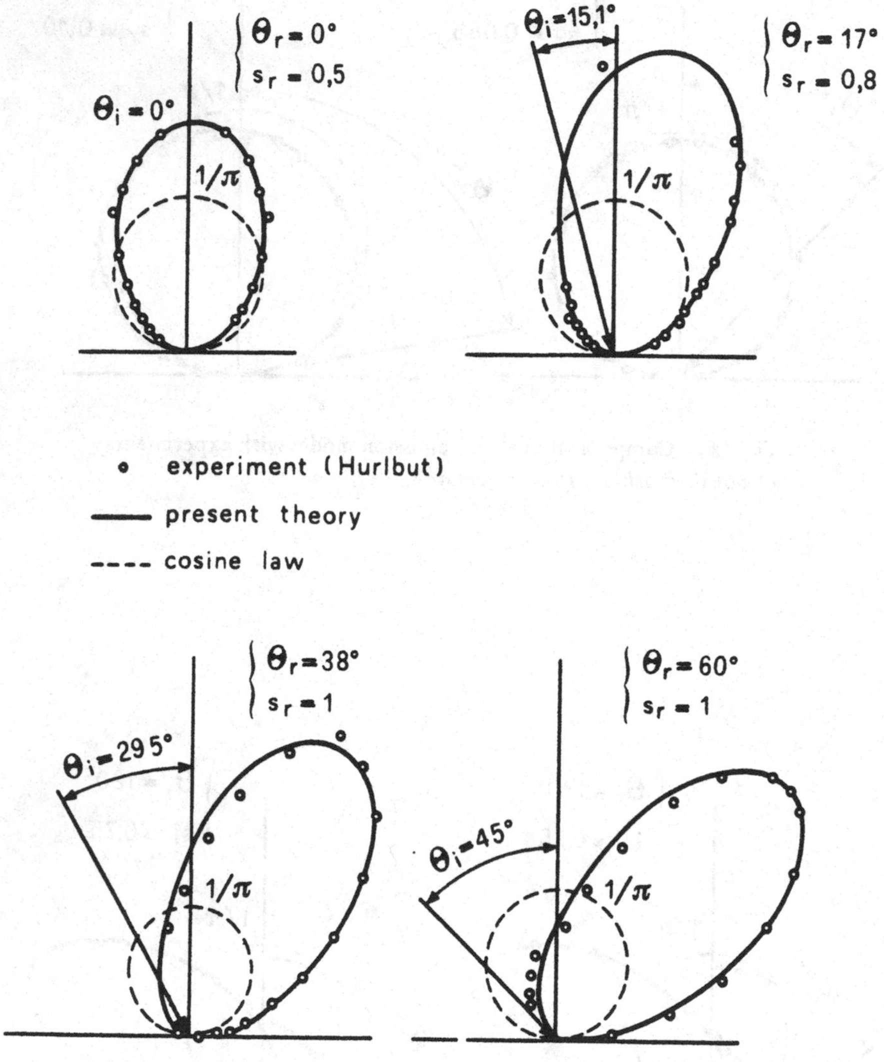

Fig. 18 – Comparison of the re-emission model with experiments for *cleaved lithium fluoride crystals*

4. Interaction Models – Impulsive Models – Rainbow Scattering

The basic problem in the study of the gas-surface interaction phenomenon is to correlate the velocity distribution function $g_i(\vec{V})$ of the incident molecules with the analogous velocity distribution function $g_r(\vec{V})$ of the re-emitted ones. That is to determine a functional relation of the form:

$$g_r(\vec{V}) = \mathscr{L}\,(g_i(\vec{V})) \tag{4.1}$$

where $g_i(\vec{V})$ is generally to be considered known, whereas the operator $\mathscr{L}$ must be determined on the basis of physical, chemical . . . criteria, taking into account above all of the surface force fields and of the consequent behaviour of the molecules on the surface. Some mathematical expressions of this operator have been proposed, as for instance the following *linear* one: (see for instance Kogan [9]):

$$g_r(\vec{V}) = \int_{\vec{V} \times \vec{n} < 0} W(\vec{V}', \vec{V})\, g_i(\vec{V}')\, dS(\vec{V}') \tag{4.2}$$

In addition some particular symmetry properties for the nucleous $W(\vec{V}', \vec{V})$ have been assumed.

From the physical point of view a very greater interest is presented by theories which, although they do not give completely the expression of this operator, allow to predict some particular or general properties of re-emitted molecules through a detailed dynamical study of the motion of the molecules interacting with the surface. As pointed out for instance by Goodman [10], *in this study the basic parameters of the problem must be taken into account* not only in their numerical values, but also in the fundamental aspect that *they characterize different physical regimes,* and that in these different physical regimes *different types of theories are applicable.* We report here the *dimensionless parameters distinguished by Goodman* in the already quoted paper [10].

$$\epsilon_{\omega} = E_o / h\,\omega_m \tag{4.3}$$

where E_o is the energy of the incident molecule, h the Plank constant, and ω_m the maximum modal frequency of the solid. This parameter *indicates whether or not quantum effects are expected to be important.* We do not investigate here this important question in detail.

(4.4)
$$\nu = \frac{2\pi V_o}{r_o\,\omega_m}$$

where r_o is a characteristic length of the gas atom-solid interaction potential, which is assumed to be the following Lennard-Jones 6-12 potential:

(4.5)
$$v_{ij}/\epsilon = (r_o/r_{ij})^{12} - 2(r_o/r_{ij})^6$$

The significance of ν is that ν^{-1} is the number of "lattice vibrations" the solid makes while the incident gas atom travels a characteristic interaction length. This number determines whether or not the interaction among the solid atoms themselves are important. *If $\nu \gg 1$ the solid atoms are effectively stationary during the gas-surface scattering process,* and the solid atom-solid atom "springs" may be ignored, that is, the surface atoms behave as free particles; *if $\nu \ll 1$, however, the solid atoms make many vibrations during the scattering process,* and the presence of the springs in the solid is very important.

(4.6)
$$\epsilon_W = E_o/W$$

where W indicates the *potential well* surrounding the surface, as we will explain shortly. The significance of this parameter is clear: if $\epsilon_w \gg 1$, the presence of the attractive well W is safely ignored and a *purely repulsive gas atom-surface atom potential* may be used; if $\epsilon_W \ll 1$, however, *trapping and hopping effects will predominate* and the presence of the well is of the utmost importance.

(4.7)
$$\mu = M_{gas}/Ms_{surface} = \text{mass ratio}$$

This parameter *gives an indication as to whether single collisions or multiple collisions* of gas atoms *striking relatively* heavy surface atoms, and models which assume single collisions to simplify the interaction will be reasonable. Therefore the interaction times will be relatively short, and the solid atom-solid atom springs will be less important. If we have $\mu \gg 1$, we have heavy gas atoms striking relatively light surface atoms, and single collision models fail, because many collisions with light surface atoms are needed to reverse the component of momentum of a heavy gas atom normal to the surface. *Interaction times will now be relatively long,* and the solid atom-solid atom springs will be more important.

On the basis of the ranges of the values assumed by the above parameters the following *different physical regimes* are fundamentally distinguished:

- thermal scattering
- structure scattering
- penetration scattering
- sputtering

Suitable theories are applicable to the different regimes, and the literature on this subject is very copious. One of the most interesting points is that the basic ground of these theories is *the assumption that the solid surface* schematically indicated in fig. 19 together with a gas atom trajectory *is surrounded by a potential well* W, as shown in fig. 20.

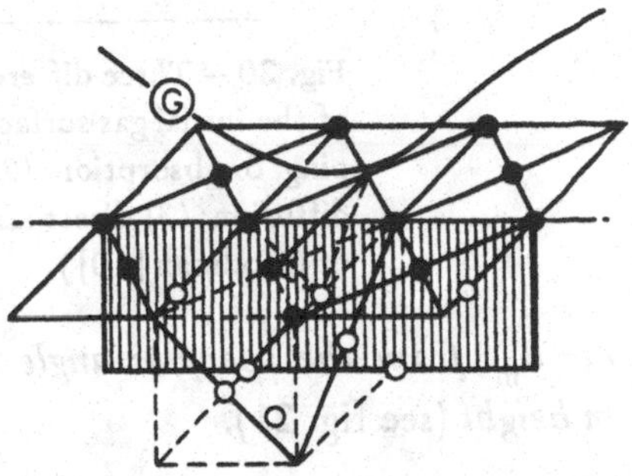

Fig. 19 – Schematical representation of surface lattice together with a gas atom trajectory (from a figure of Hurlbut [11])

The incident gas atom indicated in the upper part a) of the fig. 20 may undergo three different types of collisions as indicated in the lower part b):

(1) trapping, or absorption
(2) hopping, or surface diffusion
(3) direct scattering, that is without sitting time on the surface.

In Goodman's review [10] both the theoretical and experimental research on the matter are accurately reviewed and discussed with particular emphasis on the study of the re-emitted velocity distribution function, of the *deviation* η *of its lobular*

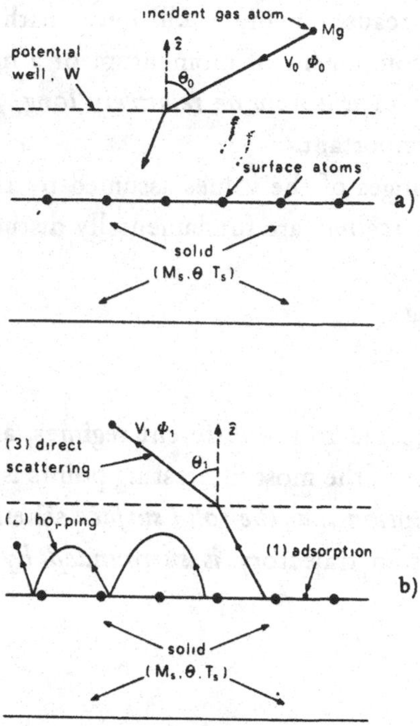

Fig. 20 – Three different types of outcome of the initial gas-surface collision: (1) trapping, or absorption; (2) hopping or surface diffusion; (3) direct scattering (from a figure of Goodman [10]).

maximum ϑ_m ***from the specular angle*** ϑ_o , of the ***width*** λ ***of the lobe at half maximum height*** (see fig. 21).

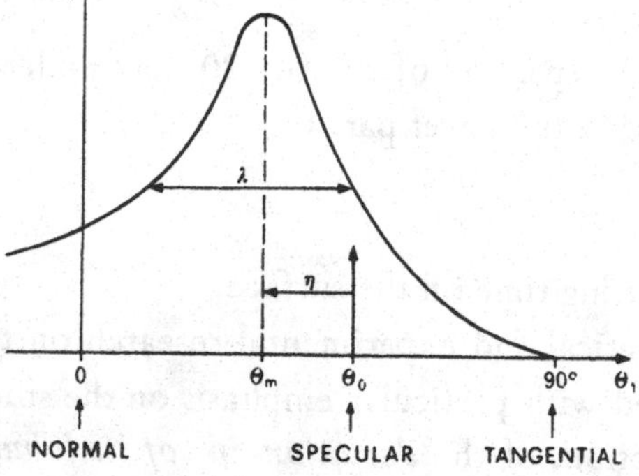

Fig. 21 – Typical monolobular velocity distribution function of re-emitted molecules (from a figure of Goodman [10])

The above short outline suggests very much *observations.* We limit ourselves to the following ones:

– **first:** I think that *the idea of the potential well* surrounding the solid surface *appears one of the most promising ideas in* the general context of *the interaction models* till now devised. *The physical ground* of this idea can be recognised in a paper of Goodman [12] where the interaction potential energy of a gas atom with a solid is obtained *by summing the 6-12 gas-atom solid-atom pairwise Lennard-Jones potential (4.5) over all atoms of a* perfect monoatomic, cubic, semi-infinite *lattice model.* The result of the calculations are presented in tables as functions of distance of gas atom from the surface.

A little time before this paper of Goodman it had been published by *Raff et alii* [13] a theoretical paper in which a previous lattice model of *Oman* et alli [14] termed the "independent oscillator lattice" was suitably extended and employed through calculation of many trajectories of interacting atoms. The research was developed using Monte Carlo techniques and carrying out numerical integrations of the 16 differential equations describing the dynamical system.

Out of this paper of Raff et alii I want only to draw *the results of the calculations of the obtained surface force field.* In fact they allow to explain the structure of the surface field forces, as shown in figures 22 corresponding to Helium beam on Nickel surface with lattice atoms fixed in their equilibrium positions. This structure corresponds to a *periodical succession of pairwise Lennard-Jones central fields,* each directed toward a surface atom. *The superposition of many central force fields periodically ordered produces the potential map with equipotential lines almost parallel to the solid surface, towards its external part. That is a one-dimensional structure in the direction normal to the surface which can be just fitted by the potential well* considered previously.

Naturally the above basic considerations would be improved also taking into account the thermal motion of surface atoms.

– **second:** It must be emphasized that the physical start-point of all interaction theories is the *surface roughness at molecular scale,* that is the fact that at this scale the surface cannot, at any respect, be considered a flat one because of the *lattice structure of the surface atoms, of their thermal motion, of the force field surrounding them,* and finally *of the absorbed atoms.*

– **third:** both theoretical and experimental evidence shown in these last years that gas-surface interaction must be considered in its *actual three-dimensional aspect,* because the behaviour of re-emitted molecules is substantially variable *in the various*

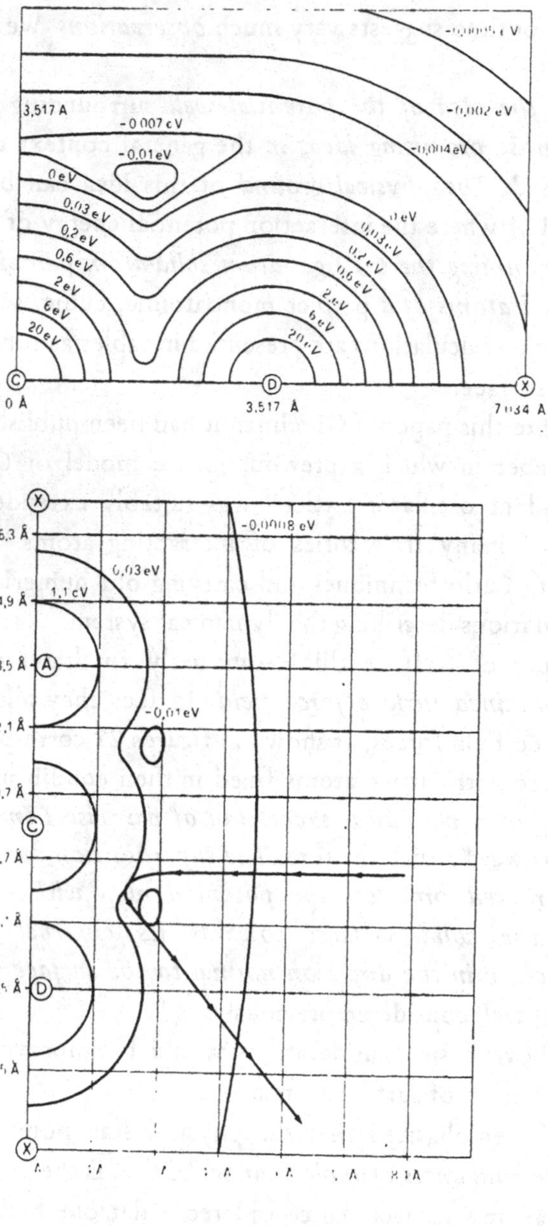

Fig. 22 – Potential map corresponding to Helium beam on Nickel surface with lattice atoms fixed in their equilibrium positions. In the lower figure it is also indicated a trajectory of a normal impinging molecule (Raff et alii [13])

azimuthal planes.

Let us now deal with the *impulsive models,* because they have particular interest in the aeorspace applications. In fact, in this case the parameter ϵ_W formula (4.6) is very high, and hence the presence of the potential well can be neglected. Well known are in the literature the "hard cube" and "soft cubes" model of Logan [15] and co-workers. From this model co-workers and myself have derived another impulsive model which presents substantial differences and seems to be more realistic than the Logan and alii one quoted above, on the basis of energetic considerations about the properties of the collisions between gas atoms and solid atoms, and furthermore as concerns the slope of the "apparent surface" presented by the solid atoms towards the outside of the solid itself.

The basic assumptions of the model are:

(i) *Every incident molecule collides only once* with a solid atom and owing to the surface force fields is scattered abroad, with variation of the velocity magnitude

(ii) *This scattering is simulated as if the solid atoms had towards the outside a rigid surface, randomly sloped,* and as if the gas molecule underwent a shock with this solid surface

(iii) *Each atom of the solid moves alternatively normally to the surface.* As regards the collisions with impinging molecules all these atoms have a common velocity W^* directed towards the outside of the solid surface and given by the Logan and Stickney paper [15] Eq. (5):

$$w^* = \frac{kT_w}{m_w \; V_i \; \cos \vartheta_i}$$

(iv) *The alternative motion of the atom keeps unchanged during the collisions*

(v) After collision the gas molecules have a velocity correspondent to a *specular reflection in the motion relative to the atom, reduced by a suitable factor* η.

Figure 23 shows schematically the model, in its bi-dimensional version particularly investigated by *Chiadò-Piat* [16], in the case of a *mono-energetic beam,* with hyperthermal velocity V_i. From the above hypothesis the following expressions for the orthogonal cartesian components u_r and w_r of re-emitted molecules velocities are obtained:

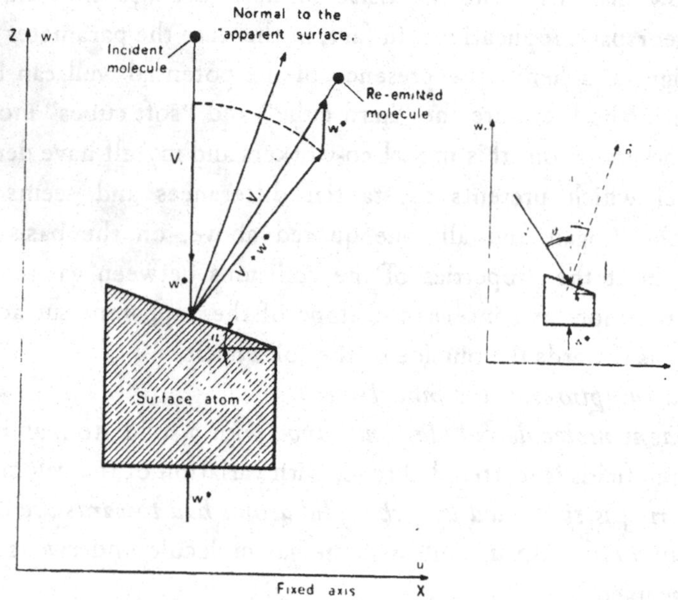

Fig. 23 – Assumed reflection law for an incident molecule, according to the impulsive interaction model. The left part of the figure refers to the normal incidence, with the details of the reflection. The right part refers to a generic incidence angle ϑ_i, without the details of the reflection.

$$
(4.8)\quad
\begin{cases}
u_r/V_i = V_r/V_i \sin\Theta = \\
\quad = \eta[\sin\vartheta_i \cos 2\alpha + (\cos\vartheta_i + \lambda)\sin 2\alpha] \\
w_r/V_i = V_r/V_i \cos\Theta = \\
\quad = \eta[(\cos\vartheta_i + \lambda)\cos 2\alpha - \sin\vartheta_i \sin 2\alpha + \lambda]
\end{cases}
$$

The parameter η simulates the effects of the dissipative forces taking place during collision. On the basis of comparisons with experiments of Stickney and Hurlbut [17] (for the comparison see Nocilla and Chiadò- Piat [18]) this parameter assumes values between 0.9 and 1 about. The parameter λ is related to the parameter w^* of the assumption (iii) as follows:

$$\lambda = w^*/V_i \tag{4.9}$$

As regards the *"apparent rigid surface, randomly sloped"* considered in the assumption (ii), it is introduced in order to simulate the fact that surface force fields in a generical point P changes as depending on the position of P in respect to the lattice centers, as already pointed out previously (see for instance fig. 22). *This fact,* which can be considered as a roughness at atomic scale, *produces on the incident molecules velocity variations different according to the exact position in which these ones impinge on the lattice.* According to this point of view we introduce a distribution function, we call $A^*(\alpha)$, for the angles relative to the slopes of these "apparent surfaces". For this distribution function the following expression is assumed:

$$\begin{cases} A^*(\alpha) = A^{**}\,[\mathrm{tg}(2k_i\alpha)] = \\ = \dfrac{2k_i}{\epsilon\sqrt{\pi}}\,(1 + \mathrm{tg}^2\,2k_i\alpha)\,\exp(-\,\mathrm{tg}^2\,2k_i\alpha)/\epsilon^2 \end{cases} \tag{4.10}$$

$$-\infty < \mathrm{tg}2k_i\alpha < \infty$$

and where the parameter k_i has the value:

$$k_i = 1/(1 - 2\,\vartheta_i/\pi) \tag{4.11}$$

The arbitrary positive parameter ϵ has the meaning of *"roughness parameter"*. For $\epsilon = 0$ all the angles α result null, and consequently there is no roughness, while for $\epsilon \neq 0$ the higher is ϵ, the higher is the variability of the inclination α, id est the deeper is the roughness.

The function $A^*(\alpha)$ is shown in fig. 24 for some values of the parameter ϵ, and for normal incidence $\vartheta_i = 0$, to which it corresponds $k_i = 1$.

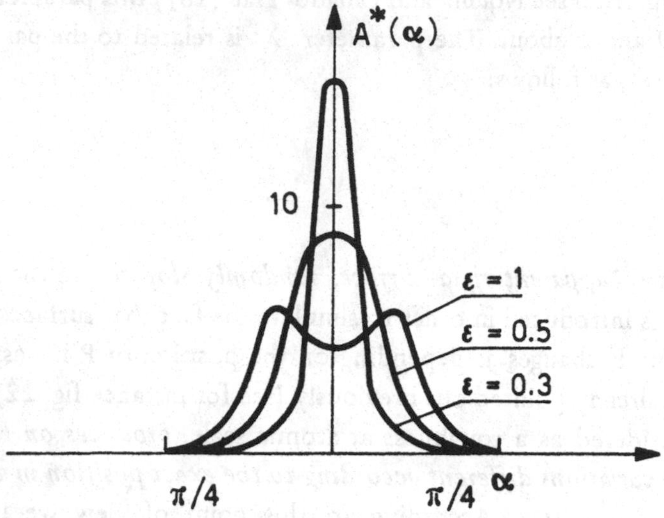

Fig. 24a – The function $A^*(\alpha)$ for some values of the parameter ϵ and for normal incidence

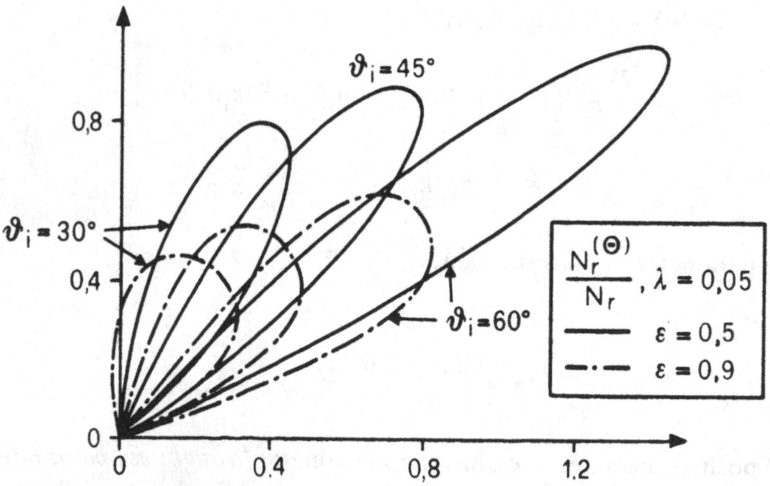

Fig. 24b – Lobes relative to the re-emitted flux according to the above bi-dimensional impulsive interaction model

The probability $dp^{(\alpha)}$ that an incident molecule collides with an atom presenting towards the outside an apparent surface with "slope" between $\alpha + d\alpha$ is:

$$dp^{(\alpha)} = A^*(\alpha)\ d\alpha \equiv A^{**}[tg(2k_i\alpha)]\ d\alpha \tag{4.12}$$

At last, the probability that a molecule is scattered towards a direction between Θ and $\Theta + d\Theta$ is:

$$dp^{(\Theta)} \equiv \frac{N_r^{(\Theta)}}{N_r}\ d\Theta = A^*(\alpha)\ \frac{d\alpha}{d\Theta}\ d\Theta \tag{4.13}$$

The re-emitted flux for unit angle in the direction Θ is:

$$\begin{aligned} N_r^{(\Theta)}/N &= A^*(\alpha)\ d\alpha/d\Theta = \\ &= 2k_i/(\epsilon\sqrt{\pi})\ (1 + tg^2\,2k_i\alpha)\ \exp(-\,tg^2\,2k_i\alpha/\epsilon^2) \end{aligned} \tag{4.14}$$

The differential ratio $d\alpha/d\Theta$ is calculated through Eqs. (4.8) after elimination of V_r. Some lobes relative to the re-emitted flux (4.14) are indicated in the fig. 24 for the following values of the parameters:

$$\lambda = 0.05\ ; \qquad \epsilon = 0.5 \text{ and } 0.9$$

and in correspondence to the following values of the incidence angle:

$$\vartheta_i = 30^\circ\ , \quad 45^\circ\ , \quad 60^\circ$$

As a *last remark* on this impulsive interaction model we observe that according to it the *molecule re-emitted in a given direction* Θ *are mono-energetic, with velocities variable for each* Θ. The dependence of this velocity V_r on the scattering angle Θ is shown in fig. 25 for the case of Argon molecular beams incident in

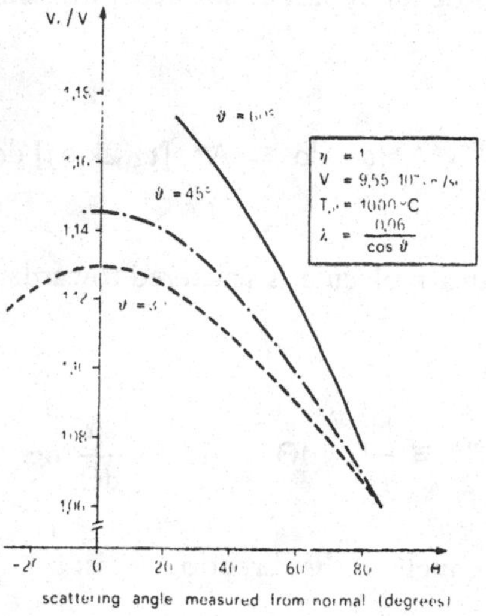

Fig. 25 – The ratio V_r/V_i versus the scattering angle Θ according to the above impulsive bi-dimensional interaction model, in the above indicated case

Platinum heated to about 1000°C, for the following angles of incidence: $\vartheta_i = 30°$, 45°, 60°, always for the following values of the parameters:

$$\eta = 1 \;; \qquad \lambda = 0.06/\cos \vartheta_i$$

The value of λ corresponds to $V_i = 0.955\ 10^5$ cm/sec.
Remarkable is the fact that for all the incidence angles the velocity V_r decreases when Θ increases from the normal. This property disagrees with the behaviour of the most probable velocity q_M and of the mean velocity $\bar{q}$, as functions of the same scattering angle Θ corresponding to the re-emission model previously described (see figs. 13, 14 and 15). In this case the maximum velocity takes place about for $\Theta = \vartheta_r$, that is in correspondence to the direction of the drift velocity of re-emitted molecules, which can be approximately considered equal to the incidence angle: $\vartheta_r = \vartheta_i$. The disagreement in this particular property is, however,

accompanied by a very good agreement of both the models as regards the "lobes" of the number flux $N_r^{(\Theta)}$.

The above bi-dimensional impulsive model *has been generalized into a spatial one* by Riganti [19]. The basic hypotheses are the same, but the slopes of apparent surfaces of the surface atoms are now depending on *two, instead of one* variable, and namely on the two angular coordinates α and β of the normal $\vec{n}_W$ to the apparent surface. These angular coordinates are assumed to vary in a suitable random way, which simulates the surface roughness. Through the calculations performed according to the flux statistics the following expression is obtained for the velocity distribution function $g_r^{(V)}$:

$$g_r^{(\vec{V})} = \frac{1}{J} \frac{2}{\epsilon \pi^{3/2}} (1 + tg^2 2\alpha) \exp(-tg^2 2\alpha/\epsilon^2) \tag{4.15}$$

where J and α are to be considered as well known functions of the angular coordinates Θ and ϕ characterizing the generical scattering direction. Obtained lobes strongly depend on the particular azimuthal plane of scattering.

Till now we considered monolobular scattering. However, in the gas-surface interaction phenomena *also multibular scattering can occur.* Experimental evidence of this fact was a very important discovery of Devienne and co-workers [20], *mainly dependent on the energy of incident beam, above 25 Ev about.* On the other hand much more recent experimental results showed the existence of *multilobular scattering, also for much minor energies, as for instance thermal ones, provided the impinging beam was contained in a suitable azimuthan plane, and the re-emitted molecules was detected also in suitable azimuthal planes,* see fig. 26.
This new kind of scattering was named *"rainbow scattering".* We report here some experimental results of *O'Keefe,* showing the very interesting *transition between monolobular to bilobular scattering for rare gases from lithium fluoride LiF* crystals in the 110 azimuth *for decreasing molecular weight of impinging gas molecules.*

Fig. 27 refers to the *same incidence angle* $\vartheta_i = 40°$, for *various rare gases.*
Fig. 28 refers to *the same rare gas neon, for various incidence angles.*
As regards this paper of O'Keefe and alii I consider very significant and revolutionary the following sentence contained in the conclusion of the paper "somewhat" disturbing is the fact *that multilobular structure has not been observed* in the scattering from crystal metals, whereas the theory, in the present state, would indicate the existence of such structure". The theory to which the Authors refer is

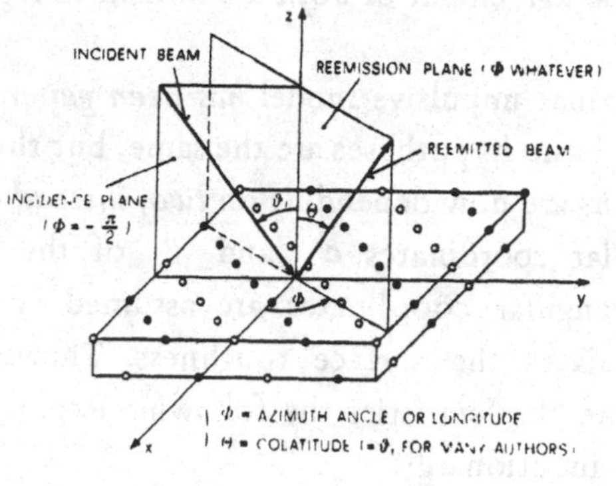

Fig. 26 – Spatial geometry for the beam-surface interaction according to the Author and co-workers Riganti and Chiadò-Piat (not all other Authors agree with this geometry)

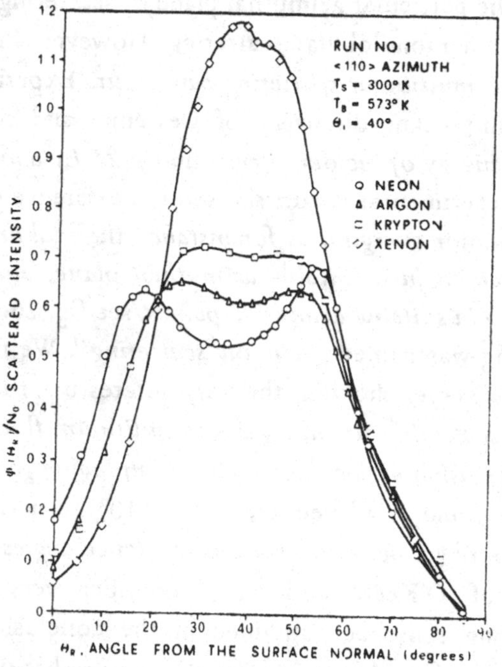

Fig. 27 – Transition between monolobular to bilobular scattering for rare gases from LiF in the 110 azimuth (O'Keefe and alii [21])

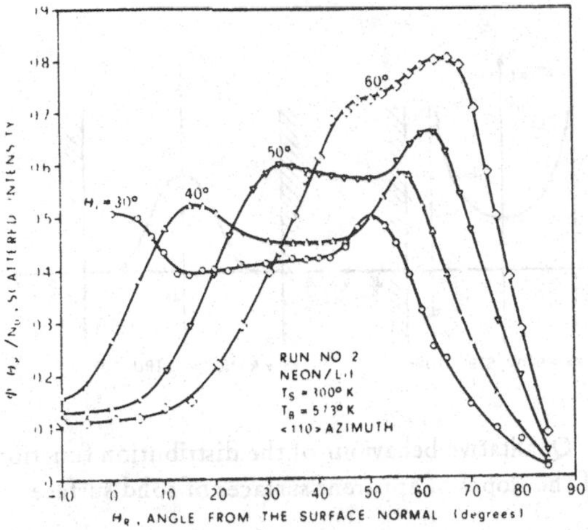

Fig. 28 – Scattered flux for Neon on LiF in the 110 azimuth for various incidence angles ϑ_i (O'Keefe and alii [21])

the rainbow scattering studied by McClure [22] (see also McClure and Wu [23]). I think that the *revolutionary* fact is *that tody it is considered "disturbing" that monolobular structure which only a few years ago was considered as fundamental in the surface interaction,* at least for moderate energies.

From a theoretical point of view we remark, as already pointed out in [24], that *the question can be related to the shape of the distribution function* A* (α) for the angles relative to the slopes of the "apparent surfaces" of the solid atoms, which was introduced in connection with the impulsive interaction model. By assuming a shape as indicated in the *left side* of the fig. 29, that is a "rainbow structure", we obtain multilobular, or *rainbow scattering;* by assuming the *right side* shape, that is a "random structure" we obtain the *classical monolobular scattering.*

In McClure's paper [22] the calculations are carried out with more accurate procedure, more satisfactory from a mechanical point of view, but the results are qualitatively the same ones.

What conclusion can be drawn on this matter? I think that both points of view of monolobate or rainbow scattering must be accepted. *The basic question is to state for which* gas-solid combinations, for which geometrical, physical, chemical, mechanical *conditions* the first one, namely *the periodical structure and bilobate flux diagrams, or the second one, namely random structure and monolobate flux diagrams must be assumed.*

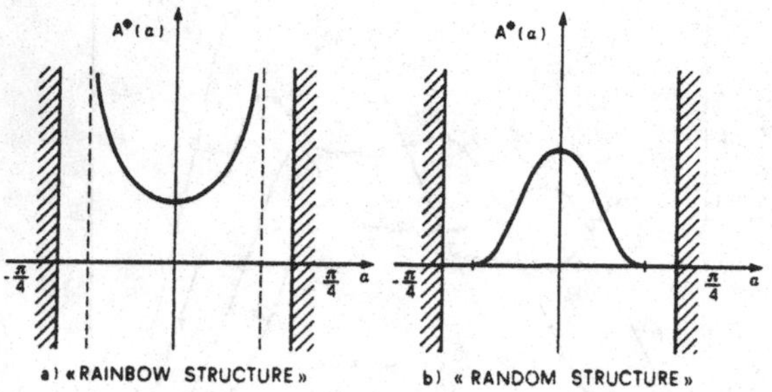

Fig. 29 – Qualitative behaviour of the distribution function $A^*(\alpha)$ of the slopes of apparent surfaces of solid surface atoms

REFERENCES

[1] KALLMANN H.K., SIBLEY W.L. — Diurnal variation of temperature and density between 100 Km and 500 Km — Space Research; pp. 270-296 (1963).

[2] FRENKEL J. — Theorie der adsorption und verwander ercheinungen — Z. Phys. 26, 117 (1924).

[3] DE BOER J.H. — The dynamical character of adsorption — Oxford University Press (1953).

[4] NOCILLA S. — Sorgenti e pozzi superficiali di molecole libere ed applicazioni. Parte I — Atti Acc. Scienze Torino, 99, 805 (1964-65). Parte II — Confronti con risultati sperimentali — ibidem, 100, 447 (1965-66).

[5] NOCILLA S. — Theoretical determination of the aerodynamics forces on satellites — Astro. Acta, 17, 245 (1972).

[6] NOCILLA S. — The surface re-emission law in free-molecule flow-3rd RGD Symp.; vol. 1, 327, Acad. Press (1963).

[7] HURLBUT F.C., SHERMAN F.S. — Application of the Nocilla wall reflection model to free-molecule kinetic theory — Phys. of fluids, 11, 486 (1968).

[8] HURLBUT F.C. — Rand Rept 339, 21-1 (1959).

[9] KOGAN M.N. — Rarefied Gas Dynamics — Plenum Press, New York (1969).

[10] GOODMAN F.O. — Review of the theory of the scattering of gas atoms by solid surfaces — Surface Science, 26, 327 (1971).

[11] HURLBUT F.C. — Gas surface interaction studies employing three-dimensional coupled lattice model — 2nd symp. jèts moléculaires, Entropie 30, 107 (1969).

[12] GOODMAN F.O. — Interaction potentials of gas atoms with cubic lattices on the 6-12 pairwise model — Phys. Rev. 164, 1113 (1967).

[13] RAFF L.M., LORENZEN J., MC COY B.C. — Theoretical investigations of gas-solid interaction phenomena, I — Jour.Chem.Phys. 46, 4265 (1967).

[13'] LORENZEN J., RAFF L.M. — idem, II: Three dimensional treatment — Jour.Chem.-Phys., 49, 1165 (1968).

[14] OMAN R., BOGAN A., LI C.H. — Theoretical prediction of momentum and energy accommodation for hypervelocity gas particles on an ideal crystal surface — 4 RGD Symposium vol. II, 396, Academic Press (1966).

[15] LOGAN R., HECK J., STICKNEY R. — Simple classical model for the scattering of gas atoms from a solid surface: additional analyses and comparisons — 5 RGD Symp., vol. I, 49, Academic Press (1967).

[16] CHIADO'PIAT M.G. — Studies on the behaviour of gas molecules scattered by a solid surface: analyses for mono-energetic beams — Entropie, 30, 103 (1969).

[17] STICKNEY R.E., HURLBUT F.G. — Studies of normal momentum transfer by molecular beam techniques — 3rd RGD Symp. 454, Acad. Press (1963).

[18] NOCILLA S., CHIADO'PIAT M.G. — Studies on the behaviour of gas molecules scattered by a solid surface: normal incidence — 6 RGD Symp., vol. 2, 1069, Academic Press (1969).

[19] CHIADO'PIAT M.G., RIGANTI R. — An impulsive tridimensional interaction model in free molecule flow — Entropie 49, 46 (1973).

[20] DEVIENNE F., ROUSTAN J. CLAPIER R. — Speed distribution of scattering molecules after the impact of a high velocity molecular beam on a solid surface — 5 RGD Symp., vol. 1, 269, Academic Press (1967).

[21] O'KEEFE D., PALMER R., SMITH J. — Rare gas scattering from LiF: III multilobular structure for neon — Gulf radiation technology, San Diego Calif., RT — 10543 (1971).

[22] McCLURE J. — Surface rainbows: a similitude between classical and diffractive scattering of atoms from cristalline surfaces" Jour.Chem.Phys., 52, 2712 (1970).

[23] McCLURE J., WU JAN — Atomic and molecular scattering from solids: I, planar scattering from one-dimensional surface arways (an extendet data set) — Scientific research laboratories D1-82-0732 (1968).

[24] NOCILLA S. — Recent theories on the interaction between molecular beams and surfaces — Entropie 49, 37 (1973).

[25] BELLOMO N. — Upper atmosphere density measurement by the Sputnik III and San Marco satellites — Astro.Acta 18, 289 (1974).

[26] RIGANTI R., CHIADO'PIAT M.G. — Drag and lift coefficient in hyperthermal free-molecule flow — T.N. 78, Ist.Mecc.Appl. Aerod.Gasd. Politecnico Torino. Also (in Italian) in: Atti 1 Congresso Naz. AIMETA, Udine (1971).

EXPERIMENTAL METHODS AND RESULTS IN RAREFIED GAS DYNAMICS

J.J. Smolderen
von Karman Institute for Fluid Dynamics

Introduction

The considerable interest in rarefied gas dynamics, which developed during the late fifties and the sixties, as a result of spectacular space programs, has led to the study of a great number of theoretical and experimental problems concerning flows of neutral and ionized gases, and to a rich harvest of original results. It is well known that the pace of these activities has been slowing down considerably in recent years, for obvious reasons connected with the de-emphasis of the space effort, and many important research groups who contributed strongly to rarefied gas dynamics have reconverted to somewhat more mundane areas. It appears worthwhile to look back at the developments of the last decades and summarize and evaluate the result obtained, so as to help orientate the research that is going at present, albeit at a reduced pace.

The area of rarefied gas dynamics is easy to define, from a theoretical point of view. It involves all gas flow phenomena for which the usual continuum Navier Stokes equations are not valid, at least in some important region of the flow. Flows in which shock waves occur are not included if these shocks have a thickness small enough compared to typical dimensions of the flow field. Quantitatively, the criterion of rarefaction is defined through the Knudsen number, Kn, ratio of a typical molecular mean free path to a typical length scale for the flow. If this length scale is clearly defined a priori, which is not often the case, then the Knudsen number is inversely proportional to the density of the gas, but also a function of temperature. Von Karman has shown that the Knudsen number is proportional to the ratio of Mach number to Reynolds number, the proportionality factor being close to unity. Of course, this relation does not solve the essential problem of the definition of a suitable length scale, which now occurs in the Reynolds number. The Chapman Enskog theory which relates the Navier Stokes equations to the Boltzmann equation of kinetic theory indicates that the length scale to be chosen must be based on the gradients occurring in the flow. This scale may depend on the Mach number as well as on the geometric length scale so that the Knudsen number may be a rather complicated function of Mach number, geometrical dimensions, density and temperature but it increases with decreasing density, all other parameters being equal.

A rarefied gas flow is defined by the fact that the Knudsen number based on local gradients is not very small compared to unity except if the regions where the Knudsen number is not small are thin layers with negligible thickness (thin shocks, Knudsen layer thickness negligible compared to boundary layer thickness). If one remembers that the mean free path in air is of the order of 10^{-5}cm in normal conditions, it is easy to see that rarefied gas flows will exist only at pressures which are a small fraction of a thousandth of the atmospheric pressure.

From the point of view of experimental facilities and instrumentation, the above definition appears to be somewhat too restrictive, because typical difficulties connected with high viscous effects and low signal levels already occur at higher pressure levels, at which most flows may still be considered as essentially continuum especially if the velocities are not too high. The important influence of flow velocity results from the Mach number dependence of the Knudsen number.

In the present series of lectures, we will concentrate on experimental research on rarefied flows of electrically neutral, non reacting gases.

Aerospace activities suggested many problems of rarefied gas dynamics, among which the most important were flows over re-entering vehicles, with particular emphasis on kinetic heating, drag and energy balance of satellites, rocket operation at extreme altitudes and in space, and the resulting exhaust plume, design problems of space simulators and their instrumentation, high atmosphere sounding devices, production of gas clouds at extreme altitudes, etc.

There are, of course, other fields for which the study of rarefied gas flows is relevant, such as astrophysics, the design of vacuum systems for chemical processing or particle accelerators, etc.

Many of these problems of considerable practical importance indicated the need for a more thorough understanding of fundamental aspects of rarefied gas dynamics such as gas surface interactions, intermolecular forces, the foundations of kinetic theory, as well as basic flow phenomena such as the Knudsen layer, the structure of shock waves, the free expansion of a gas into vacuum, flow through pipes, orifices and around bodies of simple geometric shapes (spheres, cones, thin flat plates).

We will attempt to describe, first, the present status of development of devices for flow production and stress how these devices differ from conventional wind tunnels.

The specific instrumentation for rarefied flow studies will then be described, and it will be shown that measurement of pressures, forces, heat fluxes, etc. require special technology because of the low level of signals. An important breakthrough in this field was represented by the development of electron beam techniques for the measurement of density, temperature, and even velocity distribution function. Of course, the nature and results of experiments depend crucially on the available instrumentation. We may, in fact, say that in the field of experimentation with neutral rarefied gas flows, the main limitations are due to instrumentation and not to the flow production systems.

A second part of the presentation will be devoted to the description of selected results, and the emphasis will be on experiments pertaining to basic flow phenomena such as Knudsen layer, shock structure, flow near the leading edge of a flat plate, flows around bodies of simple geometric shapes and free jets. For most of these flows, there exist some theoretical results, usually based on approximation solutions of the Boltzmann equation or one of the model kinetic equations. The experimental results might therefore lead to a check of these solutions or of the validity of the kinetic model equations. Some of the experiments were specifically designed to verify the theoretical results, and such a verification is certainly needed in view of the facts that the validity of the approximations used is not always obvious, and that it is usually impossible to calculate bounds for the errors introduced by such approximations.

It should be noted, also, that some of the basic flows mentioned involve boundaries and therefore the interaction of gas molecules with solid surfaces.

There is still a considerable lack of detailed information on this interaction and experiments may therefore be helpful in checking the validity of simple models used to represent this interaction.

I. PRODUCTION OF RAREFIED GAS FLOWS

The technology involved in the production of uniform low density gas flows at high or low speeds differs significantly from conventional wind tunnel technology and there are some obvious changes of emphasis in optimization.

Large conventional tunnels require considerable installed powers, especially at high speeds, so that power economy aspects are essential. This consideration led to the near universal utilization of intermittent facilities for high speed flow production. Viscous effects, on the other hand, are of importance only in so far as they may influence efficiency.

The technology of low density facilities, on the contrary, is dominated by viscous effects which may strongly affect the flow quality. Power efficiency of the system is usually irrelevant, the power needed being sometimes of the same order of magnitude as the power needed to supply the electronic instrumentation. Viscous effects also determine the performance and capital cost of the pumping system for high speed flows.

Intermittent operation at low densities has been used only in exceptional cases, because of difficulties connected with the very low response time of pressure measuring systems, as described below, but this argument may loose some weight as a result of the introduction of electron beam probe techniques.

The main types of rarefied flow production devices will now be described, namely:

High and low speed wind tunnels
Shock tubes
Molecular beams
Whirling arms, rotating cylinders and disks, etc.

A. High Speed Low Density Wind Tunnels

During the last twenty years, the main emphasis in the low density field was on high speed and even hypervelocity flows as a result of the requirement of the space programs. Most of the facilities built during this period were of the continuous operation type and driven by large vacuum pumps, the pumping system depending on the volume flow to be handled and on the pressure level at the pump inlet.

The mass flow in the tunnel is entirely defined by the Mach number, pressure and dimensions required at the test section, but the conditions at the pump inlet will depend significantly on the diffuser efficiency, as discussed below.

Four pumping systems have been widely used to drive low density wind tunnels, namely

1) Oil diffusion pumps
2) Cryopumps
3) Mechanical pumps
4) Steam ejectors

The oil diffusion pumps used for wind tunnel operation are very large versions of the well-known laboratory diffusion pumps based on the entrainment of low pressure gas by means of several stages of layered vapor jets. The pumps are backed by mechanical (usually vane type) exhausters. Some of these pumps use an oil vapor ejector (booster) as a last stage to increase pumping efficiency at higher inlet pressures. Units with a maximum pumping speeds of 10^3/s or higher are readily available. The useful pressure range at which the nominal can be maintained is between a fraction of a micron of mercury (millitorr) and a few microns. Fig. 1 shows a typical pumping curve which indicates the considerable drop in pumping speed at higher pressure.

The minimum pressure is conditioned by the vapour pressure of the oil at the operating temperature and high pumping speeds may be maintained at pressure levels of 10^{-4} torr and below through the use of special oils and lower heating rates.

The oil diffusion pump appears to be the most economical system, from a capital cost point of view, at least at medium pressures ($\sim 10^{-3}$ torr), but some maintenance and contamination problems, as well as uneven operation characteristics have sometimes been encountered.

The cryopump system is based on gas condensation on surfaces cooled at liquid helium or hydrogen temperature. This pumping action will therefore be effective for all gases except non condensables, down to pressures far below those needed for low density wind tunnel operation. The upper pressure for operation, however, is dependent on two factors:

a) The cooling rate resulting from the evaporation of liquid helium or hydrogen, which is related to the power of the liquefyer;

b) The total area of the cooled plates. Continuous condensation of the tunnel gas on these plates will create a thermal barrier and therefore raise the temperature at the surface. The running time at a given pressure level or the pressure level for a given running time will therefore depend on the area of the cooling surfaces.

The main limitation of this type of pumping system is the high cost of the helium or hydrogen liquefyer per unit of cooling power. Important developments in the design of such devices had been announced several years ago but the new Helium liquefyers have not been used widely.

Although the cost of a cryopumping system tends to be quite high, most large wind tunnels operating at extremely low pressure (10^{-4} torr and below) have used such a system, which is characterized by extreme cleanliness.

Mechanical pumps of volumetric type, such as root blowers, etc., have also been used for low density wind tunnel applications but their cost is significantly higher than that of an oil diffusion system of similar characteristics and their maintenance problems tend to be heavier. Here the minimum pressure of operations is conditioned by leakage effects so that these pumps cannot be operated far below 10^{-3} torr. The pumping speed is, of course, constant for all pressures and the upper pressure limit is therefore conditioned by power limitations or heating problems.

Turbomolecular pumps are designed to operate at free molecule flow conditions and are therefore limited to pressures below 10^{-3} torr, but their pumping speeds are on the low side for wind tunnel applications.

A few low density wind tunnels have used steam ejectors, operating in the 10^{-3} torr pressure range. This is a very low efficiency device which requires quite a large steam generating plant, and is justified only if such a plant is already available, because of its high capital cost.

All supersonic or hypersonic tunnels use some kind of convergent-divergent nozzle to accelerate the flow from rest to supersonic speeds, as all other techniques, such as acceleration through rotating machinery, have proved deficient in flow quality or too complicated or expensive. However, when low density operation is considered, the design of the nozzle raises severe problems as a result of severe and often dominant viscous interaction problems.

In conventional wind tunnels, boundary layer thicknesses are of the order of a percent of typical dimensions, at most, and a very simple boundary layer correction is usually sufficient to cancel flow non uniformities due to viscous effects. Low density wind tunnels, operating at pressures 10^{-2} to 10^{-6} times lower, will exhibit boundary layers ten to thousand times thicker and the corresponding correction will range from important to dominant. Two-dimensional nozzles, with their greater length and corner and cross flow effects in the boundary layers, are, of course, inapplicable to low density tunnels and axisymmetric nozzles have been universally used (The focussing effect which has been often named as a possible difficulty for such nozzles, has, in fact, never been observed).

But even optimized axisymmetric nozzles cannot be expected to solve the viscous problems at the very low pressures, as the boundary layer will invade the whole test section, unless the diameter of the nozzle is made very large, which might require a prohibitive pumping system.

It has been proposed to reduce the boundary layer thickness by suction through the nozzle walls but it appears that the additional pumping capacity needed may not be much smaller than the one required to pump a larger nozzle.

A better solution seems to be the use of wall cooling and, in extreme case, the condensation of the walls. The technique has been used with success using nitrogen cooled walls.

If a conventional nozzle is used, then a diffuser is generally added to obtain efficient pressure recovery. This pressure recovery plays a large role in the design of the pumping system, as a larger pumping inlet pressure corresponds to a lower pumping speed requirement for given mass flow and therefore to a lower cost of the pumping system.

Usually, the performance of a diffuser is compared to normal shock recovery, i.e. the ratio for the actual pressure at the diffuser outlet to the pressure behind a normal shock wave at the test section Mach number. This ratio, which can be larger than one for well-designed conventional tunnel diffusers, which was found to be very low (a few percent) in the early low density wind tunnel. More recently well-designed diffusers, using second throats, were found to yield values of pressure recovery not much below the normal shock recovery.

Free Jet Tunnels

The above-mentioned difficulties connected with thick nozzle wall boundary layers and poor diffuser efficiency have led to the use of the free jet expansion technique proposed and introduced simultaneously by several groups in the early sixties. This technique is based on the expansion flow from a stagnation chamber through a circular orifice or short sonic nozzle into a large reservoir in which the pressure is maintained at a level several order of magnitude below that of the stagnation chamber. This flow is entirely similar to the exhaust plume flow from a rocket flying at extreme altitudes, and is bounded by a barrel like stream surface, with maximum diameter much larger than the diameter of the orifice or nozzle outlet if the pressure ratio is very large. The flow may, of course, be computed by the theory of characteristics when viscous or non continuum effects are neglected and if the shape of the sonic surface is given. A barrel shaped axisymmetric shock wave will develop inside the boundary surface, as a result of the recompression after initial overexpansion at the lip of the orifice, which is further enhanced by the compression waves caused by the reflexion of the expansion waves on the free boundary. This barrel shock tends to focus on the axis of the jet and thereby creates a disk like, quasi normal terminal shock. The central flow is therefore completely surrounded by a single shock surface and one may therefore expect to reach nearly normal shock recovery based on the maximum Mach number of the flow, which occurs on the axis, immediately upstream of the terminal shock. The whole flow downstream of this shock is usually found to diffuse to subsonic conditions without further shocks.

Calculations by Owen and Thornhill for Mach number distribution on the axis, and a great number of experimental surveys have led to the conclusion that the flow within the barrel shock can be well approximated by a source flow issuing from a fictitious source located close to the centre of the orifice. A more accurate representation of the flow, proposed by Ashkenas and Shenman, involves the dependence of the density for given distance to the centre on the angle of the radius vector with the axis. The dependence is well represented by a $\cos^2(\theta / \theta_0)$ law.

The free jet concept led to very economical low density wind tunnel operation down to quite low pressures, but it has an obvious limitation because of the lack of uniformity of the flow, which is essentially conical. Both Mach number and flow direction being non-uniform, one has to work with very small models to obtain fully quantitative data. Nevertheless, the free jet tunnel has proved to be a very useful tool for semi-quantitative or qualitative studies and for the study of basic flows.

Note on MHD Acceleration and Heaters for Hypervelocity Tunnels

The production of hypervelocity (i.e. high stagnation temperature) flows at very low density is extremely difficult, not only because of nozzle design problems, but mainly because of non-equilibrium effects. These effects are due to the fact that transformation of vibrational or electronic excitation energy into translational energy through molecular collisions is a very inefficient process and it takes many thousands of collisions to obtain equilibrium between these energy modes. Now, in the hot stagnation chamber of a hypervelocity tunnel, a considerable fraction of the energy is stored in the vibrational mode, which has to be converted to kinetic energy of translation during the expansion. The number of collisions experienced by a molecule during its travel through the nozzle may be quite insufficient to obtain equilibrium in the test section. Such a lack of equilibrium makes the interpretation of the data very difficult.

The solutions have been proposed to overcome this difficulty namely M.H.D. accelerations and high frequency heating of the flow after accelerations to moderate supersonic speed. Faraday, Hall and electrodeless M.H.D. accelerators have been proposed and tested in low density high speed wind tunnels, but their performance have been quite disappointing, only a marginal increase in the energy of the flow being observed.

The high frequency or electric arc heating of a supersonic flow led to the usual difficulty of non-equilibrium flow, as most of the energy introduced was under the form of vibrational or electronic excitation and could not be converted to translational energy in the nozzle.

Heaters for low density wind tunnels must handle very low mass flows (few grams per second) and the main problem in design are the limitations in temperature of solid materials. The graphite heater, initially proposed and built by the Bogdonoff group in Princeton, seems to represent an optimum temperature in excess of 2000°C being achieved easily in non-oxydizing gas flows. The design is quite simple and has been widely used. The gas is let to a screw-like helicoidal groove machined in a graphite cylinder, which is pressed into an external graphite cylinder. The whole graphite structure is heated by a high intensity electric current.

B. Low Speed, Low Density Wind Tunnels

Low speed, low density wind tunnels have received much less attention than their high speed counterpart and only few attempts have been made to produce uniform low speed rarefied flows to study basic problems such as the Knudsen layer structure or applied problems such as high altitude probe design and calibration. The pumping speed required is generally low and diffuser efficiency irrelevant in such application, but the achievement of reasonable flow uniformity is not so easy.

The convergent ducts with smooth contour and large contraction ratio used in conventional tunnels would have prohibitive length and usually lead to an exit velocity distribution close to the Poiseuille parabolic distribution.

These considerations led to the idea of uniform flow generation by gas leaking through a flat wall of uniform porosity. Of course, the porous part of the wall must be limited and edge effect will develop. However, a calculation based on the Stokes flow equations indicate that a rather extended central uniform region will occur, which can be optimized through the use of suitable non-uniform distribution of porosity.

Itrehus has shown that the non Maxwellian velocity distribution, composed of a half-range Maxwellian representing the effusive molecules and of a distribution of molecules scattered back towards the plane relaxes to a Maxwellian after a distance of the order of one mean free path. Preliminary tests and calculations indicate that uniform flows with velocities up to sonic may be obtained by this technique.

II. MEASUREMENTS IN RAREFIED GAS FLOWS

In this section, a description will be given of measurement systems which are typical for rarefied gas flows and are not straightforward extrapolations of conventional systems. However, typical difficulties occurring in conventional measurements, due to the low density level, will be discussed. These difficulties are most severe in the case of pressure measurement. A modulation method, used to handle the small signals occurring in low density measurements will also be described.

Free Molecule Probe

According to our definition of rarefied gas flows, the Knudsen number based on overall flow dimensions must not be very small. Therefore, if a probe can be made with dimensions very small compared with the global flow scale then the Knudsen number based on the probe dimensions could be made considerably larger than unity. Such a situation may be quite useful, for it must be kept in mind that calculations of free molecule effects usually reduce to simple integrations, and that probe disturbance to the incoming flow may be strongly reduced.

In fact, if the probe Knudsen number is large enough, one will be able to deduce certain moments of the flow velocity distribution from the measurement of a pressure, a heat transfer rate or a force. Unfortunately, the moments so obtained are usually half-range moments (i.e. moments obtained by integration over a portion of velocity space only) and the results are often difficult to interpret or to correllate with theoretical results, unless detailed information is available concerning the expected velocity distribution (e.g. Maxwellian velocity distribution).

In some cases, the interpetation of the measurement will involve interaction of gas molecules with solid surface. Unfortunately, this phenomenon is not entirely understood, and the available experimental data is usually inaccurate and far from complete. This represents a serious limitation for probes involving such gas-surface interactions.

The simplest free molecule probes do not in fact involve gas-solid interaction effects and are based on pressure measurement. The latter must be considered as unfavourable, in a sense, as pressure measurements are usually quite difficult at the pressure levels encountered in rarefied gas dynamics. The earliest free molecule orifice probes were studied and developed at the University of Toronto Institute of Aerospace, by Patterson et al.

The equilibrium pressure which develops inside the orifice probe is conditioned by the balance of the incoming molecular number flux and the effusive flux, these two fluxes being independent in the free molecule regime. At given temperature, the effusive flux is proportional to number density and therefore to pressure in the probe. As a result, the probe pressure will be a measure for the incoming number flux, which is the simplest half range moment of the distribution function of the external flow, i.e.

$$\iiint_{(\vec{\xi}.\vec{n})<0} (\vec{\xi}.\vec{n})\ f(\xi)\ d\underline{\xi}$$

where $\vec{n}$ denotes the external normal to the plane of the orifice.

Patterson has suggested to rotate the probe around an axis perpendicular to the flow, and thereby measure the incoming flux for several orientations of the normal $\vec{n}$ with respect to the flow direction. Ratios of the pressures recorded for several orifice plane orientation are related to the speed ratio $U/\sqrt{2RT}$ of the flow, for an equilibrium Maxwellian distribution. Corrections have been calculated for the case of slightly non Maxwellian distributions, such as the Chapman-Enskog velocity distributions resulting from small flow non uniformities.

It is possible to combine the output of two orifice probes to obtain the measurement of significant full range moments of the velocity distribution. For instance, if two orifice probes are arranged so that the first orifice faces the flow and the second is facing in the opposite direction, then the pressure difference will be proportional to

$$\iiint_{\vec{\xi}.\vec{n}_1<0} |\vec{\xi}.\vec{n}|\ fd\xi - \iiint_{(\vec{\xi}.\vec{n}_2)<0} |\vec{\xi}.\vec{n}|\ fd\xi = \iiint (\vec{\xi}.\vec{n}_1)\ fd\xi$$

where $\vec{n}_1, \vec{n}_2$ are the normal vectors to both orifices, and n represents the number density of the flow.

Reynolds et al. have used this property to measure the velocity distribution in a Knudsen layer.

It is important to note that the condition of a large value for the Knudsen number based on the probe diameter (and therefore on orifice diameter) is not sufficient to allow for a simple interpretation of the probe output. The thickness of the wall of the probe at the orifice must be small compared with the orifice diameter, and the latter diameter must be small compared with the internal probe diameter. If these geometrical conditions are not satisfied, then corrections must be applied or the probe must be precalibrated. Another geometric effect which has received little attention up till now, occurs when the probe is located close to a boundary or immersed body. In this case, molecules scattered from the probe may be rescattered by the body to the orifice. This effect has been investigated recently at VKI.

Impact probes, in which the orifice is replaced by a short, thin tube, have also been used widely (Hughes). The ratio between the pressure in the probe chamber connected to the tube and the number flux as obtained by a simple orifice probe is a strong function of speed ratio and of the angle of tube axis with flow direction. The dependence has been calculated by Hughes (UTIAS Report No. 103).

Some other free molecule probes have been proposed, in particular to overcome the difficulties connected with pressure measurments at low densities.

The hot wire techniques, based on the conventional hot wire anemometer probes, are interesting in that the Knudsen number based on wire diameter can be made large even at medium degrees of rarefaction, a fact which is used systematically in the classical Pirani vacuum gauge. A hot wire signal would be connected to the energy flux from the flow to the wire but the relationship involves the energy accommodation coefficient of the wire and this is found to be quite variable for a given probe, so that frequent calibrations are required. The design of the calibration system represents a serious problem in itself. A further difficulty in the use of hot wires at low densities results from the fact that the convective cooling effect due to the flow does not dominate conduction and radiation, as is the case in the conventional flow applications. One cannot expect, therefore, that the wire temperature will be even approximately uniform, and this further complicates the calibration and the interpretation of the results. At best, one can expect only semi-quantitative information from low density hot wire probes.

Force probes, based on the measurement of drag on small bodies of simple geometries (disks, spheres, etc.) have also been proposed. Again, the signal may be related to some moments of the distribution function, but the relation is very complex and involves the gas-solid interaction (in particular normal and, eventually, tangential momentum accommodation).

The results can be exploited only if the incoming flow exhibits a Maxwellian velocity distribution. On the other hand, small force measurements are not easier than low pressure measurements, unless special techniques such as flow modulation, are used. These considerations explain why force probes have not been extensively used.

Pressure Measurements at Low Densities

It is well-known that accurate absolute and differential pressure measurements are quite difficult at the levels of, say, 10 μHg and below. The main sources of difficulties are the following:

a) Outgassing effects
b) Very low response times
c) Thermal effects.

We will briefly describe these effects which dominated the field of rarefied gas dynamic measurements until the advent of electron beam techniques. The first two effects are essentially due to the smallness of the free molecule mass flows which develop in elongated ducts subjected to given pressure differentials. This is due to the large numbers of collisions of individual molecules with duct walls and the resulting low probability of passage. In the millitorr (μHg) pressure range of interest for rarefied gas dynamics, the result is a very long (10^2 s or more) time for pressure transducer, especially if the dead volume of the latter is large. Optimization of duct geometry may be of some help here.

The second effect, outgassing, is due to the emission of gas absorbed or trapped in the pressure measuring system. This will lead to non negligible pressure differentials which change with time and result in drifts of the transducer signal. The effect may be reduced but not suppressed by elementary precautions such as the use of clean surfaces, the reduction of parasitic cavities and the pumping down of the tunnel pressure for extended periods (several hours) before starting the test.

The last effect, thermal transpiration, is important in pressure systems exhibiting large temperature differentials, such as would occur in high stagnation temperature wind tunnels. The effect can best be explained at the hand of the simple example of two vessels maintained at different temperature levels, say T_1 and T_2 and communicating through an orifice with diameter small compared to the main free path. Steady state will exist when the molecular fluxes through the orifice in both directions will be equal. These fluxes are proportional respectively to $\rho_1 \; \bar{c}_1$ and $\rho_2 \; \bar{c}_2$ where ρ is the density and $\bar{c}$ an average molecular velocity in the direction normal to the orifice plane. Now, $\bar{c}$ is proportional to the square root of temperature and, according to the equation of state, ρ is proportional to the ration p/T. Flux balancing will therefore require that

$$p_1 / \sqrt{T_1} = p_2 / \sqrt{T_2}$$

so that the pressures at equilibrium will not be equal, their ratio being equal to the square root of the temperature ratio.

Concerning the pressure transducers themselves, one can say that they are very sensitive and therefore fragile and expensive instruments. Most of the commercially available pressure transducers operating in the range of interest for rarefied gas dynamics are based on the membrane displacement principle, the membrane displacements being usually recorded electrically as a variation of capacitance. The small changes of capacitance involved require high frequency A.C. Wheastone bridge measurements. Some of the transducers use the self balancing principle, whereby the membrane is always restored to its equilibrium position by an automatically controlled electrostatic force. It does not seem that this additional complication is really warranted in practice. Typical characteristic for membrane transducers are a linear response up to 1 torr and a sensitivity of the order at 10^{-2} to 10^{-3} m torr (μHg).

The Electron Beam Fluorescence Techniques

The development of the electron beam fluorescent probe, proposed for the first time for local gas density measurements by Schumacker and Gadamer in 1958, must be considered as a significant break-through in experimental rarefied gas dynamics. The usefulness of the technique has been further enhanced by the work of Müntz et al. who showed that local rotational and vibrational temperatures could also be measured and that one could even deduce the velocity distribution of the gas from Doppler shift of an emission line.

The electron beam fluorescent probe represents a near-ideal flow diagnostic device in the sense that it provides highly localized (~ 1 mm³) data practically without interfering with the flow, because of the low mechanical coupling between electrons and gas molecules. In addition the response time of the device is limited only by transducers (photo multiplier) and electronic amplifiers and is therefore very short for density measurements (μs). The probe can accordingly be used to study highly unsteady flows (shock wave and sound propagation) and does not require long running time facilities as do the pressure measurements.

The density probe, which is the simplest type of electron beam application and does not require highly sophisticated spectroscopic equipment, unfortunately yields only rather limited information on the flow properties. In low speed flows, the density variations although relatively small may often be directly related to pressure (quasi isothermal flows). In high speed flows, the density is not always of direct interest and it is generally not possible to directly connect density with pressure or temperature because of the highly non isentropic, non equilibrium nature of the rarefied gas flows of interest. However, density measurement may be very useful to verify theoretical results (shock structure) or obtain semi quantitative or qualitative information (upstream influence, etc.).

The principle of the electron beam fluorescence technique is quite simple. A thin (one or two millimeters diameter) focussed beam of medium energy electrons (10 to 50 KV) generated by a conventional electron gun (TV type oxyde cathode or, preferably, longer life, higher stability tungsten cathode) is sent through the low density gas. The molecules of the gas are excited by the beam electrons and may emit photons when returning to the ground state. This emission is in the visible spectrum and the path of the beam may be observed visually. If this spontaneous emission were the only process of transition of the excited molecules to ground state, then, for a given electron beam intensity, the radiation intensity emitted would be proportional to the number of excited molecules and therefore to the gas density. However, there are competing mechanisms for the transition to the ground state, the most important of which is de-excitation by intermolecular collisions (quenching collisions). The relative importance of the spontaneous emission process and of collisional transition is expressed by the ratio of collision time to the characteristic time for spontaneous emission (of order 10^{-8}s. and constant for a given emission process). This ratio is inversely proportional to the density level, and the quenching collision process will therefore be dominant at high densities. This yields the upper limit of density level at which the electron beam technique will be useful, which turns to be in the neighbourhood of 1 torr (1 mm Hg). The technique must therefore be considered as a typical low density measurement system.

The lower limit of applicability depends only on the sensitivity of the light intensity measurement, which is mainly limited by photomultiplier tube noise. The sensity will therefore usually be better for steady state measurements, especially if some type of modulation procedure is used. The accuracy of density measurements is very good and usually of the order of 1% .

The measurement of gas temperature by electron beam methods is based on spectroscopic measurements of rotational or vibrational line intensities, which yield information on the population of the various rotational and vibrational levels. If one assumes that the levels are in equilibrium, i.e. that the population is represented by a Boltzmann distribution, then the corresponding temperature can be calculated from the measurement.

The temperature measurement is obviously much more difficult than the density measurement, not only from the instrumental point of view, because a high resolution spectrometer must be used, but also because a much higher emitted intensity is needed. The latter requirement results from the fact that intensities of individual lines must be measured, rather than a total intensity and that intensity losses occur in the spectrograph. Short response time measurements are therefore difficult. Furthermore, the interpretation of the line intensity distribution is not easy as it involves the complex process of excitation by high energy electrons.

The resulting accuracy has been estimated at 2% for rotational temperature and 15% for vibrational temperature.

Measurement of the velocity distribution of the gas has also been proposed and carried out by Müntz and his group. Such measurements are based on Doppler shift and therefore require the presence of a strong, sharp line in the radiated spectrum, which exist only for certain gases, such as helium, for instance. A detailed spectroscopic analysis of the Doppler spread of the line, due to the molecular velocities, must be carried out, the quality of definition of the velocity distribution being related to the bandwidth of the spectral analysis. The results yield integrals of the distribution function of the type

$$F(\xi_x) = \iint f(\xi_x, \xi_y, \xi_z)\, d\xi_y\, d\xi_z$$

where ξ_x is the component of velocity in the direction of the optical beam going into the spectrometer.

High resolution measurement of the distribution function require narrow bandwidth and therefore sufficiently high total radiation intensities for given sensitivity of the photomultiplier and are thus possible only at rather high density levels and the corresponding response times tend to be high.

Of course, it is also possible to deduce the bulk velocity of a rarefied flow from the Doppler shift measurement, but it is clear that the accuracy of such a measurement will depend strongly on the speed ratio, because of the error on the computation of the average velocity will be a certain fraction of the typical random velocity. Therefore, the accuracy will be low for low speed flows (small speed ratios) but very good for hypersonic flows (high speed ratios). The accuracy on the estimation of the random velocities or translational temperature has been estimated at 1% .

The spatial resolution of an electron beam measurement depends essentially on the beam diameter. While this diameter can easily be decreased under 1 mm by careful electrostatic focussing, a broadening of the fluorescent region must be expected as a result of beam scattering by gas atoms and molecules. This effect can be reduced by increasing the beam acceleration voltage (up to 50KV) and by decreasing the distance between gun and measurement point or by the use of drift tubes (in which a sufficiently low pressure is maintained).

Another problem lies in the accurate determination of the actual electron beam intensity.

The electron beam probe has been widely used for the study of high speed flow and shock structures. The most successful studies of the structure of propagating shocks have been based on absorption measurements where the density was deduced from the total scattered intensity of the whole beam, a method applicable only to one-dimensional flow configurations, and if considerable data reduction is acceptable, to axisymmetric flows. The advantage of the method, however, is its much greater sensitivity (integration of effect over a long path), which is important if short response time ($<1\ \mu s$) are to be attained, as is the case for the measurement of the structure of a shock moving in a gas at rest in a shock tube.

Here are a few suggestions for extension of the use of the electron beam which do not require the use of very sophisticated spectrometric techniques:

- Study of diffusion of a tracer gas, using a specific strong emission line of this gas, well separated from background gas spectrum.
- Measurement of density gradients by measuring A.C. signal from photomultiplier, the beam being oscillated over a small region by sinusoidal electrostatic deflection (V.K.I.).
- Measurement of the density upstream of a small disk perpendicular to the flow.

This would yield information on the mass (or number) flux ρu, at least in high speed flow, but the interpretation of results unfortunately requires data on the normal accommodation coefficient.

- Measurement of acoustic wave propagation at very low density (wavelength of order of or smaller than the mean free path). An absorption measurement might be applied in the case of plane waves.

Flow Modulation Techniques

Flow modulation techniques for rarefied gas dynamic measurements were proposed and used at V.K.I. to improve the sensitivity of low level measurements in the presence of noise and drifts and to obtain high accuracy differential measurements. The techniques are based on the use of "lock-in" amplifier systems, which are capable of extracting a very small signal from a considerable noise background, if the signal is of very well defined frequency. This latter condition will certainly be met if the signal of interest is known (or forced) to be synchronous with a sinusoidal reference signal. The lock-in amplifier uses tuned A.C. amplifier stages but its main rejection of noise results from the use of a phase sensitive output rectifier which is "locked" to the reference signal. This rectifier, coupled with a low press recorder system effectively produces an output signal proportional to the Fourier component of the input signal with respect to the reference signal, that is, to the integral

$$T^{-1} \int_0^T \text{(input signal)}.\ \text{(reference signal)}.\ dt$$

for very large values of the sampling time T. The resulting band pass of the system may then be made very narrow, even compared to the performance of the best passive filters, while maintaining a perfect stability.

The principle may be used in connection with low density measurement if a method can be found whereby the flow effects are modulated with a given, reference frequency. This is comparatively easy if a high speed, free jet tunnel is used, because the full flow may be chopped periodically by means of a disk with window, rotating in front of the orifice in the stagnation chamber. In this manner, all the signals from transducers (force pick-up, electron beam, cold wire, etc.) will exhibit a modulated component at the chopper frequency which will represent the effect of the periodically interrupted flow, while all other effects (vibrations, drifts) will be represented by components scattered over a wide frequency spectrum.

The selection of the chopping frequency raises some problems. It must, of course, be chosen low enough so that the influence of the transients due to flow establishment and to probe response time effects can be made negligible (this condition is not easily satisfied in the case of low speed flows which require a rather long establishment time and of pressure measurement, because of long response times).

On the other hand, the frequency should be high enough so that drift effects may be cancelled without the use of a prohibitively long measurement time. The measurement time is, of course, determined by the low pass filtering applied to the output of the phase sensitive rectifier. A modulation frequency of 15 Hz was used successfully at V.K.I.

Note that the flow establishment time for low speed flows may be too long for the use of such a frequency, but it is also possible to modulate by the motion of a screen which will be alternately put in front of or removed away from the probe.

The modulation technique has also been used successfully at V.K.I. to measure small flow disturbances produced by bodies, such as the far field or the interference of two parallel strips in near free molecule flow (first collision effects).

Detailed measurements were made of the upstream influence in a hypersonic flow along a flat plate. In these tests, the plate was replaced by a flat disk sector rotating around an axis perpendicular to the flow direction and the disk. Curvature of the leading edge and transients due to the passage of the corners of the sector were shown to be negligible. The fluctuating signal from the photomultiplier of an electron beam system was recorded by the lock-in amplifier system controlled by a reference signal in phase with the rotation of the disk. This fluctuating signal results from the flow disturbance created when the sector was exposed to the flow. Because of the high noise and drift rejection capability of the system, extreme sensitivity was achieved and disturbances were detected at points farther than ten mean free paths upstream of the disk leading edge.

III. EXAMPLES OF BASIC EXPERIMENTS IN RAREFIED GASDYNAMICS

The examples to be presented are not intended to give a complete survey of the history and present development of experimental rarefied gas dynamics. The selection was made so as to exhibit typical low density measurement techniques and particular emphasis was put on tests on basic flow phenomena which have been the subject of extensive theoretical analysis. Many experiments dealing with more complex flow configurations, or yielding integrated quantities such as forces, heat fluxes etc., which are obviously of considerable practical importance, have been omitted in this presentation.

1. Shock Structure Measurements

A great number of experiments have been performed on shock structure and the advent of the electron beam probe provided an ideal tool for this type of research. Both steady state normal shocks, produced by placing a suitable "shock holder" in the flow of a rarefied high speed wind tunnel and propagating normal shocks produced in a low density shock tube, have been studied.

Most of the tests were aimed at obtaining the profile of some physical quantity (pressure, recovery temperature or density) in the shock. Early experiments used free molecule probes (hot wires, orifice probes) but the results were rather inaccurate and difficult to interpret (Talbot and Sherman, 1956).

Measurements of the density profile were first attempted by Hansen and Hornig (in 1960), using the optical reflectivity principle.

Most of the recent results on density profiles were obtained using electron beam densitometry. The results published by B. Schmidt in 1969 appear to be the most complete and accurate. Schmidt used an electron beam attenuation measurement (which integrates the density effect) in the low density shock tube at the California Institute of Technology. The density profiles of shocks up to Mach numbers up to 8 in Argon exhibited signficant deviation from profiles calculated from the Navier-Stokes equations (up to 20%). The agreement with the predictions of the bimodal approximation of Mott-Smith was fair at medium Mach numbers.

A comparison was also made, for a shock at $M = 8$ in Argon, with calculations made by Bird using the Monte Carlo direct simulation method. The agreement was found to be almost perfect ($< 1\%$ error).

The density profiles, which represent the evolution of the simplest moment of the distribution function through the shock, are not extremely sensitive to the assumptions made to obtain the approximate theoretical results. Profiles for higher order moments (temperature, stresses, etc.) would therefore yield much more conclusive information about the validity of the various approximations used in shock structure calculations.

Müntz and Harnett (1969-1971) were the first to attempt a direct measurement of the distribution functions, using a spectrometric analysis of the Doppler shift of the radiation induced by an electron beam. The tests were carried out on a steady shock at $M = 1{,}59$, generated by a shock holder placed in the nozzle of the University of California-Berkeley low density wind tunnel.

Unfortunately, the shock Mach number used is too small to expect spectacular deviations from the quasi-equilibrium Chapman-Enskog solution, but the results published are very promising.

The Doppler spread of a well defined emission line of helium were measured in directions perpendicular and parallel to the plane of the shock. This yields the following integrated distribution functions:

$$F(\xi_x) = \iint f(\xi_x, \xi_y, \xi_z)\, d\xi_y\, d\xi_z$$

$$F(\xi_y) = \iint f(\xi_x, \xi_y, \xi_z)\, d\xi_x\, d\xi_z$$

where the x- and y-axis are respectively perpendicular and parallel to the plane of the shock.

Each point of a velocity distribution curve was obtained after 30 seconds of photon counting, which led to a statistical scatter of about 1% .

The main cause of error was estimated to be in the monitoring of the current distribution in the electron beam, which was measured by an independent photomultiplier tube. This resulted in a scatter by the order of 1.5% of the *maximum intensity* corresponding to the maximum value of the partially integrated distribution function. Therefore, the relative error on the values of the distribution function near the wings of the distribution could be much larger (15% for values of F equal to $0.1\, F_{max}$). The error in the evaluation of temperature was of the order of 2% and the error on the bulk velocity of the order of 3.5% .

As expected, in view of the rather low shock Mach number, the density and bulk velocity profiles agreed, within experimental errors, with the profiles predicted by the Navier-Stokes equations.

In their first report (1969), Müntz and Harnett indicated the existence of a significant discrepancy with the prediction of the Chapman-Enskog theory for the profiles of distribution half-widths in both the directions parallel and perpendicular to the flow. The discrepancies were much larger than the estimated error in measurement, especially for the distribution of the molecular velocity component parallel to the flow.

This interesting conclusion suggested a finer analysis of the deviations of the measured distribution functions from the prediction of the Chapman-Enskog quasi-equilibrium theory (which leads to the Navier-Stokes equations for the macroscopic observables). The results were published in 1971. Although the deviation from equilibrium in the weak shock, characterized by the ratio of viscous stress to pressure, never exceeds 12% in the shock, significant quantitative deviations from the Chapman-Enskog non-equilibrium correction term were observed. The correction term to the Maxwellian distribution deduced from the experiment agreed qualitatively with the Chapman-Enskog prediction however.

These measurements must, of course, be considered as a "tour de force" because of the small magnitude of the non-equilibrium effects occurring in such a weak shock. It would be very interesting to perform similar measurements on a much stronger shock, but no attempts have been made, so far, mainly because of considerable difficulties in obtaining a steady, uniform high Mach number flow with sufficiently large dimensions.

2. Translational and Rotational Freezing in Underexpanded Jets

The rarefied flow in the free expansion of a gas jet from an orifice into a near vacuum has many applications in the fields of high altitude rocket operation, high speed flow production in low density wind tunnels and generation of molecular beams.

A simplified spherical source flow has been the subject of many theoretical studies (Hamel and Willis, 1966; Edwards and Cheng, 1965; etc.).

The hypersonic part of the flow in the jet is rather well approximated by a source flow, but there is very little information on how the difference between pure source flow and jet flow might affect the theoretical results. Considerable rarefaction effects on the temperature distributions are predicted by these theories.

In fact, if the Knudsen number based on the gradients in the hypersonic part of the flow is sufficiently large, the theory predicts a completely different behaviour for the "parallel temperature" and "perpendicular temperature" i.e. for the moments

$$\frac{1}{n} \iiint f\, c_x^2\, d\xi_x\, d\xi_y\, d\xi_z, \quad \frac{1}{2n} \iiint f(c_y^2 + c_z^2)\, d\xi_x\, d\xi_y\, d\xi_z$$

where the local x-axis is in the direction of the (radial) flow. c_x, c_y and c_z are the components of the random molecular velocities.

The equilibrium theory of isentropic expansion predicts that the two moments are equal and decrease to zero according to a $r^{-4/3}$ law in a source flow. The calculations based on suitable approximations to the Boltzmann equation indicate that the parallel temperature must tend to a non zero asymptotic value ("freezing" of the longitudinal temperature) while the lateral temperature decreases to zero but less rapidly than predicted by the isentropic law. As a result, the velocity distribution function must exhibit a strong "ellipsoidal" character.

Of course, my approximations must be made to the full Boltzmann equation in order to obtain these theoretical results. The Hamel-Willis theory is valid only for Maxwell molecules, although an arbitrary coefficient is introduced in the collision frequency to allow for some kind of matching with actual gas properties. The theory assumes that the hypersonic part of the expansion is rarefied and uses the hypersonic approximation for moments suggested by Hamel. This approximation leads to a closed set of differential equations for the low order moments and, in particular, for the parallel and transversal temperatures.

Experiments on this type of flow therefore appear to be very interesting, especially if a direct measurement of distribution function or parallel and transversal temperatures could be made because the deviation from equilibrium predicted by theories is considerable. (Note that density and bulk velocity distributions are trivial in this case).

Müntz (1967) has used the electron beam method for distribution function measurement. The results were in good qualitative agreement with theory. A large difference (50 %) was observed between the parallel and perpendicular temperatures at the station farthest away from the orifice and the distribution function presented a typical "ellipsoidal" behaviour. The tendency for the parallel temperature to freeze was clearly exhibited although freezing occurred earlier than predicted by theory. The perpendicular temperature approximately followed the isentropic law.

Most of the differences observed with respect to the Hamel Willis theory could be attributed to the fact that the jet flow is not a pure spherical source flow.

It should be noted also that the discrepancy with the equilibrium law is represented by a factor five at the station of highest rarefaction, which clearly indicates the magnitude of molecular effects in these flows.

A free expanding argon jet has also been studied experimentally by the Princeton group (Abuaf, Anderson, Andres, Fenn and Miller), who used a completely different technique to obtain data on the distribution function. The method is based on the skimmer principle for the generation of molecular beam from hypersonic flows. The skimmer is a kind of conical orifice probe, the orifice being at the apex of the cone. The cone angle and thickness are small so that no normal shock is produced upstream of the orifice. The interior of the conical skimmer is connected to a high vacuum chamber. This device is expected to extract a free molecule jet from the hypersonic flow. The jet propagates into the vacuum chamber a collisionless molecular beam and its velocity distribution may be analyzed using typical molecular beam instrumentation.

The simplest way to measure lateral velocity distribution is to measure the angular spread of the jet far enough of the orifice, using standard molecular flux detector (usually some kind of ionization chamber).

The distribution of longitudinal velocity may be obtained by a time of flight measurement. The beam is periodically chopped by a rotating disk with window and the time history of the detector signal is recorded on an oscilloscope. This yields information on the velocity distribution of the molecules in the beam.

The accuracy of the longitudinal velocity distribution may be considerably improved by the use of a standard multidisk velocity selector which allows steady state measurement of each point of the distribution function.

The results of the tests are again in general qualitative agreement with the prediction of the approximate kinetic theories, the behaviour of longitudinal and lateral temperatures being entirely different and the first one exhibiting a clear tendency to "freezing". However, the quantitative behaviour of the lateral temperature is in disagreement with the theories in that it decreases more rapidly than the isentropic temperature. The discrepancy might be due to deviation from pure course flow, skimmer interaction or a systematic error, but has not yet been explained.

Rotational temperature freezing in polyatomic gases is an effect similar to the translational freezing (longitudinal temperature freezing) but occurs at somewhat higher density levels, because of the higher average number of collisions needed to obtain equipartition between translational and rotational modes.

The effect has been studied experimentally by Marrone (1967) using the electron beam fluorescence method. A spectral analysis is made of the intensities of the various rotational emission lines, from which the population of the rotational levels is deduced and a rotational temperature calculated under the assumption of rotational equilibrium.

The results clearly indicate a tendency to freezing and are in fairly good agreement with a theory developed by Repetski and Mates (1971).

3. Hypersonic Rarefied Flow in the Vicinity of the Leading Edge of a Thin Flat Plate

The hypersonic flow in the neighbourhood of the leading edge of a thin flat plate is of fundamental importance for the study of high speed laminar boundary layers and occurs in many practical applications. It is well known that the influence of the local flow near the leading edge may be felt over considerable distances downstream. Viscous interaction effects are large in this region even at high Reynolds numbers. There exists considerable doubt as to the applicability of the Navier-Stokes formalism to the leading edge flow for large Mach number at any Reynolds number if the leading edge is sufficiently sharp.

This is therefore a very important flow problem, but its theoretical treatment is quite difficult because of the large non linear effects involved. Treatments based on the Navier-Stokes equations have been given by Oguchi (1967), Shorenstein and Probstein (1968), Rudman and Rubin (1968), Chow (1967) and Cheng et al. (1969).

Gas kinetic analysis have been proposed by Charwat (1961) and later by Hamel and Cooper (1969) (first and second collision approximation valid in the immediate neighbourhood of the leading edge) by Kogan and Degtyarev (1965) and Vogenitz et al. (1970) using the Monte Carlo direct simulation method, and, more recently, by Huang et al. (1970-1973) using the BGK and the ellipsoidal statistical models in conjunction with a discrete velocity ordinate numerical integration method. All these approaches, with the exception of the Monte Carlo simulations, are based on drastic approximations for which error estimations appear to be out of question.

Experiments in this area are therefore of crucial importance. Measurements of density distributions near the leading edge, using electron beam densitometry have been performed at Princeton by Mc Croskey et al., at the N.P.L. by Lillicrap and Berry and at V.K.I. by De Geyter et al., the latter using a modulation technique.

These density distribution measurement yield important information about the upstream influence due to scattering of molecules upstream of the leading edge and about the formation of the merged shock layer.

Experiments with high Mach number flows ($M > 6$) and high ratio of stagnation to plate temperature (hyperthermal case) are most interesting. In these situations, the mean free path of the molecules scattered by the plate into the flow is much smaller than the mean free path for free stream molecules colliding with scattered molecules, the ratio being of the order of the speed ratio S_∞ of the free stream. This explains why values of surface fluxes such as normal pressure and heat transfer are much larger near the leading edge than predicted by free molecule theory (Hamel and Cooper, 1969).

Lillicrap and Berry were able to obtain density profiles near the plate by injecting the electron beam through a pressure balanced drift tube leading to a hole in the model. Their result indicates upstream influences on density of 25% at 1 free stream mean free path and of 5% at 3 mean free paths.

The downstream density rise ("shock strength") in the lateral direction, plotted against distance along the plate normalized with respect to free stream mean free path agrees very well with the Monte Carlo calculations of Vogenitz near the leading edge of the plate. No agreement was found with the theoretical results of Huang and Hwang (1970).

Other measurements using electron beam densitometry in hypersonic flows have been reported by Hickman (1970) for a wedge and by Berlin and Scott (1971) for a disk perpendicular to the flow. In the latter case, the highest Mach number was about 6 and the Knudsen number based on disk diameter about 0.1. A considerable upstream influence was observed as a result of the large thickness of the strong shock. No detailed theory exists for the flow conditions occurring in these tests.

4. Experiments on Couette Flow and Heat Transfer in Plane Parallel and Cylindrical Geometries

A great number of experiments on drag of rotating cylinders and of heat conduction between parallel plates or co-axial cylinders, have been performed at low density levels since the early investigations of Knudsen at the end of the last century. Detailed data about velocity, density and temperature profiles have been published only recently (e.g. the work of Reynolds at V.K.I., on the Kramers Problem, 1974).

Alofs et al. have published two papers (1971) reporting density distribution measurements by electron beam probe. The first paper deals with cylindrical Couette flow in the transition regime at a Mach number close to unity. This is not a pure experiment, in the sense that the density variations are due to centrifugal and temperature effects and do not yield direct information on the velocity distribution. Large relative discrepancies were noticed between the observed relative density variations ($< 1\%$) and those predicted by the Navier-Stokes solution, even taking account of slip and Burnett correction terms ($\sim 20\%$) at the lowest Knudsen number (~ 1). No gas kinetic treatment has been given for this flow configuration.

The second paper by Alofs et al. deals with the more fundamental problem of a rarefied gas placed between parallel plates with widely different temperatures. Although most theoretical gas kinetic treatments of the parallel plates thermal problem involve linearization and are therefore not expected to be valid for plate temperature ratios very different from unity, Liu and Lees (1961) and Lavin and Haviland (1962) have given approximate solutions valid for the non linear case. These solutions are based on moment methods using the free molecule regime type discontinuous velocity distribution function proposed by Lees. No estimation of the accuracy of such a treatment is possible.

At the highest Knudsen number of the tests with helium, $Kn = 0.4$, the tests results agree fairly well with the Liu-Lees theory and already diverge strongly from the predictions of continuum theories.

THE STRUCTURE OF PLANE SHOCK WAVES

Wladyslaw Fiszdon
Institute of Fundamental Technological Research

1. Introduction

Shock waves are the most conspicuous, often encountered, phenomena of gas dynamics connected with the non-linear character of propagation of disturbances [1–6]. A shock wave occupies a relatively narrow region of the flow field and is characterized by rapid changes of the hydrodynamic and thermodynamic, HT, quantities describing the state of the gas.

In the case of an ideal gas (without any dissipation effects) this narrow region is reduced to a surface of discontinuity, which is a very convenient idealized model for theoretical considerations. However, under real physical conditions, the gas passing through a shock wave transforms its primary state of equilibrium, by means of energy and momentum exchange in intermolecular collisions, into a new one. This transformation occurs over a finite time within a finite width.

The problem of the shock wave structure is to find a quantitative description of this transition region. As the physical processes of collisions determine the character of the flow in this region, the Boltzmann (B.) equation seems to be the most appropriate one for its full mathematical description. It is well known that the solution of the full B equation in the most general case is so far impossible. We must look first for means of simplifying the problem from different aspects. Some of them will be recalled below.

During the *collisions* many different *molecularr processes* connected with the possible *physical states* of the *particles* can take place. The possibility of excluding different physical processes in the investigation of a given problem depends on the *characteristic time involved i.e. the time necessary to reach a state of equilibrium* after an initial disturbance. These characteristic times, called also relaxation times, for the different physical processes, can be arranged in the following sequence, taking into account their respective values:

$$\tau_c \ll \tau_t < \tau_r \ll \tau_\nu \ll \tau_{dis} \sim \tau_{ion}$$

Here a range of temperature below 10^4 K and pressures below 10 atm was assumed and for this range the collision time τ_c is a few orders of magnitude smaller than the translational relaxation time τ_t. The rotational relaxation time τ_r is of the same order or one or two orders of magnitude larger than τ_t. The vibrational relaxation times τ_ν and particularly the dissociation and ionization relaxation times τ_{dis}, τ_{ion} are many orders of magnitude larger than the others. In the considered temperature range radiation effects can be neglected.

Taking advantage of the different orders of magnitude of the characteristic relaxation times a separation of the above physical processes is realized. For shock wave structures of monatomic molecules, in which we will be mostly interested in, only translational relaxation effects are relevant.

To facilitate our task and understanding of the above wave structure and also to build up a clear picture of its mathematical analysis and its physical character we will consider the simplest idealized case which from the macroscopic point of view is a steady one-dimensional process. To be able to do so it is necessary to assume that the whole flow field, including the shock wave, which is perpendicular to the macroscopic flow velocity, is stationary.

In these conditions B. equation for the case of a one-dimensional stationary flow without external forces is:

$$(1) \qquad c_x \frac{\partial f}{\partial x} = Q(f,f) = \int (f' f_1' - f f_1) g\, b\, db\, d\epsilon\, dc_1$$

where $Q(f, f)$ is the collision integral, $g = c - c_1$ is the relative velocity, ϵ is the angle between the plane of the *trajectory* and a fixed plane in space, b is the impact parameter and the distribution functions usually appearing in the collision integral are denoted by f with appropriate subscripts and superscripts.

The boundary conditions necessary to complete the formulation of the problem are in this case, without gas solid surface interaction effects, very simple. For a reference frame fixed to the shock wave, the gas at infinity upstream and downstream is in equilibrium described by corresponding Maxwellian distribution functions

$$(2) \qquad f_i = n_i \left(\frac{\chi_i}{\pi}\right)^{3/2} \exp\{-\chi_i[(c_x - u_i)^2 + c_y^2 + c_z^2]\}, \ (i = 1,2)$$

where:

$$\chi_i = \frac{m}{2kT_i}, \quad f_1 = f(x \to -\infty), \quad f_2 = f(x \to +\infty),$$

$c = (c_x, c_y, c_z)$ is the molecular velocity, x is in the direction perpendicular to the shock wave, k is Boltzmann constant u_1, n_1, T_1, u_2, T_2 are the HT quantities at infinity upstream and downstream, gas velocity in the x direction, number density and temperature respectively.

As known the moments of B equation, deduced using the collision invariants, give the basic conservation laws of mass, momentum and energy:

$$\begin{cases} \dfrac{\partial(\rho u)}{\partial x} = 0, & \dfrac{\partial}{\partial x}\left[\rho u\left(\dfrac{u^2}{2} + e\right) + uP_{xx} + q_x\right] = 0, \\ \dfrac{\partial}{\partial x}(\rho u^2 + P_{xx}) = 0, & \end{cases} \tag{3}$$

where e is the internal energy, P_{xx} the x-component of the stress tensor and q_x the heat flux in the x direction and the density $\rho = mn$, m is the mass of the particle.

At $\pm\infty$ where the gradients of the HT quantities are zero the above conservation laws lead to the following relations between the macroscopic flow parameters which are to the well known Rankine-Hugoniot (RH) conditions

$$\begin{cases} n_1 m u_1 = \rho_1 u_1 = \rho_2 u_2 = n_2 m u_2 = \mathcal{M} \\ p_1 + \rho_1 u_1^2 = p_2 + \rho_2 u_2^2 = \mathcal{P} \\ h_1 + 1/2\, u_1^2 = h_2 + 1/2\, u_2^2 = \mathcal{H} \end{cases} \tag{4}$$

where h is the enthalpy, and $\mathcal{M}$, $\mathcal{P}$, $\mathcal{H}$ are constants corresponding to mass flux, momentum flux and energy flux respectively.

The HT parameters entering in the expression for the distribution function at $\pm\infty$ (2) must satisfy the above compatibility conditions (4). It is worth recalling here that for real physical processes, beside the conservation laws, the additional thermodynamic requirement concerning the increase of the specific entropy must be satisfied

$$\delta\gamma \geqslant 0 \tag{5}$$

As no solution of the full B equation has been obtained so far – even in this simple case, except using the Monte Carlo numerical method, recourse must be sought to approximate methods of solving the problem of the shock wave structure (sws). Fortunately mostly we are interested in a limited number of moments of the distribution function and not in the function f itself which provides us with a more detailed description of the flow than the minimum necessary for its present physical interpretation. This is an additional justification for seeking approximate solutions giving the correct lower moments, corresponding to the HT quantities.

For the selection of the approximate methods for an adequate description of the shock wave structure the following physical considerations should be recalled: At Mach numbers (M) close to one, $M - 1 \ll 1$, the shock waves are weak, and the departure from the state of equilibrium is small. The shock wave thickness, as can be seen on fig. 1, is many mean free paths thick as M approaches unity. At higher M the departure from equilibrium is large, the shock wave is only a few mean free paths thick and the gradients of the HT quantities become very large.

Thus we can expect that approximate theories based on different assumptions will be appropriate for different ranges of M. In particular, for low M, theories based on small perturbations from equilibrium and continuum gas descriptions may be satisfactory whereas for high M it is essential to use approximations allowing for a large departure from equilibrium.

Very many approximate theories have been proposed and used during the last seventy or so years, for the determination of the shock wave structure (sws). As so far there is no approximate theory that proved itself to be most convenient and suitable for our problem, we will present a selection of the existing ones in four groupings named: continuum gas theories, kinetic models of B collision integral, kinetic models of the distribution functions, Monte Carlo methods.

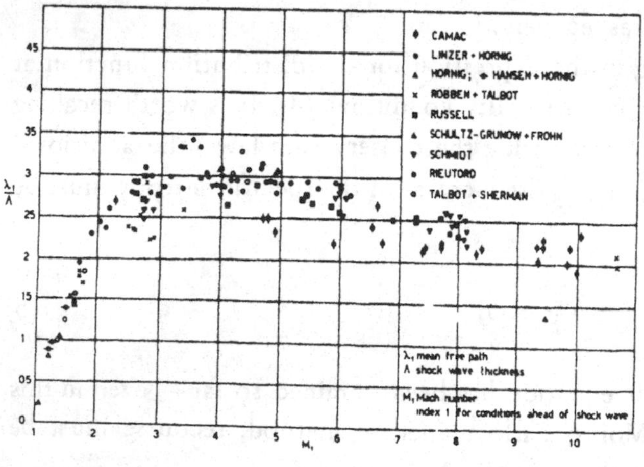

Fig. 1. Variation of inverse shock wave thickness with Mach number

II. Continuum Gas Theories

Without discussing the arguments used by the sponsors and opponents of the relevance of the continuum and the kinetic gas theories [1,5,7,9,10] in the sws. problem we will consider the hydrodynamic Navier Stokes (NS) equations as resulting from the B equations by means of the Chapman Enskog (Ch.E.) procedure [8] and hence essentially valid for small departures from equilibrium. The transport coefficients appearing in the NS equations can be determined either from the intermolecular potentials or from macroscopic measurements of the coefficients of viscosity and heat conductivity.

In the hydrodynamic NS approximation of B equations the stress tensor and the heat flux vector correspond to the simple linear Newton and Fourier relations

$$\left\{\begin{aligned} P_{xx} &= P - \frac{4}{3}\mu' \frac{\partial u}{\partial x} = P - \tau_{xx} \\ q_x &= -\varkappa \frac{\partial T}{\partial x}, \end{aligned}\right. \tag{6}$$

where $4/3\,\mu' = 4/3\,\mu + \mu_b$, μ and μ_b are the shear and bulk viscosity coefficients, $\varkappa$ is the coefficient of thermal conductivity, they are essentially functions of T. The integral of (3) with the above relations follows immediately

$$\left\{\begin{aligned} &\rho u = \mathcal{M} \\ &p + \rho u^2 - \frac{4}{3}\mu' \frac{\partial u}{\partial x} = \mathcal{L}\,\mathcal{M} \\ &h + \frac{u^2}{2} - \frac{4}{3}\frac{\mu'}{\rho}\frac{\partial u}{\partial x} - \frac{\varkappa}{\rho u}\frac{\partial T}{\partial x} = \mathcal{E}, \end{aligned}\right. \tag{7}$$

$\mathcal{E}$ and $\mathcal{L}$ are new constants.

For a perfect gas, which we will consider,

$$h = c_p T, \quad p = R\rho T = (c_p - c_\nu)\rho T, \quad \gamma = \frac{c_p}{c_\nu} \tag{8}$$

where c_p, c_ν are specific heats with constant pressure and volume respectively, the above set of equations (7) becomes:

$$(9)\quad \begin{cases} \dfrac{4}{3}\dfrac{\mu'}{\mathcal{M}}\dfrac{\partial u}{\partial x} = \dfrac{RT}{u} + u - \mathcal{L} = \dfrac{M(T,u)}{\mathcal{M}} \\[2ex] \dfrac{\varkappa}{\mathcal{M}}\dfrac{\partial T}{\partial x} = \dfrac{RT}{\gamma-1} - \dfrac{u^2}{2} + \mathcal{L}u - \mathcal{E} = \dfrac{L(T,u)}{\mathcal{M}} \end{cases}$$

R is the gas constant.

Eliminating in the above equations T a single equation for u follows:

$$(10)\quad \begin{cases} -\dfrac{4\varkappa}{3R\mathcal{M}^2}\dfrac{\partial}{\partial x}\left(\mu' u \dfrac{\partial u}{\partial x}\right) + \left[\left(\dfrac{4}{3}\dfrac{\mu'}{(\gamma-1)\mathcal{M}} + \dfrac{2\varkappa}{R\mathcal{M}}\right)u - \dfrac{\varkappa\mathcal{L}}{R\mathcal{M}}\right]\dfrac{\partial u}{\partial x} = \\[2ex] = \dfrac{\gamma+1}{2(\gamma-1)}u^2 - \gamma\mathcal{L}u + \mathcal{E} \equiv \dfrac{\gamma+1}{2(\gamma-1)}(u-u_1)(u-u_2) \end{cases}$$

The identity on the r.h.s. above results from the condition that at infinity upstream and downstream of the shock wave all gradients vanish and the gas velocities are u_1 and u_2 respectively. G.I. Taylor [11] obtained an approximate solution of (10) for the case of very weak shock waves $(u_1 - u_2)/u_1 \ll 1$ i.e. $M_1 - 1 \ll 1$, M_1 is the Mach number upstream for a shock fixed coordinate system. Comparing the orders of magnitude of terms with second and first derivatives, their ratio is of the order of λ_1/Λ and hence very small for the case considered. Here λ_1 is the mean free path upstream estimated from $\mu_1' \sim \rho_1 \bar{c}_1 \lambda_1$ with $c \sim u$ the thermal velocity and Λ is the shock wave thickness. Thus neglecting the first term in (10) and assuming that the coefficient of $\partial u/\partial x$ can be taken as constant, with values of the quantities in the square bracket corresponding to conditions far upstream (with subscript 1) equation (10) becomes then a first order differential equation, which can be easily solved, giving the velocity structure

$$(11)\quad \frac{x}{A} = \log\frac{u_1 - u}{u - u_2}, \quad \text{or} \quad \frac{u - \dfrac{u_1-u_2}{2}}{u_1 - u_2} = -\frac{1}{2}\tanh\frac{x}{A},$$

where:

$$A = \frac{2}{(u_1-u_2)(\gamma+1)}\frac{\mu_1}{\rho_1}\left[\frac{4}{3}\frac{\mu_1'}{\mu_1} + \frac{1}{P_r}\left(\gamma - \frac{1}{M_1^2}\right)\right], \quad P_r = \frac{c_p\mu}{\varkappa}$$

P_r is the Prandtl number

Using (7) i.e. $\rho u = \rho_1 u_1 = \rho_2 u_2$ the density structure follows immediately from (11):

$$\frac{\rho}{\rho_1} = \frac{1 + e^{x/A}}{1 + \rho_1/\rho_2 \, e^{x/A}}, \quad \text{with} \quad \frac{\rho_2}{\rho_1} = \frac{(\gamma+1)M_1^2}{(\gamma-1)M_1^2 + 2} \quad \text{from RH conditions} \tag{12}$$

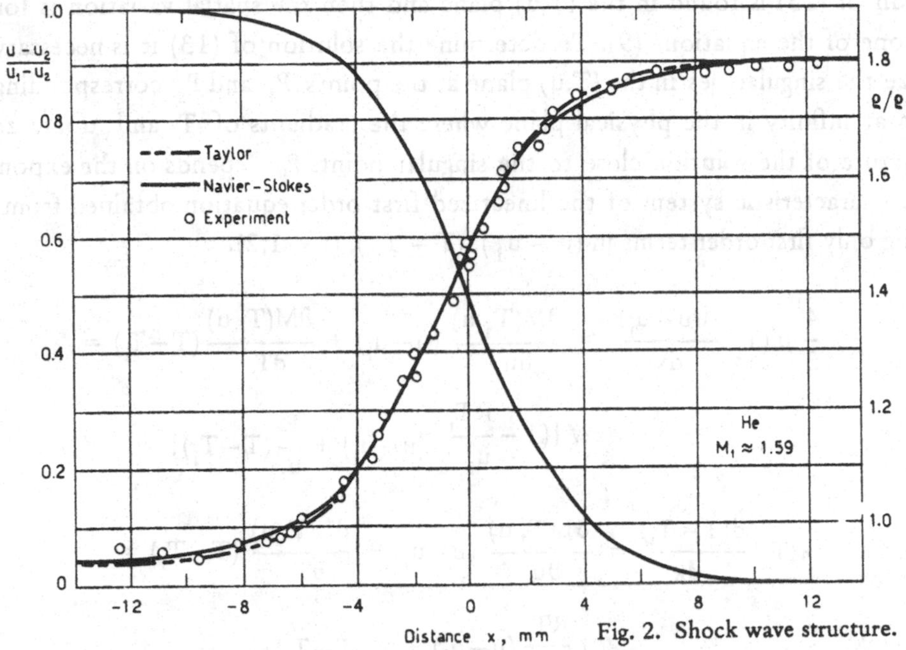

Fig. 2. Shock wave structure.

This structure is compared to fig. 2 with the results obtained from the full solution of the NS equation and with experimental results of Muntz and Harnett, all at M = 1.59. The agreement with the simplified solution is remarkable. Besides neglecting the highest order term, the main deficiency of the above approximation is the assumption concerning the constancy of the coefficients in (10) and in particular of viscosity and heat conductivity. It should be noted that taking the ratio of the two equations (9)

$$\frac{\kappa}{4\mu'/3} \frac{\partial T}{\partial u} = \frac{\frac{RT}{\gamma-1} - \frac{u^2}{2} + \mathscr{L} u - \mathcal{E}}{\frac{RT}{u} + u - \mathscr{L}} \tag{13}$$

as $\kappa/\mu' = \kappa(T)/\mu'(T)$ is practically independent of the temperature the sws in the

(T, u), (T, ρ) or (ρ, u) plane will not be affected by the law of variation of the transport coefficients with temperature. However the sws in space coordinates will be strongly affected by the temperature dependence of μ' and $\varkappa$.

Exact numerical solutions of the NS equations have been obtained by Gilbarg and Paolucci [7] and are discussed by Smolderen [13] and others. Essentially the solution of (13) is found in the (T,u) plane and then the spatial variation is found from one of the equations (9). To determine the solution of (13) it is necessary to analyze the singularities in the (T,u) plane at the points P_1 and P_2 corresponding to points at infinity in the physical plane where the gradients of T and u are zero. The nature of the solution close to the singular points P_i depends on the exponent of the characteristic system of the linearized first order equation obtained from (9) keeping only first order terms in $(u - u_i)$, $(T - T_i)$, $(i = 1,2)$.

$$\frac{4}{3}\mu'(T_i)\frac{d(u-u_i)}{dx} \simeq \frac{\partial M(T_1 u)}{\partial u}(u-u_i) + \frac{\partial M(T_1 u)}{\partial T}(T-T_i) =$$

$$\mathscr{M}\left[\left(1-\frac{RT_i}{u_i^2}\right)(u-u_i) + \frac{R}{u_i}(T-T_i)\right]$$

(14)

$$\varkappa(T_i)\frac{d(T-T_i)}{dx} = \frac{\partial L(T,u)}{\partial u}(u-u_i) + \frac{\partial L(T,u)}{\partial T}(T-T_i) =$$

$$\mathscr{M}\left[\frac{RT_i}{u_i^2}(u-u_i) + c_v(T-T_i)\right]$$

Hence for solutions of the form $e^{\zeta_i x}$ the exponents are roots of the following characteristic equation in ζ_i

$$(15)\quad \begin{vmatrix} 1-\frac{RT_i}{u_i^2} - \frac{\frac{4}{3}\mu'}{\mathscr{M}}\zeta_i\,, & \frac{R}{u_i} \\ \frac{RT_i}{u_i}\,, & c_v - \frac{\varkappa_i}{\mathscr{M}}\zeta_i \end{vmatrix} = \begin{vmatrix} \frac{1}{4\mu/3}\frac{\partial M(T,u)}{\partial u} - \zeta\,, & \frac{1}{4\mu/3}\frac{\partial M(T,u)}{\partial T} \\ \frac{1}{\varkappa}\frac{\partial L(T,u)}{\partial u}\,, & \frac{1}{\varkappa}\frac{\partial L(T,u)}{\partial T} - \zeta \end{vmatrix} = 0$$

As the product of the characteristic roots is

$$\frac{\mathscr{M}^2 c_v}{4/3\,\mu_i'\,\varkappa_i}\left(1-\frac{1}{M_i^2}\right)$$

it will be positive upstream where $M_1 > 1$ and negative downstream, $M_2 < 1$. The sum of the roots is

$$\mathcal{M}\left[\frac{c_\nu}{\varkappa_i} + \frac{1}{4\mu'/3}\left(1 - \frac{1}{\gamma M_i^2}\right)\right]$$

which is positive upstream. Hence at P_1 the two characteristic roots are both real positive and this is then an unstable node. At P_2 the characteristic roots are real and of opposite sign this will be a stable saddle point.

The interesting configuration of the trajectories of the system of equations describing the shock wave is shown on fig. 3. There is only one trajectory joining the

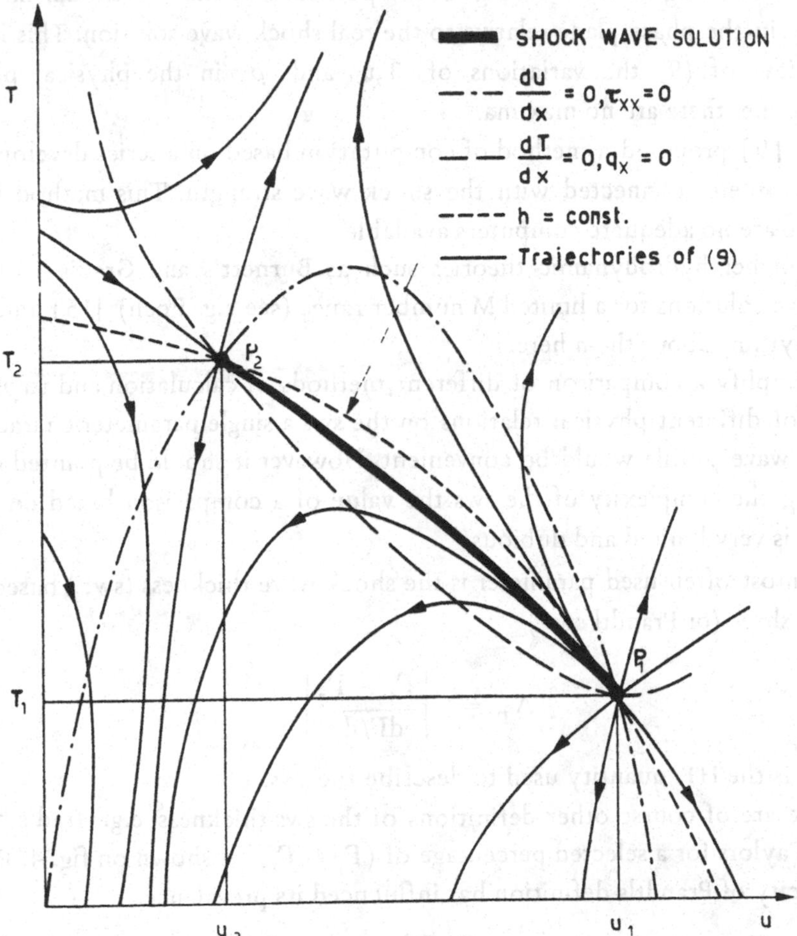

Fig. 3

points P_1 and P_2 for given $\mu'(T) \neq 0$, $\varkappa(T) \neq 0$. To obtain the solution of (13) we must start therefore from the conditions upstream, P_2, where the slope of the trajectory in the phase plane is known

$$(16) \qquad \left(\frac{dT}{du}\right)_2 = -\left(\frac{\dfrac{\partial L(T_1 u)}{\partial u}}{\dfrac{\partial L(T,u)}{\partial T} - \zeta\varkappa}\right)_2$$

here the negative characteristic root ζ for the convergent trajectory should be taken.

It is interesting to note on fig. 3 the positions of the curves $du/dx = 0$ and $dt/dx = 0$ in the phase plane relative to the real shock wave solution. This indicates that in view of (9) the variations of T,u and ρ in the physical plane are monotonic, i.e. there are no maxima.

Grad [9] proposed a method of computation based on a series development in a small parameter connected with the shock wave strength. This method is useful when there are no adequate computers available.

The higher hydrodynamic theories such as Burnett's and Grad's 13 moment method give solutions for a limited M number range (see e.g. Foch) [15] and we will not say anything about them here.

To simplify a comparison of different methods of calculation and to judge the influence of different physical relations on the sws a single parameter characterizing the shock wave profile would be convenient. However it should be pointed out that considering the complexity of the sws the value of a comparison based on a single parameter is very limited and dubious.

The most often used parameter is the shock wave thickness (swt) based on the maximum slope (or Prandtl swt)

$$(17) \qquad \Lambda_\Gamma = \left|\frac{\Gamma_1 - \Gamma_2}{d\Gamma/dx}\right|_{max}$$

where Γ is the HT quantity used to describe the sws.

There are of course other definitions of the sws thickness, e.g. Grad's based on areas, or Taylors for a selected percentage of $(\Gamma_1 - \Gamma_2)$, as shown on fig. 4. Probably the simplicity of Prandtls definition has influenced its prevalence.

This parameter is used in the non-dimensional expression λ_1/Λ_r based on the upstream mean free path λ which is the dependent variable representing different effects. As was pointed out by Smolderen [13] this arbitrary global magnitude is not adequate to estimate the validity of the NS equation. A more adequate, according to Smolderen, criterion for this purpose would be

$$Kn_u = \left(\frac{\lambda \left| \frac{du}{dx} \right|}{\sqrt{2RT}} \right)_{max} \quad \text{and} \quad Kn_T = \left(\frac{\mathcal{H} \left| \frac{dT}{dx} \right|}{T} \right)_{max} \tag{18}$$

In the case of Taylor's solution from (11) follows that the Prandtl swt

$$\Lambda_u = \frac{u_1 - u_2}{\left| \frac{du}{dx} \right|_{max}} = \frac{u_1 - u_2}{\left| \frac{u_1 - u_2}{2A} \; \frac{1}{\cosh \frac{2x}{A}} \right|_{max}} = 2A \tag{19}$$

Using the upstream Maxwellian mean free path

$$\lambda_1 = \frac{16}{5} \, \frac{\mu_1}{\rho_1} (2\pi R T_1)^{-1/2} \tag{20}$$

one gets for weak shock waves

$$\frac{\lambda_1}{\Lambda_u} = \frac{8}{5} \, \frac{(M_1^2 - 1)}{M_1} \, \frac{\sqrt{\gamma/2\pi}}{\left[\frac{4}{3} \frac{\mu_1'}{\mu} + \frac{1}{P_r} (\gamma - \frac{1}{M_1^2}) \right]} \tag{21}$$

At low M_1 this result agrees very well with results obtained from numerical calculations. To illustrate the strong influence of the law of variation of the viscosity coefficient with temperature on the shock wave thickness some of Gilbarg and Paolucci [7] results of numerical calculations of the sws thickness are shown on fig. 5.

Liepmann et al. [4] have analysed some other important properties of the sws. One of them is the variation of the total enthalpy within the shock layer. Using the total enthalpy $H = 1/2\,u^2 + h$ the conservation of energy equation can be

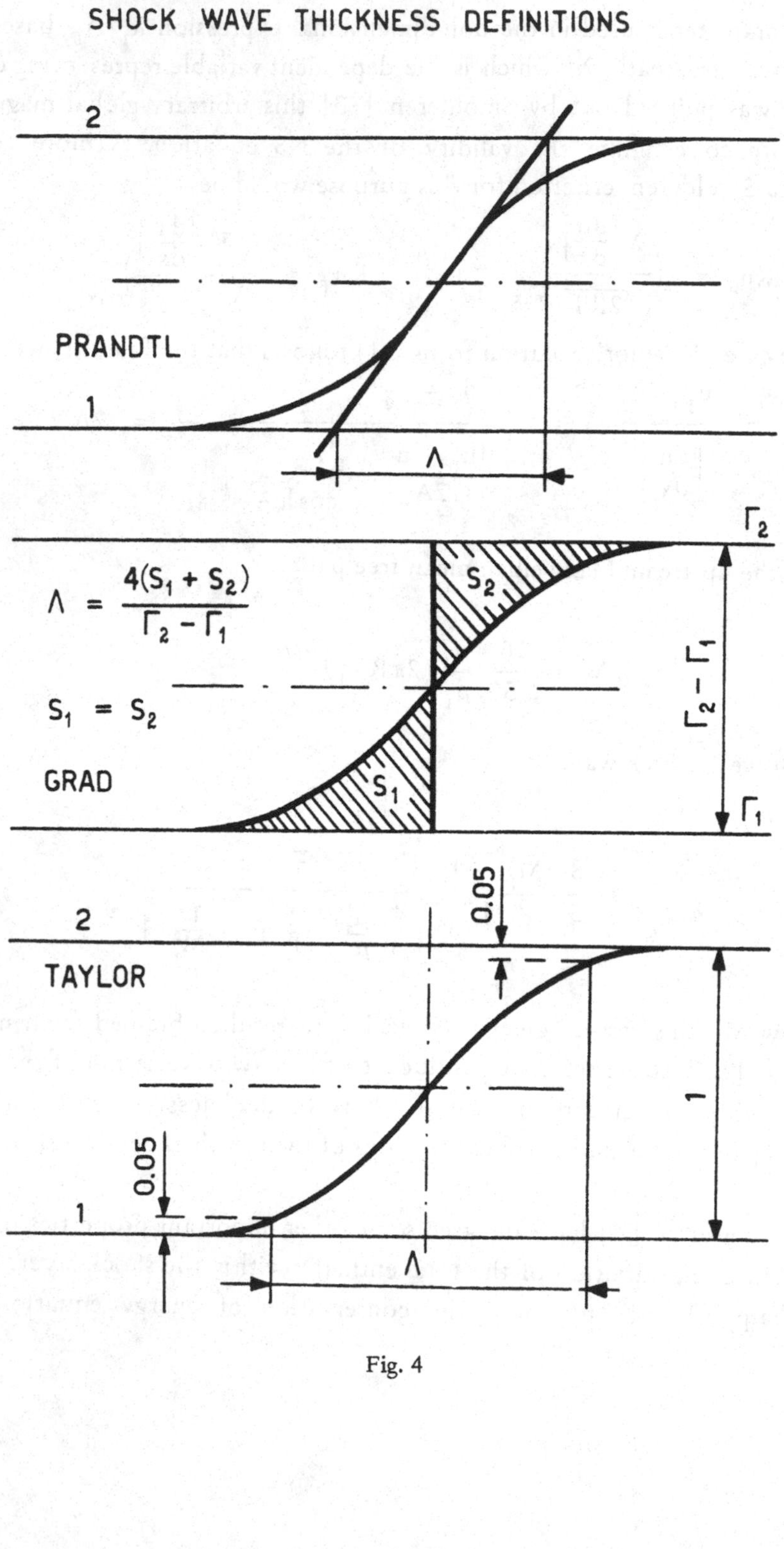
SHOCK WAVE THICKNESS DEFINITIONS
2
PRANDTL
1
Λ
Γ2
Λ = 4(S1 + S2) / (Γ2 − Γ1)
S1 = S2
GRAD
S2
S1
Γ2 − Γ1
Γ1
0.05
2
TAYLOR
1
0.05
1
Λ

Fig. 4

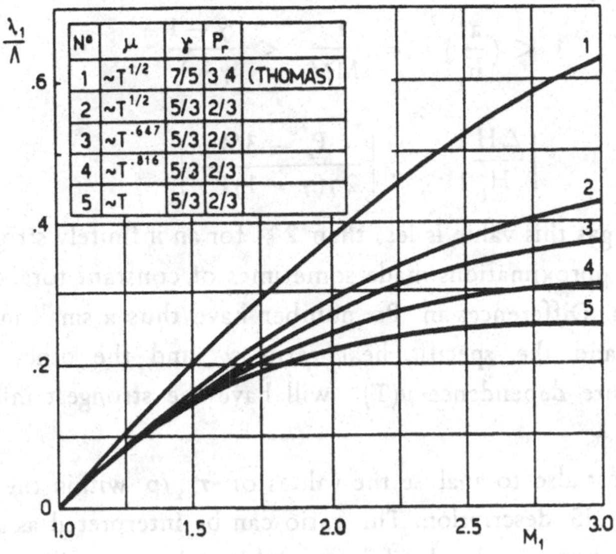

Fig. 5

written:

$$\mathscr{M}(H - H_1) = \tau_{xx}\, u - q_x \tag{22}$$

Taking advantage of the fact that P_r is close to 3/4 an upper bound of variation of the total enthalpy can be estimated as follows: At the point where $\Delta H = H - H_1$ is a maximum from (22)

$$\left|\frac{\Delta H}{H_1}\right| = \left|\frac{P_r - 3/4}{P_r}\;\frac{\tau_{xx}\, u}{\mathscr{M} H_1}\right| \tag{23}$$

Since the r.h.s. of the above equation has a small parameter $(P_r - 3/4)$, $\tau_{xx}\, u$ can be evaluated assuming a constant total enthalpy. Then from (7) the maximum value of τ_{xx} within the shock layer is

$$|\tau_{xx}\, u|_{\max} = \frac{\gamma + 1}{8\gamma}\, \mathscr{M}\, u_1^2 \left(1 - \frac{a^{*2}}{u_1^2}\right)^2 \tag{24}$$

where a^* is the velocity of sound at the point where it is equal to the gas velocity [4] $a^{*2} = 2H_1(\gamma - 1)/\gamma$. Hence:

$$\left|\frac{\Delta H}{H_1}\right| < \left|\frac{P_r - 3/4}{P_r}\;\frac{\gamma + 1}{8\gamma}\left(1 - \frac{a^{*2}}{u_1^2}\right)^2\right| \tag{25}$$

Since as given in [4]

$$(26) \qquad 1 < \left(\frac{a^*}{u_1}\right)^2 = \frac{1}{M_1^{*2}} < \frac{\gamma - 1}{\gamma + 1}$$

there follows

$$(27) \qquad \left|\frac{\Delta H}{H_1}\right| < \left|\frac{P_r - 3/4}{2\gamma(\gamma + 1)P_r}\right|$$

For a monatomic gas this value is less than 2% for an infinitely strong shock wave. This justifies the approximations made sometimes of constant total enthalpy in the NS approximation. Differences in Pr number have thus a small influence on the shock thickness and the specific heat ratio γ and the other coefficient of viscosity-temperature dependence $\mu(T)$ will have the strongest influence on this layer.

It is of interest also to analyse the values of τ_{xx}/p within the shock layer as resulting from the NS description. This ratio can be interpreted as a dimensionless measure of the gradients in the shock layer and it is the expansion parameter of the Ch.E. procedure $\tau_{xx}/p << 1$; is a sufficient condition for the validity of the NS equation [14]. For a monatomic gas this ratio is related to second moments of the distribution function ($p = 1/3\rho\,(c_x^2 + 2c_T^2)$), $\tau_{xx} = p - \rho c_x^2$, c_T velocity perpendicular to c_x

$$(28) \qquad \left|\frac{\tau_{xx}}{p}\right| = \left|\frac{2(\bar{c}_x^2 - \bar{c}_r^2)}{\bar{c}_x^2 + 2\bar{c}_r^2}\right| < 2$$

This ratio is usually considerably smaller than 2. From the NS conservation of momentum equation (7) $\mathcal{M}\,(u - u_1) + p - p_1 = \tau_{xx}$:

$$(29) \qquad \frac{\tau_{xx}}{p} = -\frac{\gamma + 1}{\gamma - 1}\;\frac{(M_1^* - M^*)\,(M^* - M_2^*)}{\dfrac{\gamma + 1}{\gamma - 1} - M^{*2}}$$

where $M^* = u/a^*$ is related to $M = u/a$, $M_1^* M_2^* = 1$, [4] :

$$(30) \qquad M^{*2} = \frac{(\gamma + 1)M^2}{2 + (\gamma - 1)M^2}$$

For an infinitely strong shock wave at the sonic point

$$(31) \qquad \left(\frac{\tau_{xx}}{p}\right)_{M_1 \to \infty} = \frac{\sqrt{\dfrac{\gamma + 1}{\gamma - 1}}\left(\sqrt{\dfrac{\gamma + 1}{\gamma - 1}} - 1\right)}{\sqrt{\dfrac{\gamma + 1}{\gamma - 1}} + 1}$$

Furthermore the largest possible value of τ_{xx}/p occurs for

$$M_1^* = \sqrt{\frac{\gamma+1}{\gamma-1}} \quad \text{and} \quad M^* \to M_1^*$$

i.e. close to the upstream end of the shock layer and is

$$\left(\frac{\tau_{xx}}{p}\right)_{max} = \frac{1}{\gamma-1} \tag{32}$$

From the last two relations it is seen that for a monotonic gas at the sonic point the stress ratio does not exceed 2/3 while upstream the maximum value of τ_{xx}/p is 3/2. The variation of the NS value of the parameter τ_{xx}/p within the shock layer is shown on fig. 6 for different values of M_1^*.

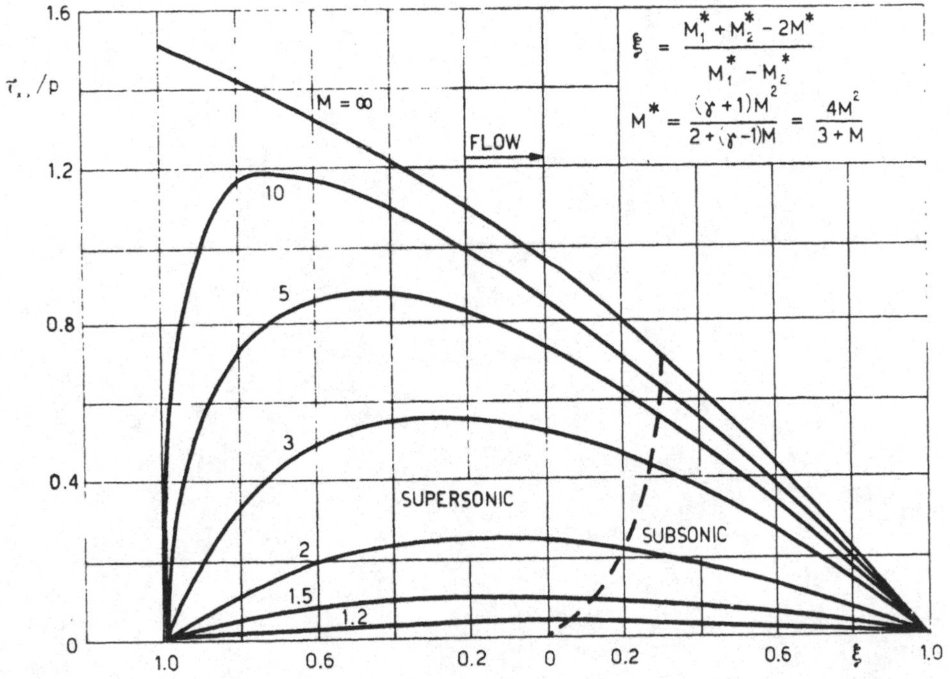

Fig. 6

As the first order distribution function of the Ch.E. procedure, from which the NS. equations result involves terms proportional to τ/p and $q/p \sim (\tau/p)M$ the difference between the upstream and downstream parts of the shock layer, from the point of view of the relevance of the NS description, are emphasized. It can be expected that the NS approximation will fail sooner in the upstream region.

However, the question concerning the range of applicability of the NS equation for the sws can only be answered by comparison with results obtained using a less restrictive approximation to the B equation and with reliable experimental results.

We will consider now the next grouping of approximate methods of solution of the B equation for the sws problem.

III. Kinetic Models of the Collision Integral

The details obtained from the full solutions of B. equation as mentioned before are much too fine and a cruder model of the collision process may be quite satisfactory for our purpose. This idea together with the great difficulties connected with the calculation of the non-linear quintuple collision integrals has led to attempts to replace them by a *collision model relation with moments which are the physical quantities of main interest.* The collision model must satisfy the same conservation laws and behave similarly close to equilibrium as B. collision integral.

Formally the collision integral can be written

$$c_x \frac{\partial f}{\partial x} = Q(f,f) = \mathscr{L}(f)\left\{\frac{\mathscr{G}(f,f)}{\mathscr{L}(f)} - f\right\} \tag{33}$$

where the non-linear "gain" and "loss" operators are:

$$\left\{\begin{aligned} \mathscr{G}(f,f) &= \int f' f_1' \, g\, b\, db\, d\epsilon\, dc_1 \\ \mathscr{L}(f) &= \int f_1 \, g\, b\, db\, d\epsilon\, dc_1 \end{aligned}\right. \tag{34}$$

and these expressions can be used only when the intermolecular collision law is such that each one of the above integrals converges separately. The simplest and most successful statistical collision model introduced by Bhatnager, Gross and Krook (BGK) [16] consists in assuming

$$\left\{\begin{aligned} &\mathscr{L}(f) = An\,, \quad \mathscr{G}(f,f) = AnF \\ &F = n(2\pi RT)^{-3/2} \exp\left[-\frac{(\bar{c}-\bar{u})^2}{2RT}\right] \end{aligned}\right. \tag{35}$$

$$f = f(x, c_x, c_y, c_z)$$

As shown by Holway [17] this is equivalent to using Shanon's maximum uncertainty principle to F considered as a probability density subject, to the condition of expressing the number density mass velocity and temperature by the usual kinetic relations. The BGK model as shown by Kogan [18] can also be considered as a rough model representation of the collision integrals inspired partly by Maxwell molecules and partly by rigid spheres [14] and $A = |c| n \lambda$ corresponds to the number of collisions. Thus in the sws problem considered B equation is replaced by the BGK model equation

$$c_x \frac{df}{dx} = An(F - f) \tag{36}$$

which can be looked upon as a relaxation equation for deviations from the local Maxwellian distribution F with the boundary conditions given by (2). Integrating (36) formally for $c_x > 0$ and $c_x < 0$ with the corresponding boundary condition, the integral relation for f is obtained

$$\begin{cases} f(x; c_x > 0) = f_+ = \int_{-\infty}^{x} \frac{AnF}{c_x}\left[\exp\left(-\int_{x'}^{x} \frac{And x''}{c_x}\right)\right]dx' \\ f(x; c_x < 0) = f_- = \int_{+\infty}^{x} \frac{AnF}{c_x}\left[\exp\left(-\int_{x'}^{x} \frac{And x''}{c_x}\right)\right]dx' \end{cases} \tag{37}$$

$An = An(x)$ or const.

where the complementary function solutions of (36) are omitted as the boundaries are at $\pm\infty$; because far away from the shock $f(\pm\infty; c) = f_{1,2}(c)$ independently of the sign of c_x and the boundary conditions will be satisfied.

Extensive numerical calculations of the shock wave structure were made by Chahine et al. [19] and Anderson et al. [20] using essentially an iteration method, starting with the NS solution as the 0 order approximation. This procedure is very convenient as for the BGK model only the first few moments are needed for computing the collision terms. As emphasized by Liepmann et al. [14] the higher iterates obtained have nothing in common with the higher order hydrodynamic approximations of the kinetic theory, because in the BGK case the form of f does not change, remaining the same as (37) and only the parameters entering into f change unlike in the Ch.E. expressions.

Chahine's numerical procedure converges after 1-15 iterations depending on the M number and once the shock wave profile is obtained the distribution function can be found from (37). Typical distribution functions across the sw (for a given transversal molecular velocity and at a given position) as the dependence of c_x on c_r are illustrated on fig. 7 for M = 10. At this M number, the strong *bimodal character* of the distribution function over a large part of the shock layer is noticeable and the connected *deep penetration* of the supersonic stream can be seen.

The variation of the total enthalpy within the sw. layer calculated using the NS and BGK models shown on fig. 8 indicates that although for $1 < M < 10$ the absolute magnitude of $(H-H_1)/H_1$ remains small, within the previously indicated limits, but the signs are opposite.

Shock wave profiles obtained for the NS and BGK models at M = 5 and M = 10 are shown for illustration on fig. 9. Similar profiles were calculated numerically for other M numbers and are given in [19], [20]. These results indicate that at M numbers below about 1.8 both profiles coincide practically. At higher M numbers the differences between the two methods grow and are particularly marked in the upstream wing of the sws. It should be also noted that the differences in the velocity and density profiles deduced using the two above mentioned models are much smaller than those of the temperature profiles. Obviously the maximum slope thickness will not show the large difference in the profile of the upstream and downstream wings of the sw and in this respect the area based Grads definition of sw thickness will be more representative. For illustration a comparison of temperature slope and area sw thicknesses calculated using NS and BGK models are shown on fig. 10.

The agreement of the NS and BGK density profile with experimental results of Muntz and Harnett [21] shown on fig. 11 is excellent.

Anderson's [20] solutions of the sws using the BGK model are quite thorough and less demanding in computer time and their main features will be outlined below. A rather essential step is the formal removal of the collison frequency term $An = \gamma$ from the relaxation BGK equation by introducing a new independent variable.

$$\tau = \frac{1}{a_2} \int_0^x \gamma(s)ds, \quad \text{with} \quad a_2 = \sqrt{\frac{kT_2}{m}} \tag{38}$$

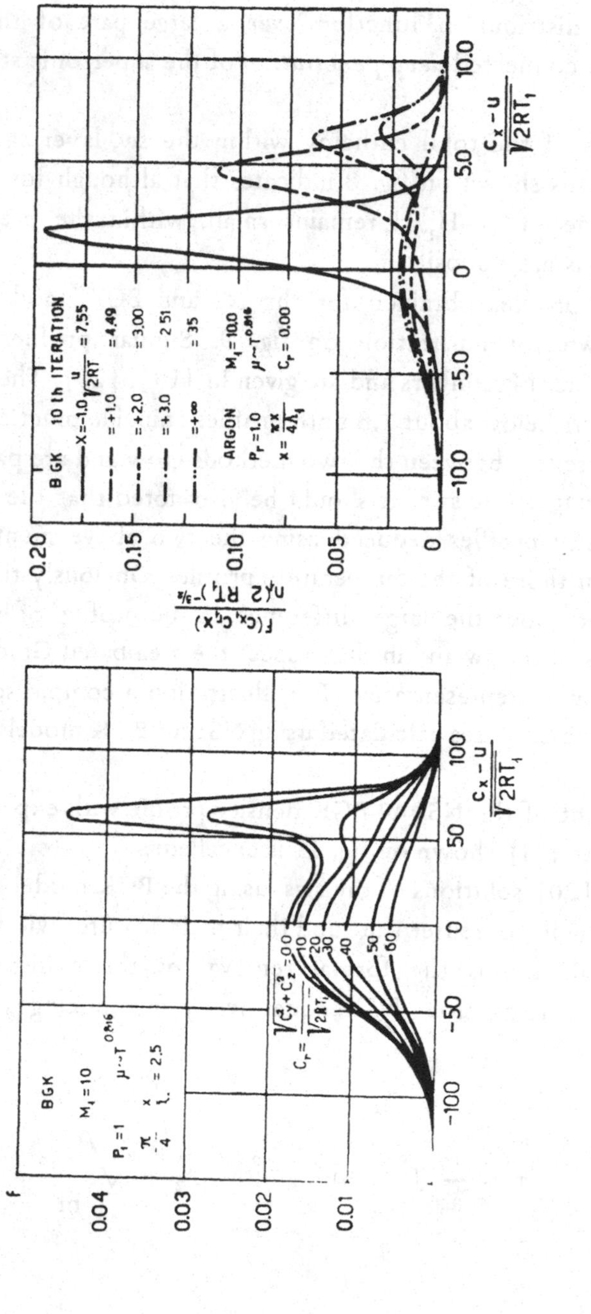

Fig. 7. (Chahine, Narasimha, 1965)

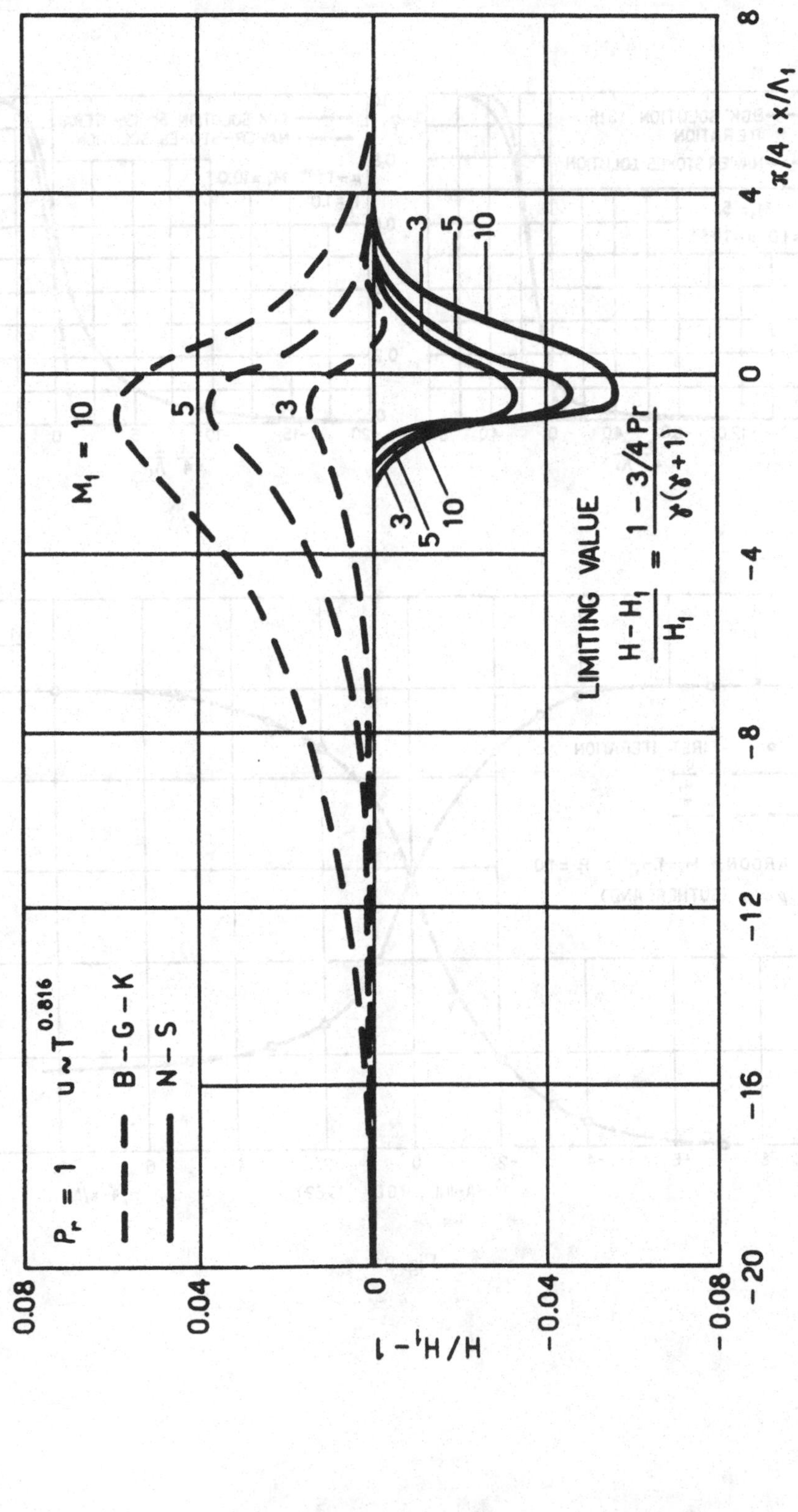
$P_r = 1$
$\mu \sim T^{0.816}$
B – G – K
N – S
$M_1 = 10$
5
3
10
LIMITING VALUE
$\frac{H - H_1}{H_1} = \frac{1 - 3/4\,Pr}{\gamma(\gamma+1)}$
0.08
0.04
0
−0.04
−0.08
$H/H_1 - 1$
−20
−16
−12
−8
−4
4
8
$\pi/4 \cdot x/\Lambda_1$

Fig. 8

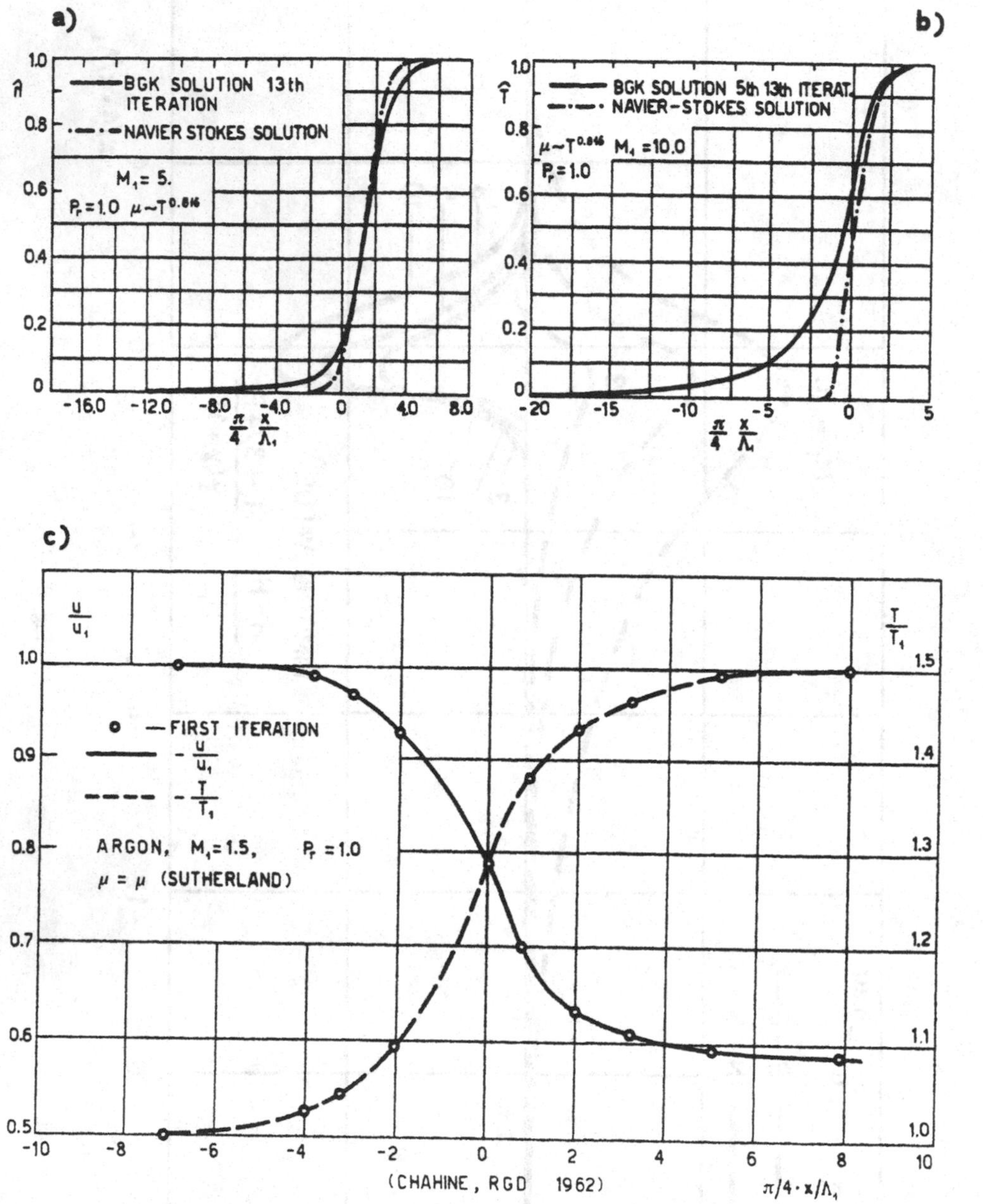
a)
BGK SOLUTION 13th ITERATION
NAVIER STOKES SOLUTION
M₁ = 5.
Pr = 1.0 μ ~ T^0.816
1.0
0.8
0.6
0.4
0.2
0
-16.0 -12.0 -8.0 -4.0 0 4.0 8.0
π/4 x/Λ₁
b)
BGK SOLUTION 5th 13th ITERAT.
NAVIER-STOKES SOLUTION
μ ~ T^0.816 M₁ = 10.0
Pr = 1.0
1.0
0.8
0.6
0.4
0.2
0
-20 -15 -10 -5 0 5
π/4 x/Λ₁
c)
u/u₁
T/T₁
FIRST ITERATION
- u/u₁
- T/T₁
ARGON, M₁ = 1.5, Pr = 1.0
μ = μ (SUTHERLAND)
1.0
0.9
0.8
0.7
0.6
0.5
1.5
1.4
1.3
1.2
1.1
1.0
-10 -8 -6 -4 -2 0 2 4 6 8 10
(CHAHINE, RGD 1962)
π/4 · x/Λ₁

Fig. 9

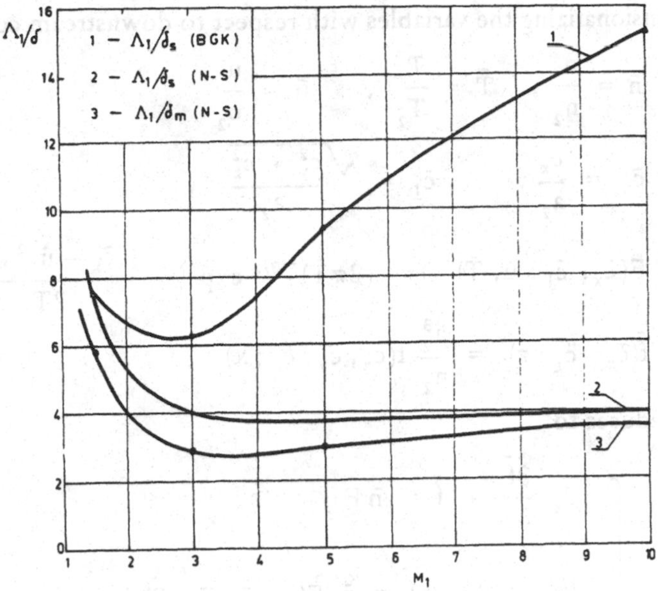

Fig. 10. (Liepmann, Narasimha, Chahine, 1962)

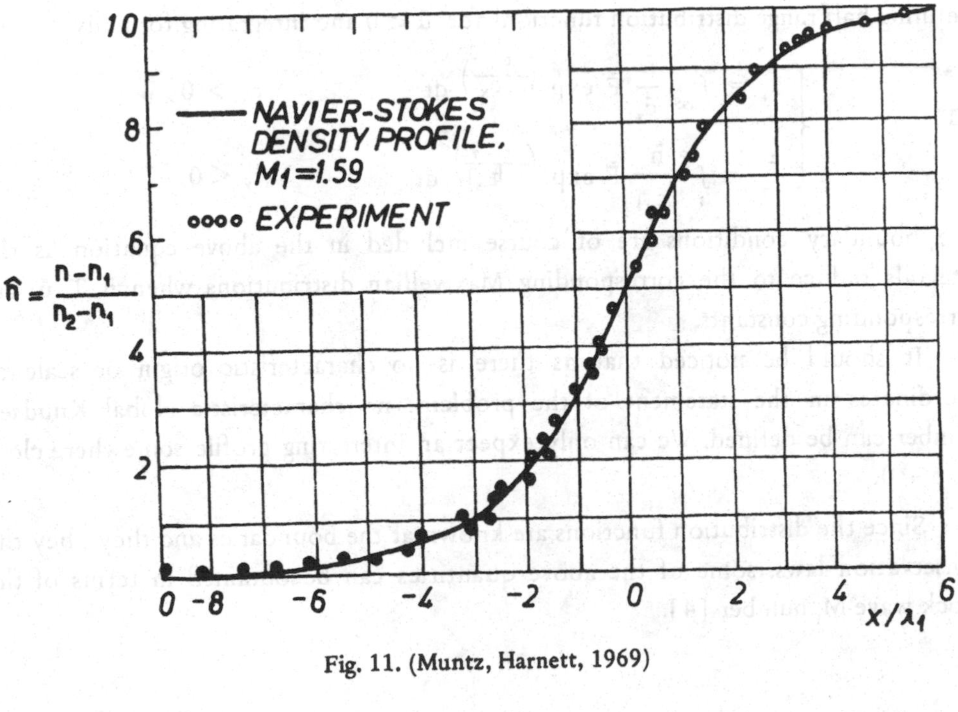

Fig. 11. (Muntz, Harnett, 1969)

Non dimensionalizing the variables with respect to downstream conditions:

$$(39)\quad \begin{cases} \tilde{n} = \dfrac{n}{n_2}\,, \quad \tilde{T} = \dfrac{T}{T_2}\,, \quad \tilde{u} = \dfrac{u}{u_2} \\ \tilde{c}_x = \dfrac{c_x}{a_2}\,, \quad \tilde{c}_\perp = \dfrac{\sqrt{c_y^2 + c_z^2}}{a_2} \\ \tilde{F}(\tilde{c}_x, \tilde{c}_\perp; \tilde{u}, \tilde{T}) = (2\pi\tilde{T})^{-3/2} \exp\{ - \dfrac{[(\tilde{c}_x - \tilde{u})^2 + \tilde{c}_\perp^2]}{2\tilde{T}} \}\,, \\ \tilde{f}(\tilde{c}_x, \tilde{c}_\perp; \tau) = \dfrac{a_2^3}{n_2} f(c_x, c_y, c_z; x) \end{cases}$$

(36) and (37) reduces to

$$(40)\quad \begin{cases} \tilde{c}_x \dfrac{\partial \tilde{f}}{\partial \tau} + \tilde{f} = \tilde{n}\tilde{F} \\ f(c_x, c_\perp; -\infty) = \tilde{n}_1 \tilde{F}(\tilde{c}_x, \tilde{c}_\perp; \tilde{u}_1, \tilde{T}_1) \\ f(\tilde{c}_x, \tilde{c}_\perp; +\infty) = \tilde{n}_2 \tilde{F}(\tilde{c}_x, \tilde{c}_\perp; \tilde{u}_2, \tilde{T}_2) \end{cases}$$

Defining half range distribution functions for $\tilde{u} \gtrless 0$ and integrating formally:

$$(41)\quad \begin{cases} \tilde{f}_+ = \int_{-\infty}^{\tau} \dfrac{\tilde{n}}{\tilde{c}_x} \tilde{F} \exp\left(-\dfrac{t-\tau}{\tilde{c}_x}\right) dt & c_x > 0 \\ \tilde{f}_- = \int_{\tau}^{\infty} \dfrac{\tilde{n}}{|\tilde{c}_x|} \tilde{F} \exp\left(-\dfrac{t-\tau}{|\tilde{c}_x|}\right) dt & \tilde{c}_x < 0 \end{cases}$$

The boundary conditions are of course included in the above equation as the integrals reduce to the corresponding Maxwellian distributions when $\tilde{n}, \tilde{T}, \tilde{u}$ are corresponding constants.

It should be noticed that as there is no characteristic origin or scale of coordinates in the statement of the problem, no characteristic global Knudsen number can be defined. We can only expect an interesting profile somewhere close to $\tau = 0$.

Since the distribution functions are known at the boundaries and they obey the conservation laws, some of the above quantities can be obtained in terms of the shock wave M_1 number [4].

$$\left\{\begin{aligned} \tilde{n}_1 &= \frac{n_1}{n_2} = \left(\frac{u_1}{u_2}\right)^{-1} = \frac{(\gamma-1)M_1^2+2}{(\gamma+1)M_1^2}, \\ \tilde{T}_1 &= \frac{T_1}{T_2} = \left\{1+\frac{2(\gamma-1)(\gamma M_1^2+1)(M_1^2-1)}{(\gamma+1)^2 M_1^2}\right\}^{-1}, \\ \tilde{u}_1 &= \sqrt{\gamma\tilde{T}_1}\,M_1, \end{aligned}\right. \tag{42}$$

and by definition $\tilde{n}_2 = 1$, $\tilde{T}_2 = 1$.

Substituting the formal integrals (41) in the definition relations of lower order HT moments the following non-linear integral equations, characterized by a single parameter M_1, are obtained:

$$\left\{\begin{aligned} \tilde{n} &= \int_{-\infty}^{+\infty} \frac{\tilde{n}}{\tilde{T}} H_1\left[\operatorname{sgn}(\tau-t)\frac{\tilde{u}}{\sqrt{\tilde{T}}}, \frac{|\tau-t|}{\sqrt{\tilde{T}}}\right]dt, \\ 3\tilde{n}\tilde{T}+\tilde{n}\tilde{u}^2 &= \int_{-\infty}^{+\infty} \tilde{n}\sqrt{\tilde{T}}\left\{H_3\left[\operatorname{sgn}(\tau-t)\frac{\tilde{u}}{\sqrt{\tilde{T}}}, \frac{|\tau-t|}{\sqrt{\tilde{T}}}\right]+\right. \\ &\qquad \left. 2H_1\left[\operatorname{sgn}(\tau-t)\frac{u}{\sqrt{\tilde{T}}}, \frac{|\tau-t|}{\sqrt{\tilde{T}}}\right]\right\}dt \end{aligned}\right.$$

The kernel function

$$H_n(p,q) \equiv \frac{1}{\sqrt{2\pi}}\int_0^\infty y^{n-2}\exp\left[\frac{1}{2}(y-p)^2-\frac{q}{y}\right]dy \tag{43a}$$

has been studied by Abramovitz [22], Anderson, Mecomber [23], Chahine, Narasimha [27]. To pass from the used τ variable to the physical space variable, once the collision frequency is specified, an additional quadrature is required.

In Andersons [20] procedure the above analytic problem is replaced by a discrete analogue and the resulting discrete problem is then solved iteratively with the integral operators replaced by a Gaussian quadrature taking into account the singularity at $t=\tau$.

To use the convenient Chebyshev interpolation, the interval $-\infty<\tau<\infty$ was mapped by Anderson into the interval $-1<z<1$ by the transformation

$$\left\{\begin{aligned} z &= \tanh\left(\frac{\tau}{\Delta}\right), \\ \tau &= \frac{1}{2}\Delta\ \ln\left(\frac{1+z}{1-z}\right) \end{aligned}\right. \tag{44}$$

The collocation points in the z domain are mapped thus into a set of points close to the origin in the τ domain. The interpolation sample points, at which the equations are satisfied, are chosen as extrema of an appropriate Chebyshev polynomial, including the points $z = \pm 1$. The parameter Δ is chosen arbitrarily so that roughly 98% of the sw. profile is for $-3\Delta < \tau < 3\Delta$. The iteration is started for convenience, with an appropriately normalized hyperbolic tangent function. After 10-15 iterations the process is stationary and convergence ceases. This behaviour is an interesting unsolved numerical problem.

Anderson's numerical results agree very well with Chahines.

One of the deficiencies of the BGK method is that the corresponding $P_r = 1$ whereas for real gases it is $0.5 < P_r < 1$ The effect of the P_r number on the sw profile as calculated from the NS description is shown on fig. 12 and is quite noticeable. This has incited the search for new models giving the possibility of introducing additional parameters. Holway's ellipsoidal model and Segal and Ferziger's [24], and Shakkow's [25] polynomial models are the results of such developments.

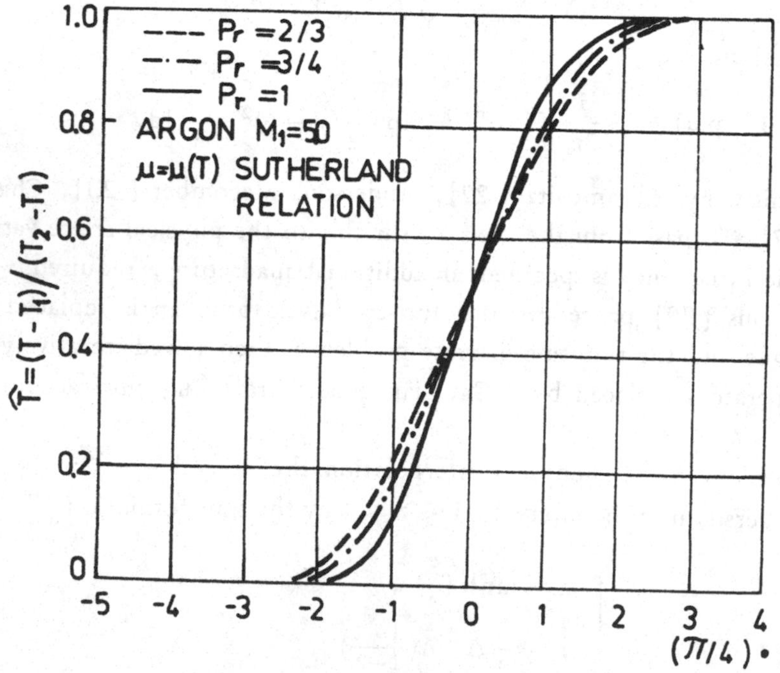

Fig. 12. (Liepmann, Narasimha, Chahine, 1952)

We will describe these models following Segals [26] unifying approach to the "higher order" BGK models.

In this approach, somewhat reminiscent of Anderson's at the beginning, the collision integral of (33) is replaced by a model operator

$$Q(f, f) \cong K(\mathcal{M}_i)\ \{\Psi(\bar{c}_1, \mathcal{M}_i) - f\} \tag{45}$$

where the average collision frequency $K(\mathcal{M}_i)$ which models the operator L(f), (34), is assumed independent of c and the function $\Psi(c, \mathcal{M}_i)$ models $\mathcal{G}(f,f)/\mathcal{L}(f)$, called "emission function", describes the average distribution following a collision. The quantities $\mathcal{M}_i$ are velocity moments of f defined as

$$\mathcal{M}_i \equiv \int \varphi_i(c)\, f(c)\, dc \tag{46}$$

where $\varphi_i(c)$ is a selected set of functions of c which are usually the HT moments n, u, T, τ_{xx}, q_x.

The main feature of the models is the replacement of the five-fold collision integral by an operator on the selected moments of the distribution function which are the quantities of physical interest.

For convenience, a similar transformation as Anderson's is used:

$$\begin{cases} dt = K(x)dx\ , \quad \text{or}: \quad t = \int_{x_0}^{x} K(x)\,dx \\ dx = \dfrac{dt}{K(t)} \quad\quad \text{or}: \quad x = \int_{t_0}^{t} \dfrac{dt}{K(t)} \end{cases} \tag{47}$$

where the dependence of K on the variable is through $\mathcal{M}_i$ and x_0, t_0 are arbitrary reference origins.

One of the additional advantages of using t, is that the shock profiles tend to an invariant form at high values of M in these coordinates.

In the above variables using dimensional notations with corresponding boundary conditions the problem is rewritten below for convenience (t is not time but the quantity defined in (47)

$$(48)\quad \begin{cases} c_x \dfrac{\partial f(c,t)}{\partial t} + f(c,t) = \Phi\,(c, m_i(t))\,, \\ f(c, x = -\infty) = f^{(o)}(n_1, u_1, T_1) = f_1^{(o)}\,, \\ f(c, x = +\infty) = f^{(o)}(n_2, u_2, T_2) = f_2^{(o)}\,, \\ f^{(o)}(n, u, T) = n(2\pi kT)^{3/2} \exp\left[\dfrac{-(c-u_x)^2}{2RT}\right]. \end{cases}$$

Formally integrating over t, in the present notation, integrals of the form (41) are obtained and denoting *the split range integral* by $⨏$ can be written as:

$$(49)\qquad f(c,t) = ⨏_{-\infty}^{+\infty} \frac{\Phi\,(c, \mathcal{M}_i(t'))}{|c_x|} \exp\left(-\left|\frac{t-t'}{c_x}\right|\right) dt'.$$

Velocity moments of the above distribution function are:

$$(50)\qquad \mathcal{M}_i(t) = \iiint dc\, ⨏_{-\infty}^{+\infty} \varphi_i(c) \frac{\Phi\,(c, m_i(t'))}{c_x} \exp\left(-\left|\frac{t-t'}{c_x}\right|\right) dt'$$

To obtain a finite set of moments of the model equation (48), and not of the full B equation, in the one dimensional sw structure case of interest the five HT non zero moments n, u, T, τ_{xx}, q_x are used. They form a closed set which corresponds respectively to the five selected quantities for φ_c :

$$1,\ \frac{c_x}{n}\,,\ \frac{mc^2}{3kn}\,,\ (c_x^2 - \frac{1}{3}c^2)/n\,,\ \frac{c_x c^2}{2n}\,.$$

Considering a set of moments $\mathcal{N}_i$ related to the five nonzero $\mathcal{M}_i$ ones corresponding to the choice $\varphi_i^* = (1, C_x, C^2, C_x^2, C_x C^2)$:

$$\begin{cases} \mathcal{N}_1 = n \\ \mathcal{N}_2 = nu \\ \mathcal{N}_3 = n(3\frac{k}{m}T + u^2) \\ \mathcal{N}_4 = n(\frac{k}{m}T + u^2 + \tau_{xx}) \\ \mathcal{N}_5 = nu(5\frac{k}{m}T + u^2 + 2\tau_{xx}) + 2nq_x \end{cases} \tag{51}$$

As moments of the collisional invariants $\psi_i^{(col)} = (1, c_x, c_x^2)$ annul the collision operator, hence also the corresponding moments of the l.h.s. of the B. equation i.e. of $c_x\, \partial f/\partial x$. Integrating this result over x, the moments of f taken with $c_x \psi_i^{(col)}$ must therefore be constants and therefore u_2, u_4 and u_5 are constants of flow. τ_{xx} and q_x vanish at the end of points and finally to determine all the $\mathcal{N}_c$ and hence all the non-zero $\mathcal{M}_i(t)$, only two equations for $\mathcal{N}_1(t)$ and $\mathcal{N}_3(t)$. Putting $\varphi_i^* = (1, c^2)$ into (50) and denoting $c^2 = c_x^2 + c_\perp^2$ the two equations are

$$\mathcal{N}_1(t) = \int_{-\infty}^{+\infty} dc_x \int_0^{\infty} 2\Pi c_\perp\, dc_\perp \, ⨍_{-\infty}^{+\infty} \frac{\Psi(c_x, c_\perp, \mathcal{M}_i(\mathcal{N}_i(t')))}{|c_x|} \exp\left(-\left|\frac{t-t'}{c_x}\right|\right) dt'$$

(52)

$$\mathcal{N}_3(t) - \mathcal{N}_4 = \int_{\infty} dc_x \int_0^{\infty} 2\Pi c_\perp^3\, dc_\perp \, ⨍_{-\infty}^{+\infty} \frac{\Psi(c_x, c_\perp, \mathcal{M}_i(\mathcal{N}_i(t')))}{|c_x|} \exp\left(-\left|\frac{t-t'}{c_x}\right|\right) dt'$$

The kernels of the above integrals are rather inconvenient for computation as they involve double quadratures over the components of c. However if the emission function Ψ has the form of Maxwellian functions multiplied by velocity polynomials the kernels can be expressed in terms of the well studied functions $H_n(p.q)$, (43a).

Making use of these functions and performing the $c_\perp$ integration after substituting Ψ for each model into (52) the following general form, used for numerical calculations is obtained:

$$(53)\quad \begin{cases} n(t) = \int_{-\infty}^{\infty} n(t')\, K_n^{MOD}(n(t'), T(t'); (t-t'))\, dt' \\ n(t)\, T(t) = \int_{-\infty}^{\infty} n(t')\, K_T^{MOD}(n(t'), T(t'); (t-t'))\, dt' + \frac{m}{3k}\left(\mathcal{N}_4 - \mathcal{N}_2^2 / n(t)\right) \end{cases}$$

The kernels K in (53) for all models are sums of lower order H_n functions and have a logarithmic singularity at $t' = t$.

The derivation of the kinetic models according to Segal and Ferziger [24] is subject to the following conditions:

1. Conservation of mass, momentum and energy.
2. The collision operator acting on the equilibrium Maxwellian distribution function must produce a null result.
3. Q is rotationally invariant.
4. Model operators act only on a small number of moments $\mathcal{M}_i$ (f) and not on f itself.
5. The collision frequency is independent of the molecular velocity **c** and is a function of $\mathcal{M}_i$. The dependence on **c** is a further possibility.
6. The behaviour of the model operators close to equilibrium should resemble that of the full B linearized operator.

Close to equilibrium

$$(54)\qquad f = f_o + \delta f, \qquad \delta f \equiv f_o \Phi(c, r)$$

where Φ is a small dimensionless quantity and f_o is a Maxwellian distribution. Hence

$$(55)\qquad \mathcal{M}_i(f) = \mathcal{M}_{i,o} + \delta \mathcal{M}_i, \qquad \delta \mathcal{M}_i = \int \delta f \Phi_i \, dc$$

and

$$(56)\qquad Q_M = Q_{M,0} + \sum_{i=1}^{N} \left(\frac{\partial Q_M}{\partial \mathcal{M}_i}\right)_{f=f_o}, \qquad \delta \mathcal{M}_i \equiv f_o Q_M(\Phi)$$

The functions Φ_i in $\mathcal{M}_i$ are chosen as the normalized Maxwell eigenfunctions $\psi_i(c)$ corresponding to the linearized B. operator [28].

Using (45)

$$f_o Q_M(\Phi) = K_o \{ \sum_{i=1}^{N} (\frac{\partial \Psi}{\partial \mathcal{M}_i})_o \ \delta \mathcal{M}_i - f_o \Phi \} = K_o \{ \Sigma (\frac{\partial \Psi}{\partial a_i})_o \ a_i - f_o \Phi \}$$

$$(\frac{\partial \Psi}{\partial a_i}) = f_o(\lambda_i^B + 1)\Psi_i(c), \quad i = 1, \ldots N \tag{56a}$$

The values of the different quantities are given on table I.

As shown by Segal and Ferzinger [24]

$$f_o Q_M(\Phi) = f_o K_o \{ \sum_{i=1}^{N} (\lambda_i + 1) a_i \Psi_i(c) - \Phi \} \tag{57}$$

Applying a unified treatment for all the models the collision frequency and the P_r are

$$K(\mathcal{M}_i) = (\frac{P}{\mu}) \frac{1}{|\lambda_4|} = \frac{5kp}{2m\mathcal{H}} \cdot \frac{1}{|\lambda_5|}$$

$$p = nkT, \quad P_r = \lambda_5/\lambda_4 \tag{58}$$

The different models can now be deduced using different values of N and λ.

The BGK model is the simplest one, with $N = 3$ the minimum allowable N. All λ_i, $i > 3$ eigenvalues of the BGK model are equal to -1. The emission function

$$\Psi^{BGK}(c, \mathcal{M}_i) = f^{(o)}(n, u, T) = n(2\pi \frac{k}{m} T)^{-3/2} \exp[-\frac{(c-u)^2}{2RT}] \tag{59}$$

The collison operator is

$$Q^{BGK} = K^{BGK} [f^{(o)}(n, u, T) - f] \tag{60}$$

with the linearized form

$$f_o Q^{BGK}(\Phi) = f_o K_o \left[\frac{n - n_o}{n_o} + \frac{c\,u}{\frac{kT_o}{m}} + \left(\frac{c^2}{\frac{2kT}{m}} - \frac{3}{2} \right) \frac{T - T_o}{T_o} - \Phi \right] \tag{61}$$

Using the Chapman Enskog procedure

$$K^{BGK} = P/\mu \,, \qquad P_r^{BGK} = 1 \tag{62}$$

The ellipsoidal model developed by Holway [17] corresponds to a bivariate Gaussian emission function

$$\Psi^E(c) \approx \exp\left[-\frac{1}{2} \sum_{i,j} \epsilon_{ij} (c_i - b_i)(c_j - b_j) \right] \tag{63}$$

which on application of the conservation relations yields

$$\left\{ \begin{aligned} &\Psi^E(\mathbf{c}, m_i) = \frac{n}{(2\pi)^{3/2}} (\det \underline{\underline{\mathcal{E}}})^{1/2} \exp\left(-\frac{1}{2} \underline{\underline{\mathcal{E}}} : cc \right) , \\ &\underline{\underline{\mathcal{E}}} = \left(\lambda \underline{\underline{\tau}} + \frac{k}{m} T \underline{\underline{I}} \right)^{-1} , \\ &Q^E = K^E (\Psi^E - f) \end{aligned} \right. \tag{64}$$

where $\underline{\underline{\tau}}$ is the stress tensor, $\underline{\underline{I}}$ is a unit tensor, $C = c - u$ and λ is an adjustable parameter. The eigenvalue λ_4 is, in this case with $N = 4$, equal to $(\lambda - 1)$ and using the Ch.E. procedure

$$P_r^E = \frac{1}{1 - \lambda} = \frac{2}{3} \,, \qquad K^E = \frac{2}{3} \frac{P}{\mu}$$

By this choice of $\lambda = -1/2$ the value of P_r is improved but $\lambda_4 = -3/2$ which compares unfavorably with the correct Maxwell molecule value of $-2/3$ (for the BGK model $\lambda_4 = -1$) and may be an explanation why the ellipsoidal model has not been successful. The polynomial model is a further attempt to extend the BGK model. The main idea of Segal and Ferziger [24] was to develop a non-linear hierarchy of models similar to the linear Gross-Jackson linearized ones.

They assume

$$\frac{\partial \Psi}{\partial a_i} = (\lambda_i + 1)\, \Psi_i(c)\, F(c.\, a_i)\,, \qquad i = 1 \text{ to } N \tag{65}$$

where the function F must satisfy the condition

$$F(c, a_i)\, |_{f = f_o} = f_o \tag{65a}$$

The choice of F is of course very important and there are two simple possibilities

$$F = f^{(o)}(n, u, T) \quad \text{and} \quad F = \Psi \tag{66}$$

The second choice was suggested by Cercignani and Tironi [30] however this choice does not satisfy physical conditions for a power of c equal or greater than 3.

A mixed choice leads to the ellipsoidal model.

The simple choice of

$$F(c, a_i) = f^{(o)}(n, u, T) \qquad \text{for all } i \tag{67}$$

was used which leads to the following emission function

$$\Psi^{Pol}(c, \mathcal{M}_i) = f^{(o)}(n, u, T)\,[1 + \sum_{i=4}^{N} (\lambda_i + 1) A_i\, \Psi_i(\frac{c-u}{\sqrt{2RT}})] \tag{68}$$

A_i and ψ_i are given on Table I. For $N = 5$ and $\lambda_5 / \lambda_4 = 2/3$ the Ch.E. procedure gives

$$P_r^{Pol} = 2/3 \quad , \qquad K^{Pol} = (\frac{P}{\mu}) / |\lambda_4| \tag{69}$$

there remains only one adjustable parameter λ_4. The main advantage is that this parameter can be adjusted without affecting P_r.

For each model the kernel function K of (53) can be deduced ; ' the coupled pair of equations was solved numerically in a similar way as Anderson used for the BGK model.

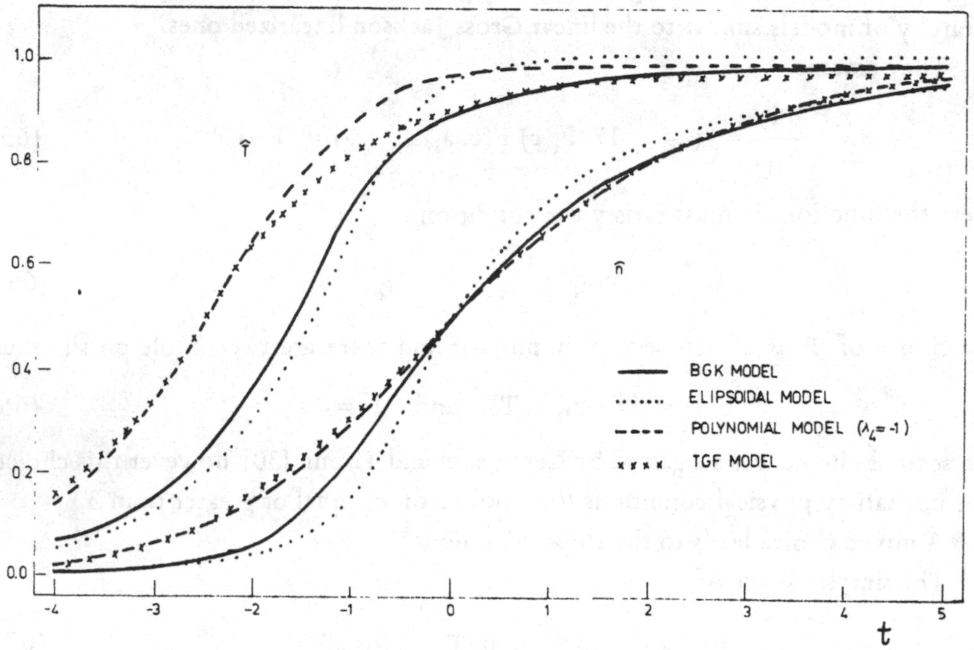

Fig. 13. (Segal, Ferziger, 1971)

Reduced temperature and density sw profiles calculated for the different models using the t coordinate are shown on fig. 13 indicating the large differences in some parts of the profile between the models.

The variation of the total enthalpy and longitudinal stress with reduced density within a shock wave calculated for different models is shown on fig. 14.

The same basic approach consisting in generalizing the collison term and the emission term of the relaxation equation (36) (48) was used by Shakkow [25] and in Zhuk, Rikov, Shakkov's paper [25] to determine the sws.

In the relaxation equation, repeated here for convenience using an obvious slightly modified notation

(70) , (36) , (48) $$c_x \frac{\partial f}{\partial x} = K(\Psi - f)$$

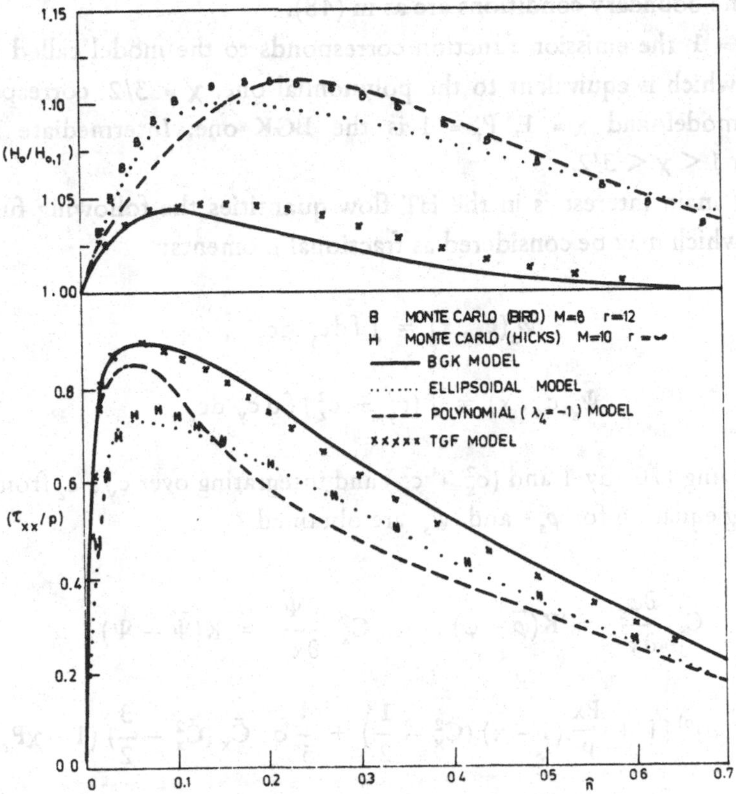

Fig. 14. (Segal, Ferziger, 1971)

the different quantities are denoted

$$[\,\bar{c}^2 = (c_x - u)^2 + c_y^2 + c_z^2\,]\left(\frac{2kT}{m}\right)^{-1}, \qquad \bar{c}_x = (c_x - U)\left(2\frac{k}{m}T\right)^{-1/2}, \tag{71}$$

and the emission function is

$$f^{(o)} = n(2\pi(k/m)T)^{-3/2}\exp(-\bar{C}^2), \quad \delta_x = \frac{1}{n}\int \bar{C}_x \bar{C}^2 f\, dc,$$

$$\tau_{xx} = m\int (c_x - u)^2 f\, dc, \quad P_x = \tau_{xx} - P, \quad c_\perp^2 = (c_y^2 + c_z^2)\left(2\frac{k}{m}T\right)^{-1}$$

The boundary conditions are as in (48).

$$\Psi = f^{(o)}\left\{1 + \frac{P_x}{P}(1 - X)\left(C_x^2 - \frac{1}{2}C_\perp^2\right) + \frac{4}{5}q_x C_x\left(C^2 - \frac{5}{2}\right)(1 - x\mathrm{Pr})\right\}, \tag{72}$$

$k = P/u$. The boundary conditions are as in (48).

For $\chi = 1$ the emission function corresponds to the model called by Shakkov "S" model which is equivalent to the polynomial one. $\chi = 3/2$ corresponds to the ellipsoidal model and $\chi = 1$, $P_r = 1$ is the BGK one. Intermediate models are obtained for $1 < \chi < 3/2$.

As our main interest is in the HT flow quantities the following functions are introduced which may be considered as fractional moments:

$$\varphi_s(c_x, x) = \int f\, dc_y\, dc_z \ ,$$

(73)

$$\Psi_s(c_x, x) = \int (c_y^2 + c_z^2)\, f\, dc_y\, dc_z$$

Multiplying (70) by 1 and $(c_x^2 + c_z^2)$ and integrating over c_y, c_z from $-\infty$ to $+\infty$ the following equation for φ_s and ψ_s are obtained

$$(74) \qquad C_x \frac{\partial \varphi_s}{\partial x} = K(\bar{\varphi} - \varphi) \ , \qquad C_x \frac{\partial \Psi}{\partial x} = K(\bar{\Psi} - \Psi) \ ,$$

$$(75) \qquad \bar{\varphi} = \varphi^{(o)} \left[1 + \frac{Px}{P}(1-x)\left(\bar{C}_x^2 - \frac{1}{2}\right) + \frac{4}{5} q_x \bar{C}_x \left(\bar{C}_x^2 - \frac{3}{2}\right)(1 - \chi P_r)\right]$$

$$\bar{\Psi} = \varphi^{(o)} T \left[1 + \frac{Px}{P}(1-x)\left(\bar{C}_x^2 - 1\right) + \frac{4}{5} q_x \bar{C}_x \left(\bar{C}_x^2 - \frac{1}{2}\right)(1 - \chi P_r)\right]$$

$$\varphi_s^{(o)} = n(\pi T)^{-1/2} \exp(-C_x^2) \ , \quad \chi = \frac{P}{K\mu} = \text{const}, \quad K = \frac{8}{5\sqrt{\pi}} \frac{nT}{\mu}$$

Equation (74) is written in nondimensional quantities referred to values upstream of the sw

$$(76) \begin{cases} n' = \dfrac{n}{n_1} \ , \quad T' = \dfrac{T}{T_1} \ , \quad c' = \dfrac{c}{(2k/m T_1)^{1/2}} \ , \quad u' = \dfrac{u}{(2k/mT_1)^{1/2}} \ , \quad x' = \dfrac{x}{\lambda} \\[2ex] f' = \dfrac{(2k/mT_1)^{3/2}}{n_1} f \ , \quad \mu = \dfrac{\mu}{\mu_1} \ , \quad \lambda_1 = \dfrac{16}{5n_1} \dfrac{\mu_1}{(2\pi m k T_1)^{1/2}} \end{cases}$$

but the dashes have not been used.

The boundary conditions are:

$$x = +\infty: \quad \varphi_s = n_2 (\pi T_2)^{-1/2} \exp(-\bar{C}_x^2), \quad \Psi_s = T_2 \varphi_s$$

$$x = -\infty: \quad \varphi_s = (\pi)^{-1/2} \exp(-\bar{C}_x^2), \quad \Psi_s = \varphi_1 \tag{77}$$

The quantities are related to the fractional moments by:

$$n = \int \varphi_s \, dc_x, \quad u = \frac{1}{n} \int c_x \varphi_s \, dc_x, \quad P_\perp = \int \Psi \, dc_x$$

$$T = \frac{2}{3}\left(\frac{P_\perp}{\eta} + \frac{1}{n} \int c_x^2 \varphi_s \, dc_x - u^2\right), \tag{78}$$

$$S_x = \frac{1}{nT^{3/2}} \left[\int c_x^3 \varphi_s \, dc_x + \int c_x \Psi_s \, dc_x - 3u \int c_x^2 \varphi_s \, dc_x + 2u^3 n - uP_\perp \right]$$

The problem of the sws is reduced to the integration of (74) with the boundary conditions (77) and of course the HT quantities must satisfy the conservation laws.

The equations (74) are integrated in an iterative procedure

$$C_x \frac{\partial \varphi_{s,n+1}}{\partial x} = K_n (\bar{\varphi}_{s,n} - \varphi_{s,n-1}), \quad c_x \frac{\partial \Psi_{s_1 n+1}}{\partial x} = Kn(\bar{\Psi}_{s,n} - \Psi_{s_1 n+1}) \tag{79}$$

using a finite difference method. For this purpose the variables were introduced

$$S = \int_{x_i}^{x} K(x')\, dx' \tag{80}$$

and (79) was written in its integral form

$$\varphi_s(c_x, s) = \varphi_s(c_x, 0)\, e^{-\frac{s}{c_x}} + \int e^{\frac{s'}{c_x}} \bar{\varphi}_s(c_x, s') \frac{ds'}{c_x}\, e^{-\frac{s}{c_x}} \tag{81}$$

and within the small interval $0 \leqslant S' \leqslant S$ the subintegral function $\bar{\varphi}$ in (81) was replaced by a linear approximation

$$\bar{\varphi}_s(s') = \bar{\varphi}_s(0) + (\bar{\varphi}_s(S) - \bar{\varphi}_s(0)) \frac{S'}{S} \tag{82}$$

The range of integration for the molecular velocity was taken

$$- 3T_2^{1/2} \leqslant C_x \leqslant 3T_2^{1/2} + u_1$$

Some results of the calculations are given on the following figure. The differences in the temperature profiles calculated for different P_r numbers are shown on fig. 15 for M = 5. The relative longitudinal stress and heat flux are shown on fig. 16 and 17 for different M numbers. The differences of the profiles for different P_r numbers are significant. The important effect for viscosity laws on the sw profile and on the relative positions of the profiles of different HT quantities is illustrated on fig. 18.

A comparison of results obtained using the continuum gas theories and the kinetic models of collision integrals with experimental results are shown on fig. 19 and fig. 20. The agreement with experimental results of the NS description is satisfactory as already stated for $M < 2$ whereas $M > 4$ the collision integral models appear to agree better with experiments.

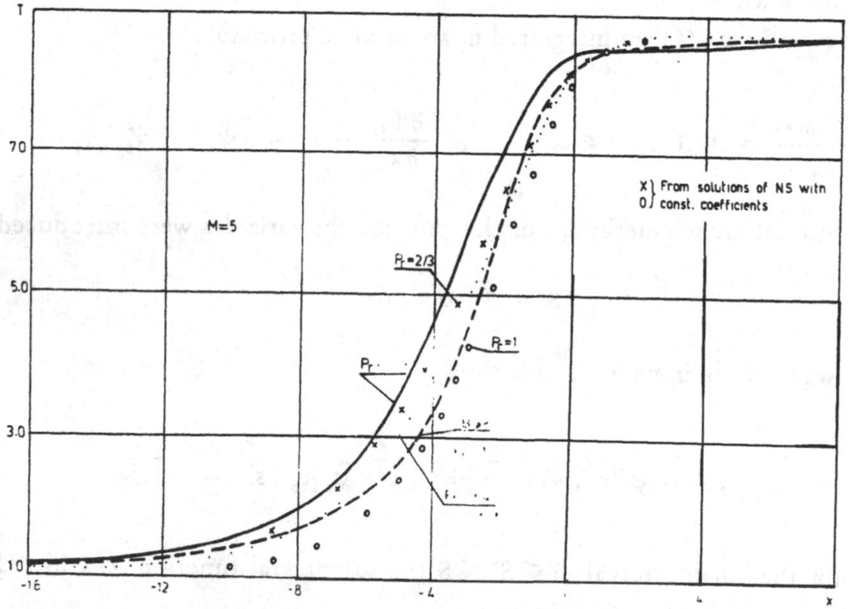

Fig. 15. (Shakhov, 1969)

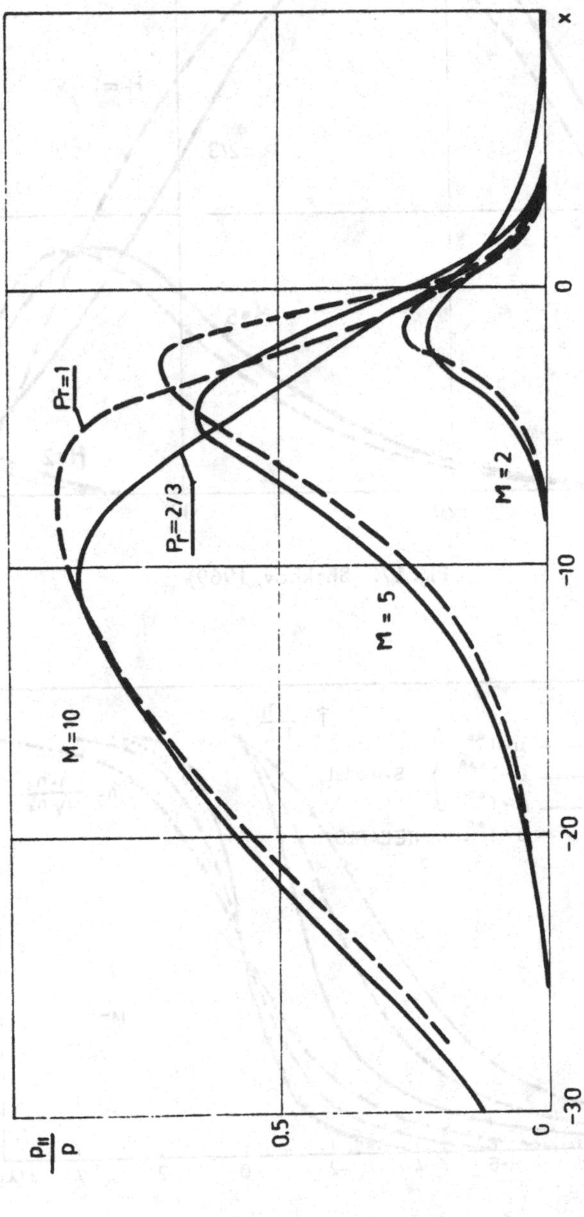

Fig. 16. (Shakhov, 1969)

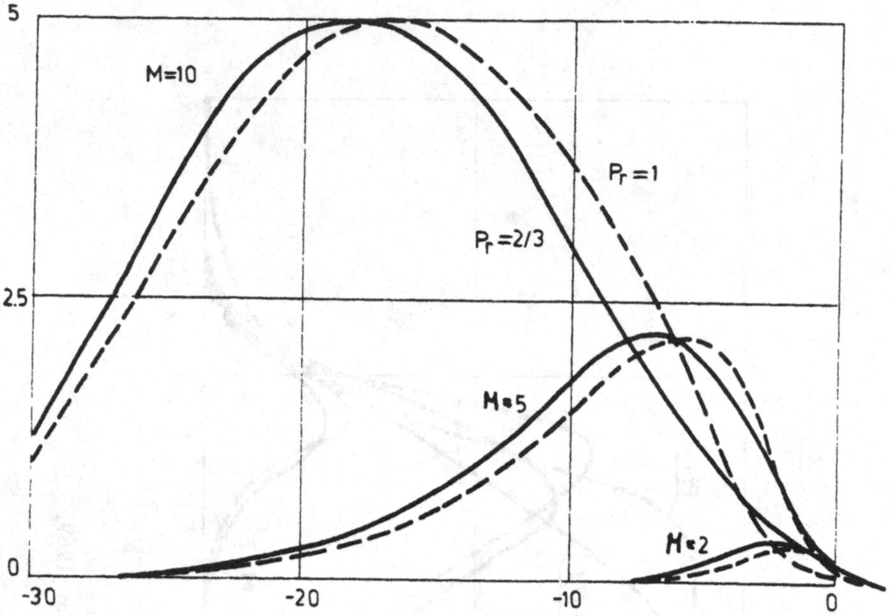

Fig. 17. (Shakhov, 1969)

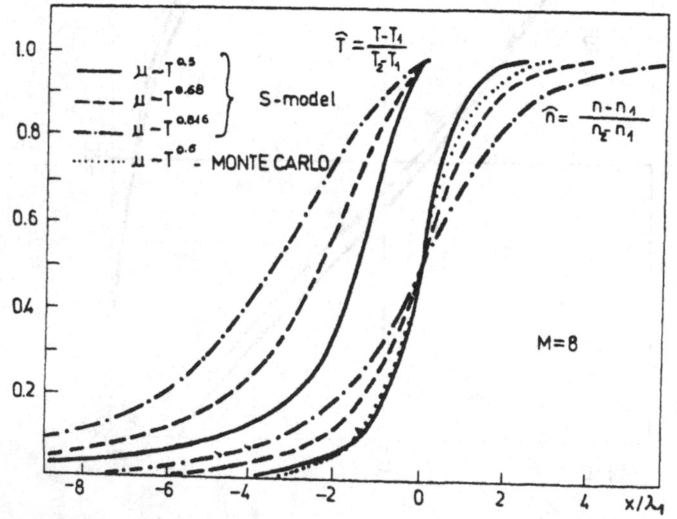

Fig. 18. (Zhuk, Rikov, Shakhov, 1973)

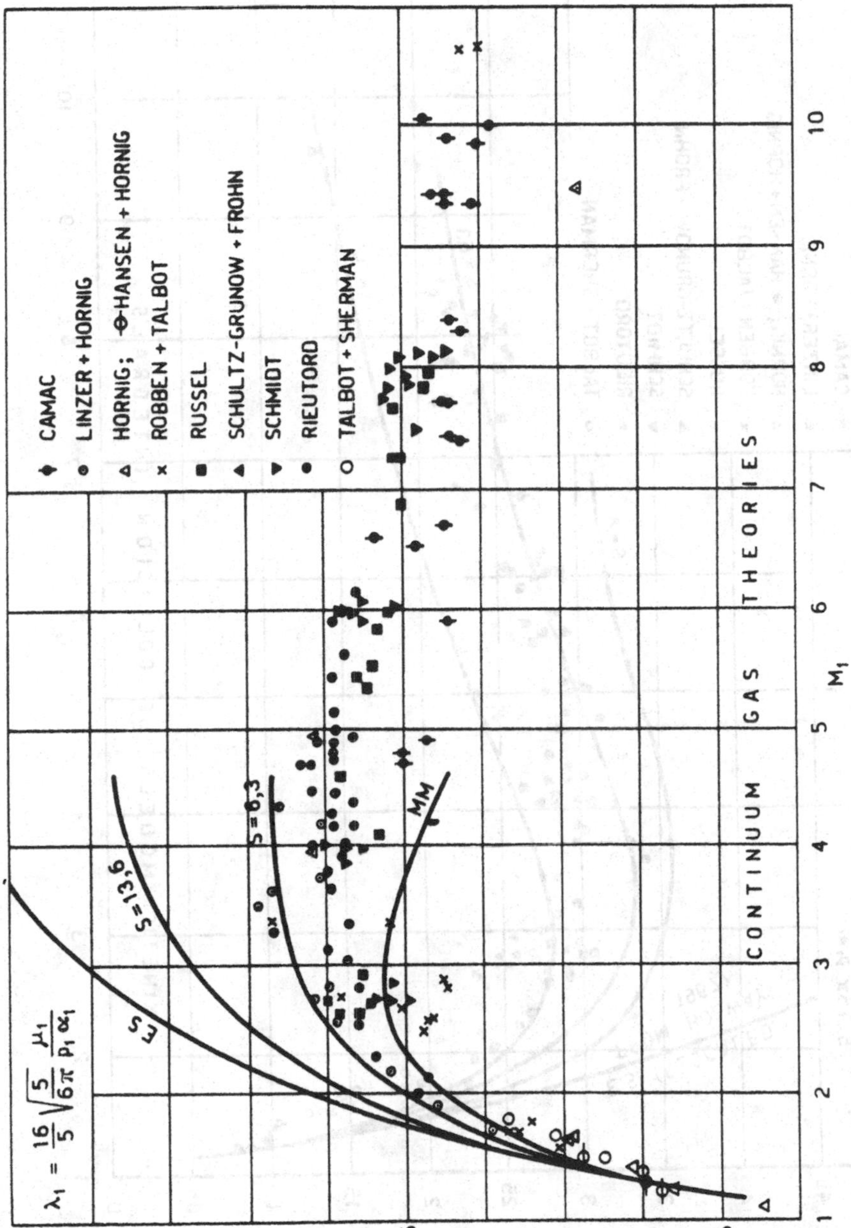

CAMAC
LINZER + HORNIG
HORNIG; HANSEN + HORNIG
ROBBEN + TALBOT
RUSSEL
SCHULTZ-GRUNOW + FROHN
SCHMIDT
RIEUTORD
TALBOT + SHERMAN
S=13,6
S=8,3
MM
CONTINUUM GAS THEORIES
M₁

Fig. 19

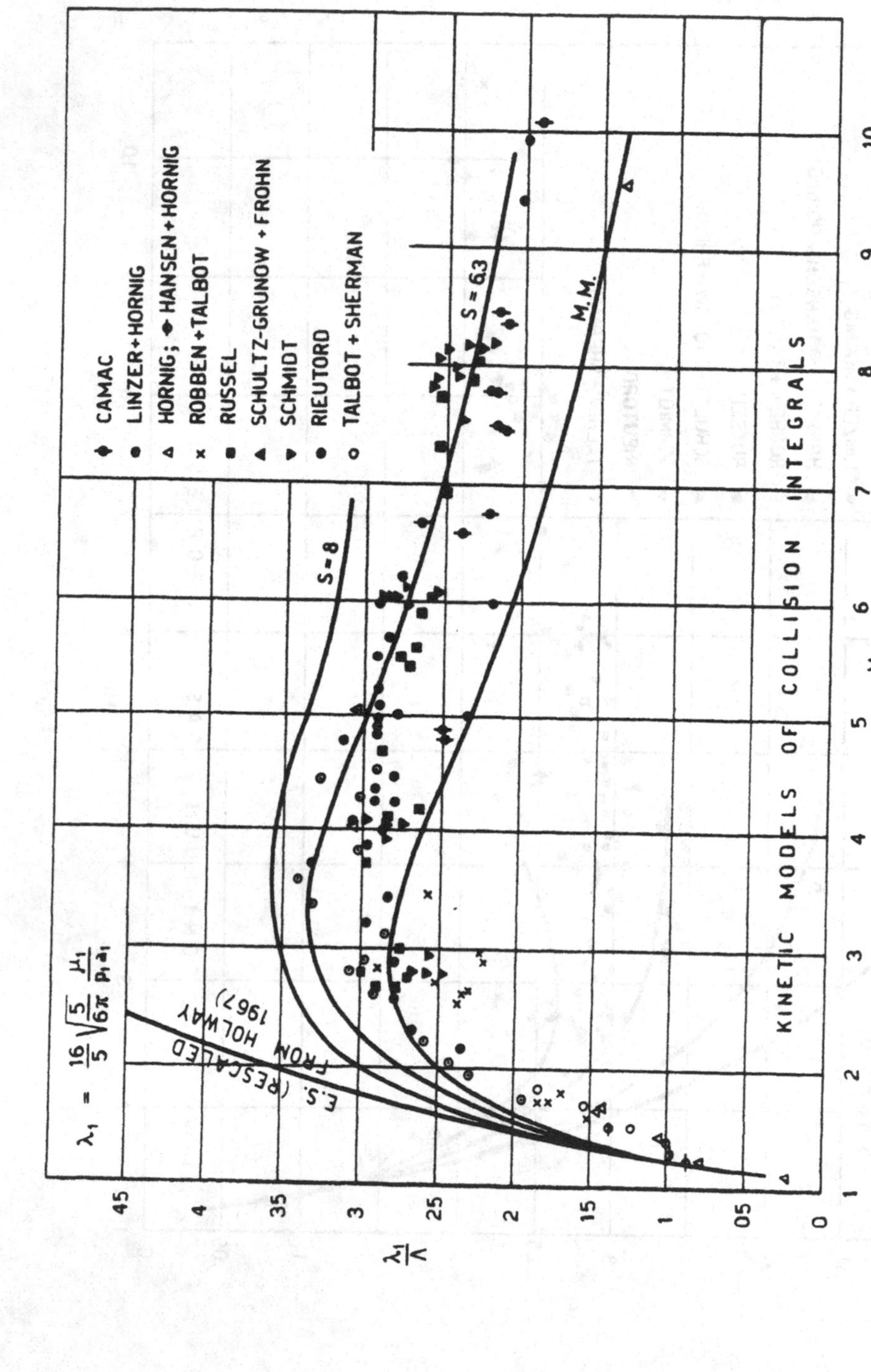
$\lambda_1 = \frac{16}{5}\sqrt{\frac{5}{6\pi}}\,\frac{\mu_1}{p_1 a_1}$
CAMAC
LINZER + HORNIG
HORNIG; HANSEN + HORNIG
ROBBEN + TALBOT
RUSSEL
SCHULTZ-GRUNOW + FROHN
SCHMIDT
RIEUTORD
TALBOT + SHERMAN
E.S. (RESCALED FROM HOLWAY 1967)
S = 8
S = 6.3
M.M.
KINETIC MODELS OF COLLISION INTEGRALS
$\frac{\lambda_1}{\Lambda}$
M_1
0
05
1
15
2
25
3
35
4
45
1
2
3
4
5
6
7
8
9
10

Fig. 20

IV. Kinetic Models of the Distribution Function

To obtain an approximate solution of the B equation for the sws problem, besides the previously described method, many approaches based on finding expressions of the distribution function based on physical grounds are used.

The first and most frequently used method was introduced by Mott Smith [31] and independently in a later publication, based on earlier work, by Tamm [32].

Mott Smith suggested that a better description function within a sw than the skewed Maxwellian $\hat{M}^*$ resulting from the Ch.E. procedure may be a bimodal distribution composed of the upstream $f_1^{(o)} = f^{(o)}(n_1, u_1, T_1)$ and downstream $f_2^{(o)} = f^{(o)}(n_2, u_2, T_2)$ distributions. Thus the proposed distribution function is made up of the sum of two known equilibrium distributions with number densities varying with position indicating their relative importance at different points within the sws.

Thus taking

$$f(c, x) = (1 - \tilde{n}(x))\, f_1^{(o)} + \tilde{n}(x)\, f_2^{(o)} \tag{84}$$

the conservation equations are satisfied if the HT quantities with subscripts 1 and 2 satisfy the Rankine Hugoniot conditions. An additional transport equation of some nonconserved function of velocity $\Phi(c)$ is required to determine the unknown function $\tilde{n}(x)$. Mott-Smith used $\Phi(c) = c_x^2$ and $\Phi(c) = c_x^3$. The one dimensional stationary transport equation is [2] [31]

$$\frac{\partial}{\partial x} \int c_x \Phi f \, dc = \int (\Phi' - \Phi) f f_1 \, g b db d\varepsilon dc_1 dc \tag{85}$$

where $f_1 \equiv f(c_1)$, $\Phi' \equiv \Phi(c')$, is the velocity of the first molecule after collision, $g = |c_1 - c|$. The integral on the r.h.s. of (85) depends on the interaction potential of the molecules. Substituting (84) in (85) as shown in [31] after rather lengthy calculations the required equation for $\tilde{n}(x)$ is obtained

$$\frac{d\tilde{n}(x)}{dx} = \frac{B}{\lambda_1} (1 - \tilde{n}(x))\, \tilde{n}(x) \tag{86}$$

where B is a constant depending on Φ, the molecular interaction law and the HT parameters and λ_1 is the $\hat{M}$. mean free path. Choosing the origin $x = 0$ as the point where $\bar{n}(0) = 1/2$ and integrating (86) yields

$$n(x) = \frac{e^{\frac{B}{\lambda_1}x}}{1 + e^{\frac{B}{\lambda_1}x}} \tag{87}$$

The total number density for a given x is

$$n(x) = (1 - \bar{n}(x)\, n_1 + \bar{n}(x)\, n_2 \tag{88}$$

which on substituting n_2/n_1 from (12) and (87) gives

$$\frac{n(x)}{n_1} = \frac{M^2 + \frac{2}{\gamma - 1} + M^2 \frac{\gamma + 1}{\gamma - 1} e^{B\,x/\lambda_1}}{(M^2 + \frac{2}{\gamma - 1})\ (1 + e^{Bx/\lambda_1})} \tag{89}$$

The Prandtl density sw thickness, $(n_2 - n_1)\, |\, dn/dx\, |^{-1}_{max}$ is easily found to be

$$\Lambda_1 = \frac{4\lambda_1}{B} \tag{90}$$

The values of B for the two moments used by Mott-Smith is shown on fig. 21 stressing the large influence of the choice of the moment equation on the calculated physical quantity. Rode and Tennenbaum [33] have made calculations of shock wave thickness for moments with different powers of c_x showing that the results for powers exceeding 6 become meaningless, see fig. 22. It is interesting also to compare the calculated sw thicknesses for different interaction laws as calculated by Muckenfuss [34] using the bimodal Mott Smith model; as seen on fig. 23 this has a very marked effect on the considered magnitude.

Tamm's approach avoids the arbitrariness of the choice of the additional moment equation by using the variational principle of finding the minimum of the functional

$$S = \int [c_x \frac{\partial f}{\partial x} - Q(f, f)]^2 \, dx \, dc \tag{91}$$

with the trial function in the bimodal form. Although Oberai [35] and Narasimha and Desphande [36] have used similar approaches, this method deserves further developments.

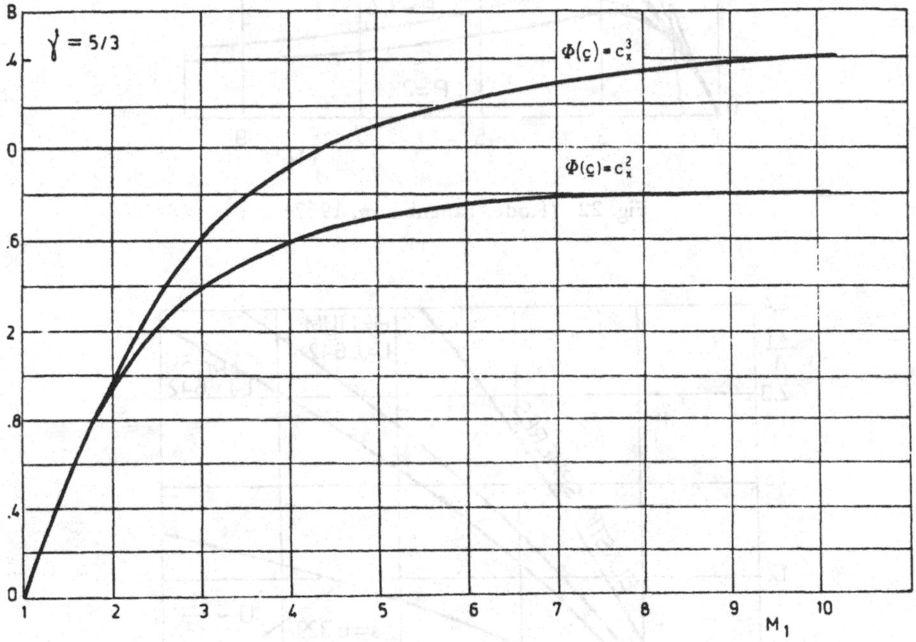

Fig. 21. (Mott-Smith, 1951)

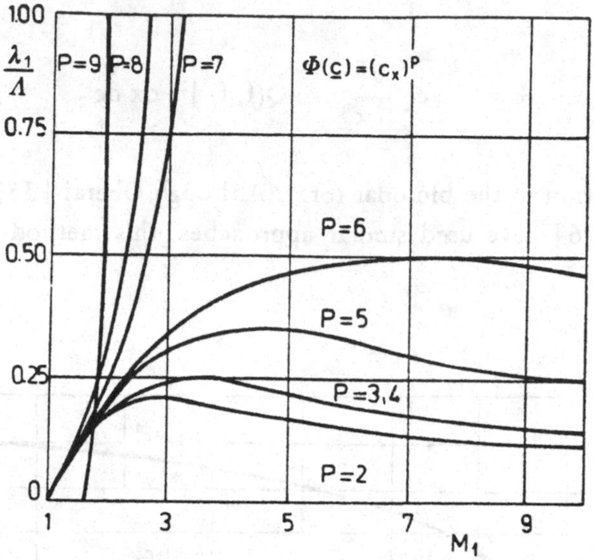

Fig. 22. (Rode, Tanenbaum, 1967)

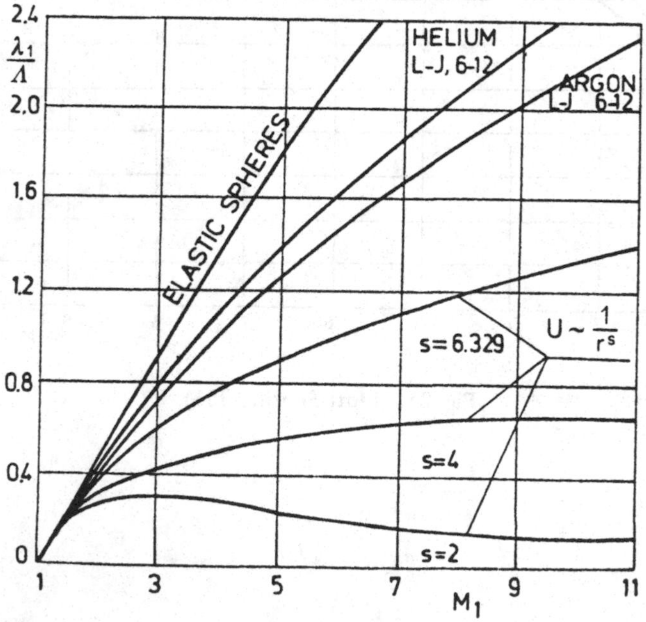

Fig. 23. (Muckenfuss, 1962)

A generalization of the MS method proposed and analysed by Salven, Grosch and Ziering [37] consists in approximating the distribution function by a linear combination of $\hat{M}$ type functions

$$f = \sum_{i}^{N} n_i(x)\, f_i(c) \tag{92}$$

where f_1 and f_2 are the MS forms and the others must satisfy certain consistency conditions. A particularly simple case is the three term approximation:

$$f(x, c) = a_1(x)\, F_1(c) + a_2(x)\, F_2(c) + a_3(x)\, F_3(c) \,, \tag{93}$$

where

$$F_\alpha(c) = (2\pi\, b_\alpha^2)^{-3/2} \exp\left[-\frac{1}{2b_\alpha^2} \left((c_x - u_\alpha)^2 + c_y^2 + c_z^2 \right) \right] \qquad (\alpha = 1, 2)$$

$$F_3(c) = (2\Pi b_3)^{-2} (c_x - u_3) \exp\left[-\frac{1}{2b_3^2} \left((c_x - u_3) + c_y^2 + c_z^2 \right) \right], \qquad b_i^2 = \frac{kT_i}{m}$$

The HT quantities with subscripts 1 and 2 correspond to upstream and downstream conditions and the boundary conditions are:

$$\begin{aligned} &a_1(-\infty) = n_1 \,, \qquad a_2(+\infty) = n_2 \,, \\ &a_1(+\infty) = a_2(-\infty) = a_3(-\infty) = a_3(+\infty) = 0. \end{aligned} \tag{94}$$

To determine the 3 coefficients $a_i(x)$ moments of the stationary one dimensional B equation will be taken, using the functions of c, $\Phi_j(c)$, leading to quations:

$$\frac{d}{dx} \int c_x\, \Phi_k\, f(x, c)\, dc = \int \Phi_k\, Q(f, f)\, dc \,. \tag{95}$$

For the assumed distribution function (93) the above equation yields

$$\sum_{i=1}^{3} I_i(\Phi_k) \frac{da_i}{dx} = \sum_{i,j=1}^{3} I_{ij}(\Phi_k)\, a_i\, a_j \,, \qquad (k = 1, 2)$$

where

$$I_i(\Phi_k) = \int c_x \, \Phi_k(c) \; F_i(c) \, dc$$

(96)

$$I_{ij}(\Phi_k) = \int (\Phi_k' - \Phi_k) \, F_i(c) \, F_j(c) \, g b db d\epsilon \, dc, dc$$

using the symmetry relation of the collision integral [23], [8]. If $\Phi_k(c)$ is one of the conserved quantities $(1, c_x, c_y, c_z, c^2)$ the $I_{ij}(\Phi_k) = 0$ and the following set of 5 equations results

$$(97) \quad \begin{cases} u_1 \dfrac{da_1}{dx} + u_2 \dfrac{da_2}{dx} + I_3(1) \dfrac{da_3}{dx} = 0 \,, \\ (u_1^2 + b_1^2) \dfrac{da_1}{dx} + (u_2^2 + 5b_2^2) \dfrac{da_2}{dx} + I_3(c^2) \dfrac{da_3}{dx} = 0 \,, \\ u_1(u_1^2 + 5b_1^2) \dfrac{da_1}{dx} + u_2(u_2^2 + 5b_2^2) \dfrac{da_2}{dx} + I_3(c^2) \dfrac{da_3}{dx} = 0 \,, \\ I_3(c_y) \dfrac{da_3}{dx} = 0, \qquad I_3(c_z) \dfrac{da_3}{dx} = 0 \end{cases}$$

Integrating (97) from $-\infty$ to $+\infty$ with the boundary conditions (94) and requiring $a_3(x) \neq 0$ there results

$$(98) \qquad I_3(c_y) = I_3(c_z)$$

and obviously the R.H. relations are recovered

$$(99) \quad \begin{cases} u_1 n_1 = u_2 n_2 \,, \\ (u_1^2 + b_1^2) n_1 = (u_2^2 + b_2^2) n_2 \,, \\ u_1(u_1^2 + 5b_1^2) n_1 = u_2(u_2^2 + 5b_2^2) n_2 \end{cases}$$

Using the above relations in (97) and dividing the momentum and energy conservation equations by the mass conservation equation the following consistency conditions result (besides (98))

$$\begin{cases} \dfrac{1}{u_1}(u_1^2 + b_1^2) = \dfrac{1}{u^2}(u_2^2 + b_2^2) = \dfrac{I_3(c_x)}{I_3(1)} , \\ (u_1^2 + 5b_1^2) = (u_1^2 + 5b_2^2) = \dfrac{I_3(c^2)}{I_3(1)} , \end{cases} \tag{100}$$

Thus in order to satisfy the conservation equations one of the upper three equations (97), say the first one, must be satisfied together with equations (98) and (100).

Since the conservation equations reduce to one differential equation among the three unknown function $a_i(x)$ two additional moment equations, one more than in the MS method, are required.

For the case of monatomic Maxwell molecules from the consistency conditions:

$$f(x, c) \cong \sum_{i=1}^{N} a_i(x)\, F_i(c) \tag{101}$$

This can easily be generalized for the case of a distribution function corresponding to

$$u_3 = \frac{1}{2u_1}\left(u_1^2 + \frac{kT_1}{m}\right)$$
$$\frac{5kT_3}{m} = u_1^2 + \frac{5kT_1}{m} - 3u_3^2 \tag{102}$$

Salwen, Grosch and Ziering [37] have made an analysis of the asymptotic behaviour of equations (96) for the case of Maxwell molecules and starting from the asymptotic value at $x = +\infty$ have made numerical calculations for three pairs of momenta of $\Phi_k(c)$ (c_x^2, c_x^3), $(c_x^2, c_x c^2)$, $(c_x^3, c_x c^2)$. The results of these trimodal calculations are compared on fig. 24 with those of bimodal calculations of the sw profile and on fig. 25 of the Prandtl shock wave thickness displaying appreciable

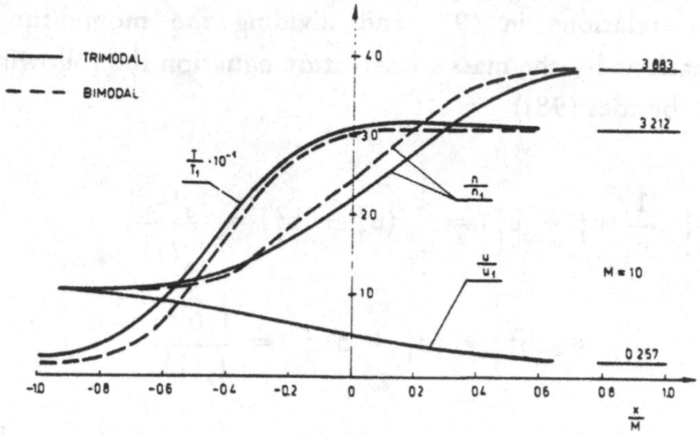

Fig. 24. (Kogan, 1966)

differences between the cases considered. In particular the results obtained at $M \approx 1$ indicate that $F_3(c)$ has been well chosen for weak shocks. For strong shocks perhaps another choice of $F_3(c)$ would be better.

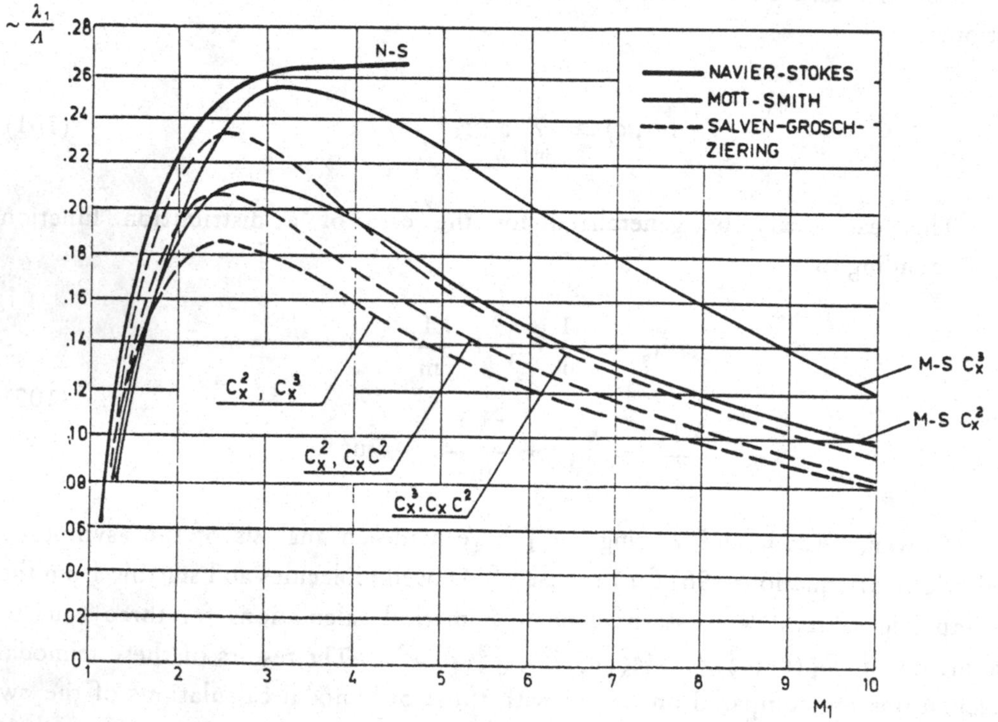

Fig. 25. (Slaven, Grosch, Ziering, 1964)

An interesting feature of the sw profile is that at higher M, a significant change in temperature occures far ahead of any significant change of density. Also the asymmetry of the sw profiles calculated using different methods is interesting. The trimodal models exhibit a larger asymmetry than the bimodal M.S. model for which the density profile is symmetric and the velocity profile shows a small asymmetry. For a quantitative measure of this feature it is necessary to introduce a general accepted definition of asymmetry.

A comparison with experimental results of the models of the distribution function is made on fig. 26 showing the different ranges of agreement and deviations of the bimodal and trimodal models.

Some other models of the distribution functions are commented in [38].

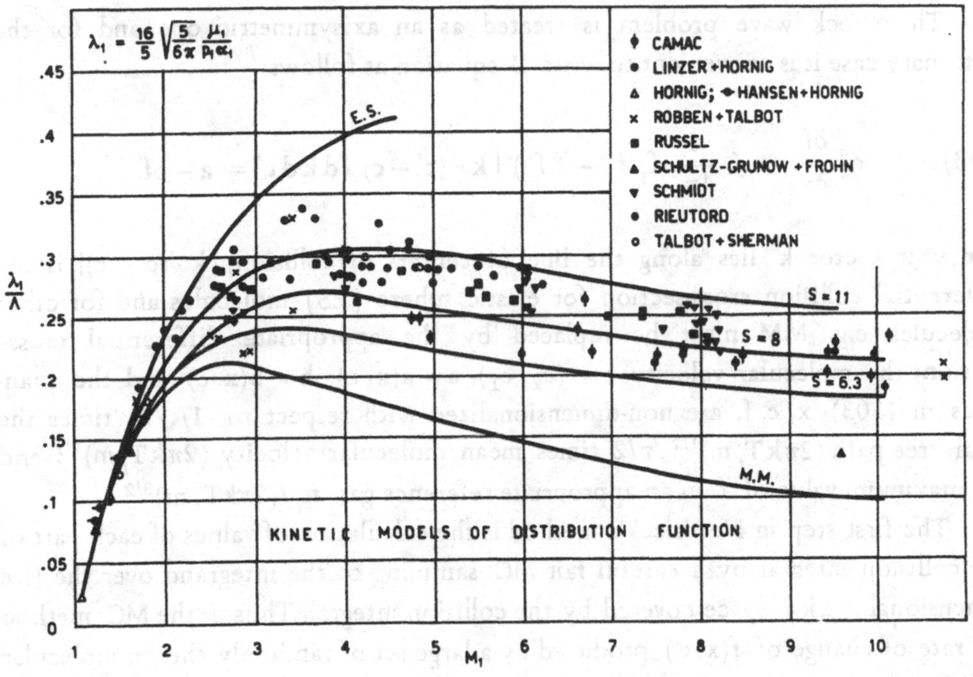

Fig. 26

V. Monte Carlo Methods

The mathematical difficulties inherent in the calculation of the B collision integral and the shortcomings of the different approximate methods have imposed and the use of the Monte Carlo (M C) methods for the determination of the sw structure. The availability of large scale computers is an additional favorable factor.

Essentially the MC method is used in two different ways: either as a tool for solving B equation or for direct simulation of the motions of molecules for the conditions corresponding to the existence of shock waves. An outline of these two applications of this stochastic computational procedure and some of the main results obtained will be given below.

Nordsieck's [39] MC method of evaluation of the B collision integral has so far yielded the largest number of results due to the efforts of Hicks, Yen et al. [40] [41].

The shock wave problem is treated as an axisymmetric one and for the stationary case it is convenient to write B equation as follows:

$$(103) \qquad c_x \frac{\partial f}{\partial x} = \int \frac{1}{4\pi}(f_1 f_1' - f f') \, | k \cdot (c' - c) | \, dk \, dc' = a - bf$$

The unit vector k lies along the line of centers of collision, $|k \cdot (c' - c)|$ is the differential collision cross-section for elastic sphere (ES) molecules and for other molecules e.g. MM must be replaced by the appropriate differential cross-section; the molecular velocity $c = (c_x, c_\perp)$, $a = a(x, c)$, $b = b(x, c)$ and the quantities in (103) x, c, f, are non-dimensionalized with respect to $1/\sqrt{2}$ times the mean free path $(2\pi k T_1 m)^{1/2}$, $\pi/2$ times mean molecular velocity $(2\pi k T_1 m)^{1/2}$ and the maximum value of f in an appropriate reference gas $n_1/(2\pi k T_1 m)^{3/2}$.

The first step in Nordsieck's method is the calculation of values of each part of the collision integral by a careful fair MC sampling of the integrand over the five dimensional (c, k) space covered by the collision integral. Thus in the MC method the rate of change of f(x, c) produced by a large set of randomly chosen molecular collisions in the c, k space is computed (for a sample of N values the random errors are proportional to $N^{-1/2}$. The two-dimensional velocity space was divided e.g. in Hicks calculations into 226 cells as shown on fig. 27 and for each discrete value of x; f, a, a − bf was evaluated at the center of each cell. Yen et al. [41] have completed very recently extensive calculations made for ES and Maxwell molecules

(MM) Nordsick's improved method with an iterative procedure starting from an initial MS distribution function at 9 positions within the shock wave. The Mach number range covered was 1.1 to 10, the nine positions used correspond to nine equal divisions of the number density n and Monte Carlo samples of 2^{13} collisions in four independent runs were generally used. This method allows the determination of the properties of the sw profile and we will recall some of them and particularly wherever data exist the influence of the intermolecular potential on these properties.

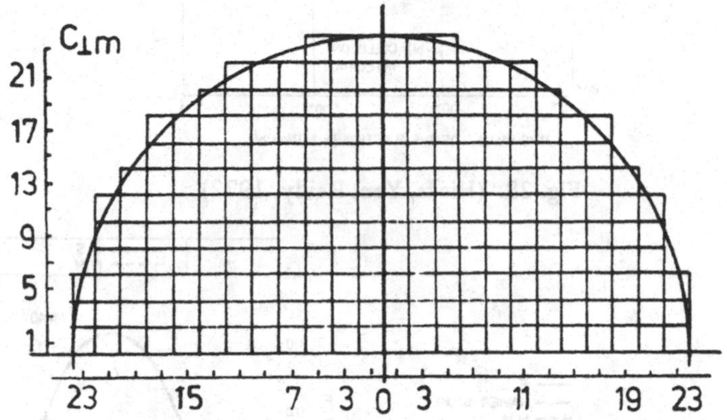

Fig. 27. (Nordsieck, Hicks, 1966)

Using as a measure of departure from local equilibrium the r.m.s. (a – bf)/(a + bf) the results are shown on fig. 28 where the large variations with M are noteworthy. It can be seen there, that the departure from equilibrium as measured by the above value is larger near the cold side for $M > 1.2$. For $M < 1,2$ it is larger at the center which indicates the strong diffusion effect of the high speed molecules moving upstream.

Most results are given as functions of reduced density

$$\hat{n} = \frac{n - n_1}{n_2 - n_1} \tag{104}$$

and not of the space variable, as using this quantity as an independent variable exposes many interesting facts and saves one quadrature operation.

The density gradient is shown for different types of molecules on fig. 29. It is seen there that the differences between the models are very large and that the position of maximum gradient shifts from the cold side to the hot side as the M number increases.

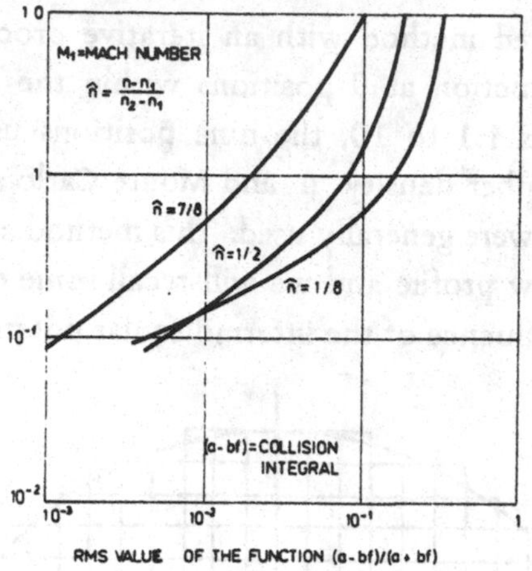

Fig. 28. (Hicks, Yen, Reilly, 1972)

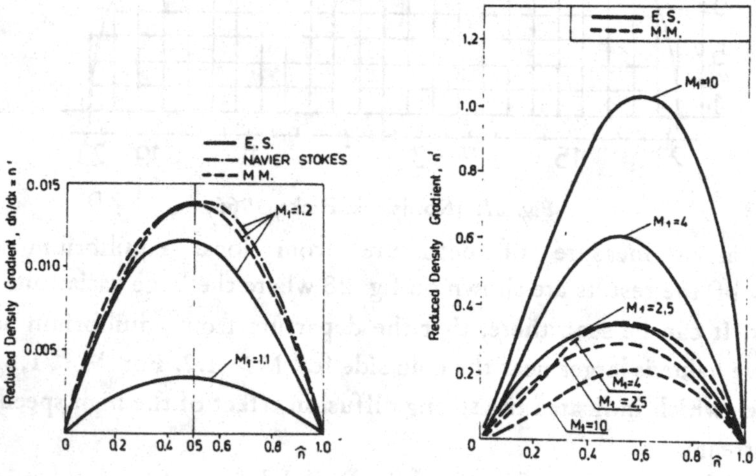

Fig. 29. (Yen, NG, 1973)

Comparing now the HT quantities i.e. moments of the distribution function

$$\mathscr{M}_i = \int f \Phi_i \, dc \tag{105}$$

$$\Phi_1 = 1, \quad \Phi_2 = c_x, \; \Phi_3 = c_x^2, \quad \Phi_4 = c_x c^2, \quad \Phi_6 = c_x^3, \quad \Phi_9 = c_\perp^2$$

the dimensionless HT quantities u, T, τ, q, related respectively to u_1, T_1, p_1, u_1^2, are expressed by the above moments as follows:

$$\left\{\begin{array}{l} u = \mathscr{M}_2/n \\ T_x = 2\pi\,[-u^2 + \mathscr{M}_3/n] \\ T_\perp = \pi\,\mathscr{M}_9/n \\ T = 1/3\,T_x + 2/3\,T_\perp\,, \\ \tau = 2/3\,n\,(T_\perp - T_x)\,, \\ q = (2\pi\,\mathscr{M}_4/\mathscr{M}_2) - 3T_x - 2\pi u^2\,, \\ q_x = (2\pi\,\mathscr{M}_6/\mathscr{M}_2) - 3T_x - 2\pi u^2\,, \end{array}\right. \tag{106}$$

Yen's and al. [41] calculations have shown that the non invariant moments of $f(\mathscr{M}_4, \mathscr{M}_6, \mathscr{M}_9)$ are nearly linear functions of the density n. As an example the variation of $\mathscr{M}_9$ vs n is shown on fig. 30.

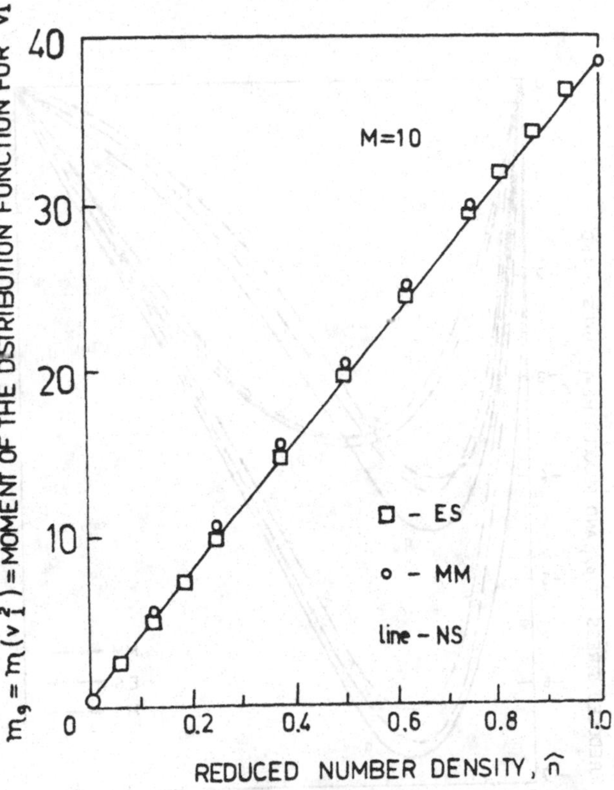

Fig. 30. (Yen, NG, 1973)

It can be observed therefrom that the moments of f that determine the HT properties are strongly coupled and that this coupling is nearly independent of the intermolecular interaction law. Hence it can be expected that the variation of the macroscopic properties with respect to n and to each other would be nearly the same for a real gas whose molecular properties are between MM and ES. It should be also noted that the MS sw predicts an exactly linear dependence of the moments from n and therefore independence of macroscopic properties with respect to n on the collision law.

The variation of the heat fluxes and stress with n is shown on fig. 31 for different models. Although the overall variation is similar in the three cases shown the differences in the upstream and downstream wings is appreciable.

The variation of the positions of maxima of different HT quantities including the BH function:

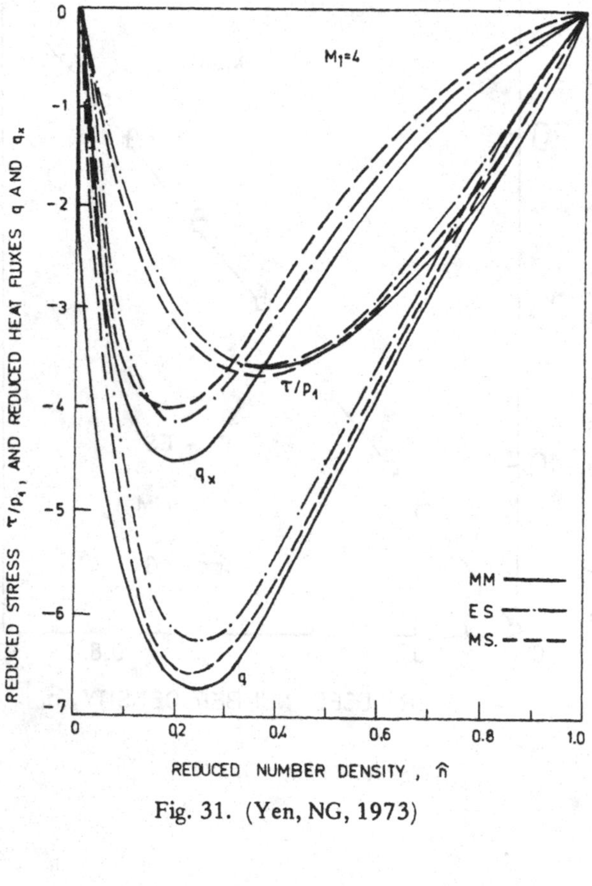

Fig. 31. (Yen, NG, 1973)

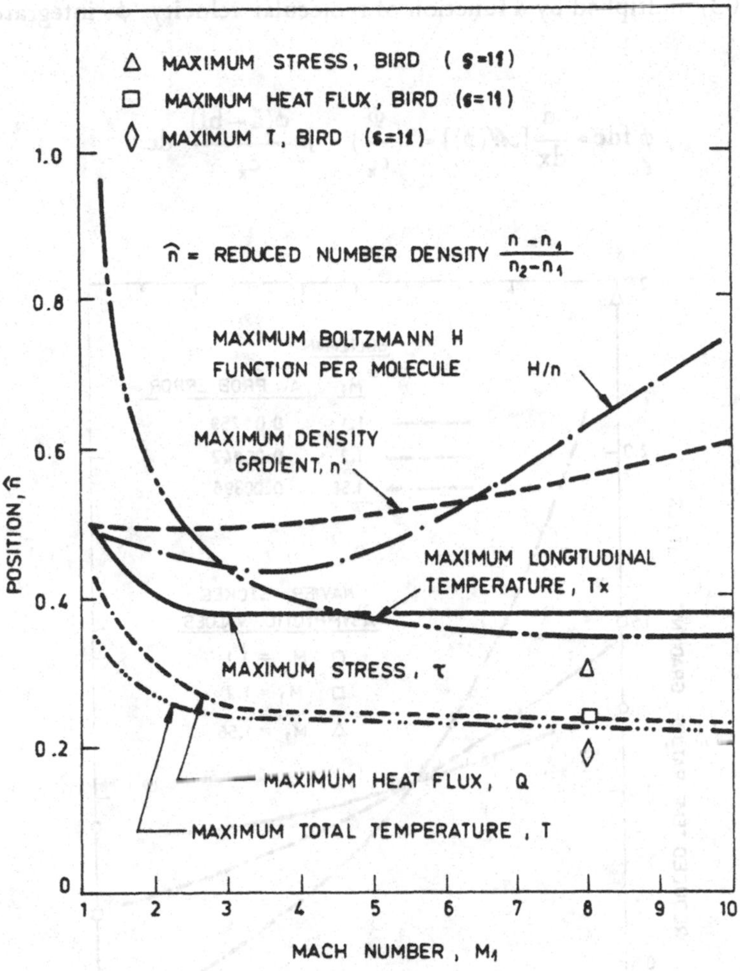

Fig. 32. (Hicks, Yen, Reilly, 1972)

is illustrated for ES molecules on fig. 32 showing at higher M small variations with M except for the longitudinal temperature T .

The variation of $dT/dn = d\{(T-T_1)/(T_2-T_1)\}/d\{(n-n_1)/(n_2-n_1)\}$ on n for various M_1 are shown on fig. 33 exhibiting the large differences particularly at lower M.

To analyse the spatial variation of the HT macroscopic quantities, the B equation (103) multiplied by a function of molecular velocity ϕ integrated over c is used:

$$\int \phi \, f dc = \frac{d}{dx}[\mathscr{M}(\phi)] = I(\frac{\Phi}{c_x}) = \int \frac{\phi(\breve{a} - \check{b}f)}{c_x} dc \tag{107}$$

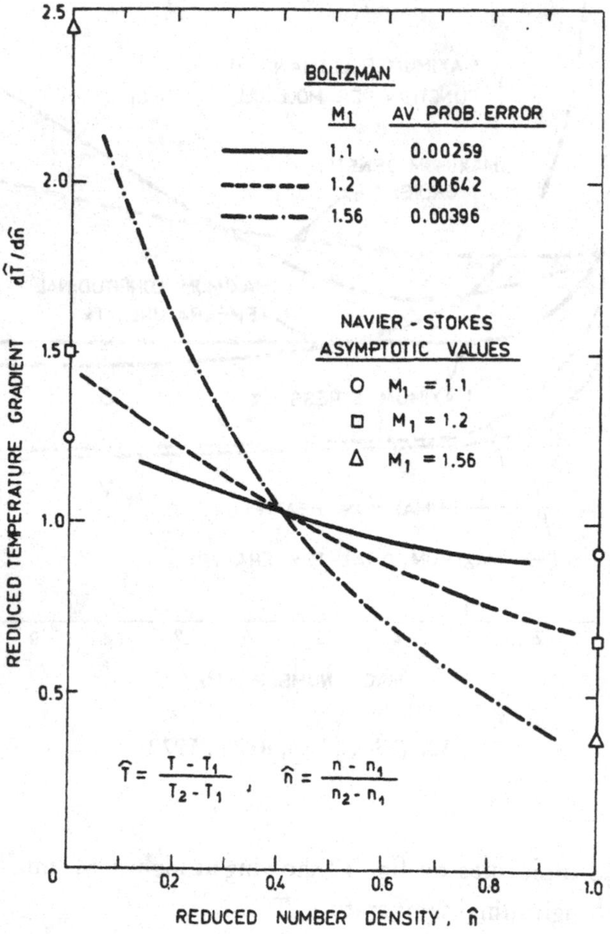

Fig. 33(a). (Hicks, Yen, Reilly)

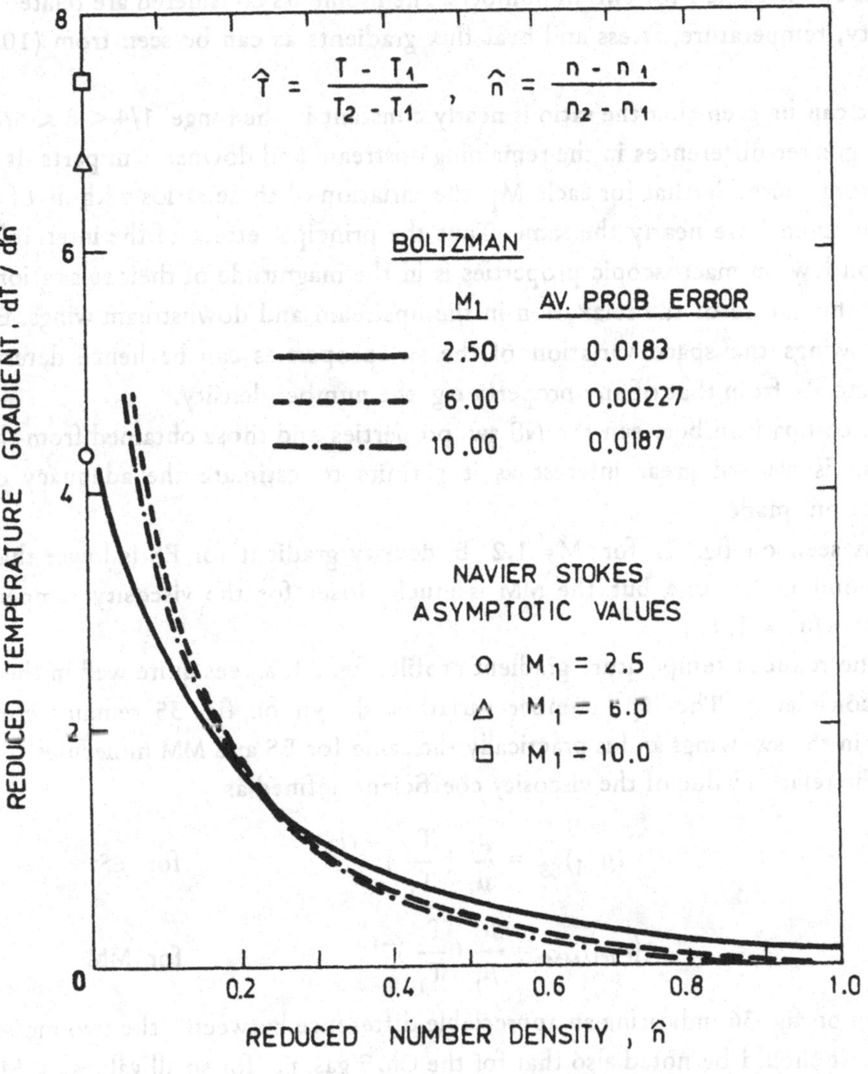

$\hat{T} = \frac{T - T_1}{T_2 - T_1}$, $\hat{n} = \frac{n - n_1}{n_2 - n_1}$
BOLTZMAN
M_1 AV. PROB ERROR
2.50 0.0183
6.00 0.00227
10.00 0.0187
NAVIER STOKES
ASYMPTOTIC VALUES
○ M_1 = 2.5
△ M_1 = 6.0
□ M_1 = 10.0
REDUCED TEMPERATURE GRADIENT, $d\hat{T}/d\hat{n}$
REDUCED NUMBER DENSITY , $\hat{n}$
0 2 4 6 8
0 0.2 0.4 0.6 0.8 1.0

Fig. 33(b)

To make a quantitative comparison of the effect of the molecular model (MM or ES) the ratio of the moments of the collision integrals of three moments are compared on Table I for two M numbers. The moments considered are related to the velocity, temperature, stress and heat flux gradients as can be seen from (105) and (106).

It can be seen that the ratio is nearly constant in the range $1/4 < \hat{n} < 5/8$ but shows greater differences in the remaining upstream and downstream parts. It is also seen from the table that for each M_1 the variation of these ratios with $\hat{n}$ of all the three moments are nearly the same. Thus the principal effect of the intermolecular collision law on macroscopic properties is in the magnitude of their relaxation rates and in the nature of the relaxation in the upstream and downstream wings. Except in the wings the space variation of the sw properties can be hence determined satisfactorily from that of any property e.g. the number density.

A comparison between the NS sw properties and those obtained from the MC solution is also of great interest as it permits to estimate the adequacy of the assumptions made.

As seen on fig. 27 for $M = 1.2$ B. density gradient for ES is lower than the corresponding NS one but the MM is much closer for the viscosity temperature relation $u/u_1 = T/T_1$.

The reduced temperature gradient profile, fig. 34, agrees quite well in the three cases considered. The P_r number variation shown on fig. 35 remains constant except in the sw wings and is practically the same for ES and MM molecules.

The relative value of the viscosity coefficient defined as

$$(\mu_{rel})_{ES} = \frac{\mu}{\mu_1}\left(\frac{T}{T_1}\right)^{-1/2} \qquad \text{for ES}$$

(108)

$$(\mu_{rel})_{MM} = \frac{\mu}{\mu_1}\left(\frac{T}{T_1}\right)^{-1} \qquad \text{for MM}$$

is shown on fig. 36 indicating an appreciable difference between the two molecular models. It should be noted also that for the Ch.E gas, i.e. for small values of $M - 1$, μ_{rel} should be one.

A very interesting result shown by Yen et al. [41] is the graphical display of the velocity distribution function shown on fig. 37 at three positions in the sw for ES and MM. The photographs show

a) the deep penetration of high speed molecules from the cold side towards the hot

i	r	l	m	λ_i^{*}	$\Psi_i(\underline{c})$	a_i^{*}	A_i^{**}
1	0	0	0	0	1	$\frac{n-n_0}{n_0}$	0
2	0	1	0, ±1	0	$\sqrt{2}\underline{c}$	$\sqrt{2}\underline{u}/(2RT_0)^{1/2}$	0
3	1	0	0	0	$\sqrt{\frac{2}{3}}(\frac{3}{2}-c^2)$	$-\sqrt{\frac{3}{2}}\frac{T-T_0}{T_0}$	0
4	0	2	0, ±1, ±2	−3/5	$\sqrt{2}(\underline{\underline{cc}}-\frac{1}{3}\underline{\underline{\delta}}c^2) \triangleq 2\underline{\underline{\overset{0}{c}c}}$	$\sqrt{2}\underline{\underline{\tau}}/2RT_0$	$\sqrt{2}\underline{\underline{\tau}}/2RT$
5	1	1	0, ±1	−2/5	$\sqrt{\frac{4}{5}}\underline{c}(\frac{5}{2}-c^2)$	$-2\sqrt{\frac{4}{5}}\underline{q}/(2RT_0)^{3/2}$	$-2\sqrt{\frac{4}{5}}\underline{q}/(2RT)^{3/2}$

* $a_i \equiv \pi^{3/2}\int e^{c^2}\phi\,\Psi_i\,d\underline{c}$ is the perturbation of the moment m_i from absolute equilibrium.

** $A_i \equiv \frac{1}{n}\int f^{(0)}\phi\,\Psi_i(\underline{C})\,d\underline{C}$ is the perturbation of the moment m_i from local equilibrium ($\underline{C} = \underline{c} - \underline{u}$).

Table I

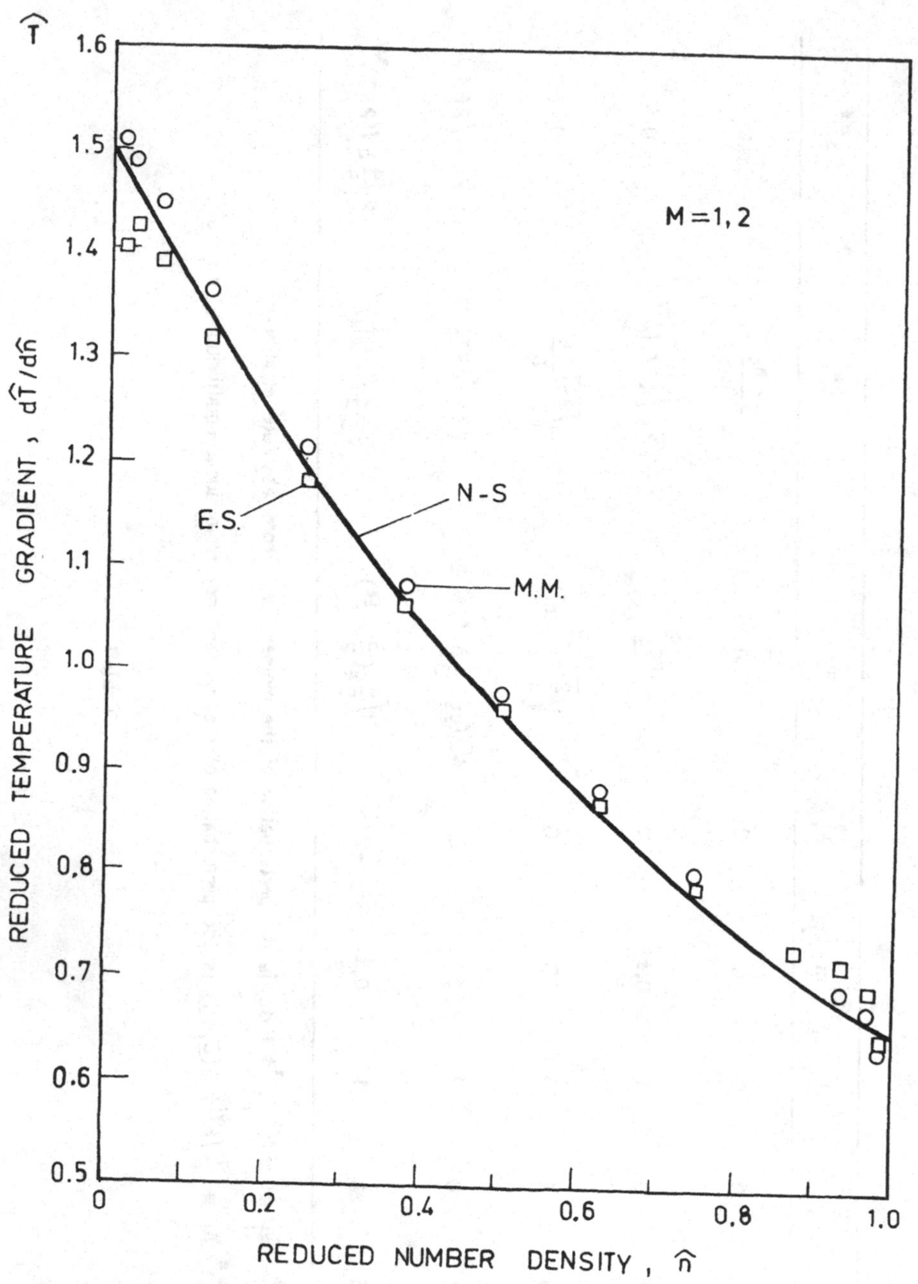

Fig. 34. (Yen, NG, 1973)

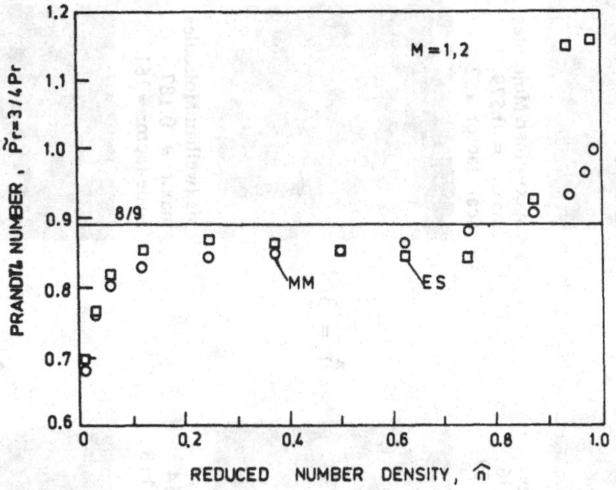

Fig. 35. (Yen, NG, 1973)

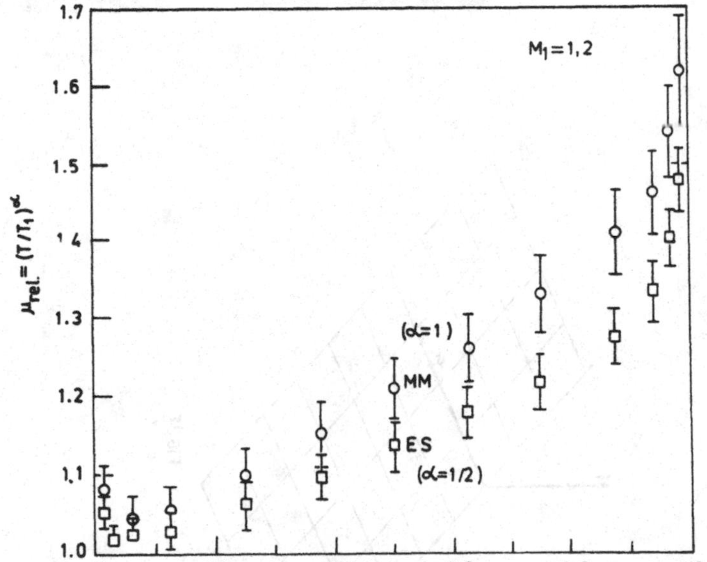

Fig. 36. (Yen, NG, 1973)

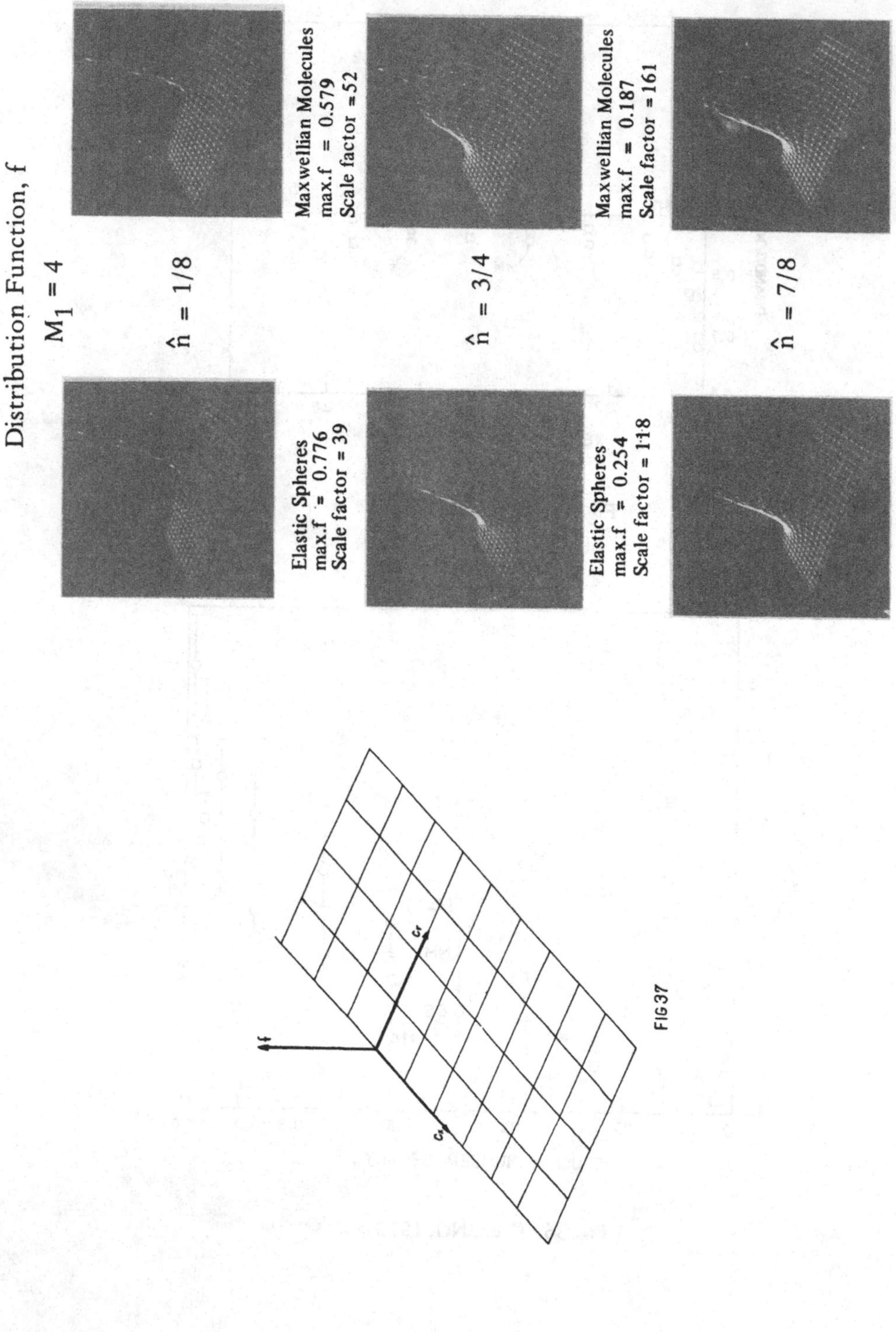
Distribution Function, f
$M_1 = 4$
$\hat{n} = 1/8$
Elastic Spheres
max.f = 0.776
Scale factor = 39
Maxwellian Molecules
max.f = 0.579
Scale factor = 52
$\hat{n} = 3/4$
Elastic Spheres
max.f = 0.254
Scale factor = 118
Maxwellian Molecules
max.f = 0.187
Scale factor = 161
$\hat{n} = 7/8$
Elastic Spheres
max.f = 0.195
Scale factor = 154
Maxwellian Molecules
max.f = 0.205
Scale factor = 146
f
FIG 37

Fig. 37

side in the case of ES which is lacking for the MM;

b) the bimodal character of the distribution functions are less pronounced in the case of MM;

c) the relaxation toward equilibrium in the downstream wing is completed earlier for MM as the distribution function in this case is much closer to equilibrium values at $\hat{n} = 3/4$ and $7/8$ than for ES.

Cheremisin [43] used also MC method for evaluating the integrals in an iteration procedure in the physical space. The position coordinate is used as independent variable. For this purpose B equation (103) is written in the integral form with superscripts in parenthesis corresponding to the iteration number:

$$
\begin{aligned}
f^{(n)}(c_x > 0, x) &= f_{01} \exp\left[-\frac{1}{c_x}\int_{L-}^{x} \check{b}^{(n-1)}(c, x_1)\,dx_1\right] + \\
&+ \frac{1}{c_x}\int_{L-}^{x} \check{a}^{(n-1)}(c, x_1)\exp\left[-\frac{1}{c_x}\int_{x_1}^{x} \check{b}^{(n-1)}(c, x_2)\,dx_2\right]dx_1 , \\
f^{(n)}(c_x < 0, x) &= f_{02} \exp\left[-\left|\frac{1}{c_x}\right| \int_{x}^{L+} \check{b}^{(n-1)}(c, x_1)\,dx_1\right] + \\
&+ \frac{1}{|c_x|}\int_{x}^{L+} \check{a}^{(n-1)}(c, x_1)\exp\left[-\frac{1}{|c_x|}\int_{x_1}^{x} \check{b}(c, x_2)\,dx_2\right]dx_1
\end{aligned}
\tag{109}
$$

where f_{01} and f_{02} denote the Maxwellian equilibrium distributions for upstream and downstream and the limits of integration at ± infinity are replaced by limits at finite distances L−, L+ sufficiently far from the sw position determined initially as the position of the gasdynamic discontinuity.

As the HT macroscopic quantities are moments of the distribution function it was possible to calculate them at each phase point (c, x) with a statistical accuracy and the integrals a,b could be determined using the MC method. Very few results are given in [43], we are showing on fig. 38 the variation of longitudinal and transversal temperature at M = 2 calculated by Cheremisin and for comparison on an adjoining figure some experimental results of Muntz and Harnett [42] at M = 1.59. The corresponding NS values have been calculated from the following relations between temperatures and stresses:

$$
\begin{cases}
T_\perp = T + \dfrac{2}{3}\dfrac{\mu}{k\eta}\dfrac{\partial u}{\partial x} \\[2ex]
T_x = T - \dfrac{4}{3}\dfrac{\mu}{kn}\dfrac{\partial u}{\partial x}
\end{cases}
\tag{110}
$$

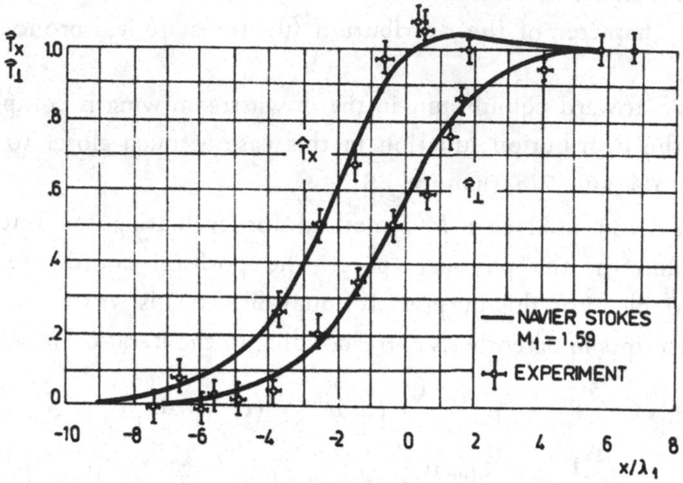

Fig. 38(a) (Muntz, Harnett, 1969)

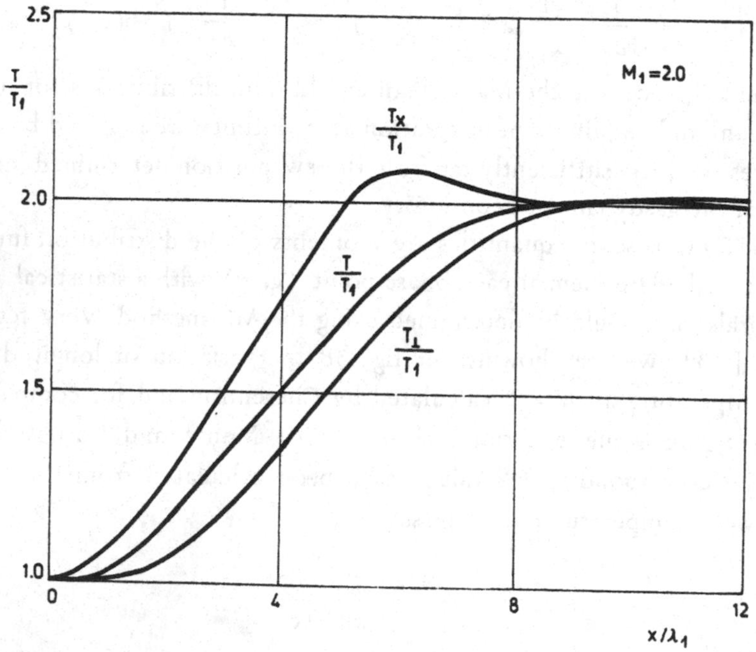

Fig. 38(b) (Cheremisin, 1970)

The qualitative correlation is very good and in both cases the overshoot of T_x appears.

In Bird's MC method, which will be outlined now, the general approach is not to solve the equations of flow but rather to simulate directly the motion of the gas particles and deduce therefrom the gas flow. It can in some respect be considered as a clean numerical experiment. The relation between the B equation and the direct simulation MC method is analysed in [45] and as described there: "The direct simulation MC method is a technique for the computer modeling of a real gas by several thousand simulated molecules". The six coordinates describing the physical state of the particle in the six dimensional velocity-postion phase space are stored in the computer and modified with time as they go through representative collisions. The molecules are followed concurrently.

The method is based on the same assumptions as those used in the derivation of the B equation. Molecular chaos is assumed and only binary collisions are considered and the range of molecular forces is assumed to be small compared with the molecular distance. The mean free path is assumed to be much larger than the molecular spacing. Of course the comparatively small sample of simulated molecules is assumed to form a valid representation of the real gas. To satisfy this requirement, in a typical application as given by Bird [44], the number of followed particles is approximately 10 orders of magnitude smaller than the number of molecules in the real gas. The fluctuations being inversely proportional to the sample size are therefore 5 orders of magnitude greater than in the real gas. This imposes a limit to applications in which the shock waves are sufficiently strong to produce a reasonable signal to noise ratio. Experience has shown that the computation procedure is stable.

A typical computation flow chart, used by Bird [4], is shown on Table II. The first step is to select and set up the initial configuration. Often this is a uniform gas in Maxwellian equilibrium or it may correspond to a state calculated using approximate results say the NS or MS method. The six coordinates of each simulated molecule are stored in the computer. The simulated spatial region of computation is divided into a number of cells of dimensions smaler than the distance over which significant flow changes are expected. The boundary conditions are the Maxwellian equilibrium distribution for upstream and downstream. After setting up the zero time configuration the molecules are allowed to move and collide among themselves. The two processes are decoupled, by computing collisions appropriate to a small time Δt_m which is much smaller than the mean collision time, ensuring thus

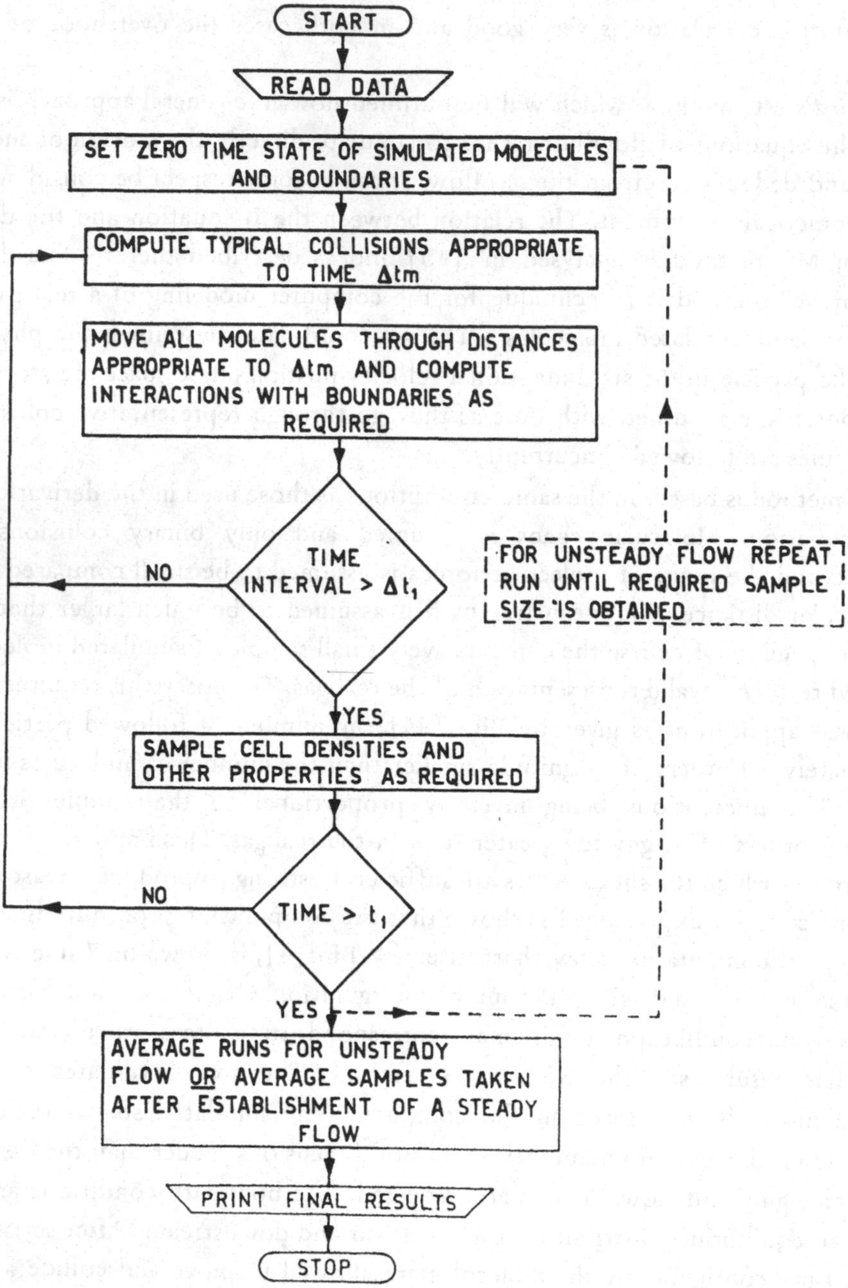
START
READ DATA
SET ZERO TIME STATE OF SIMULATED MOLECULES AND BOUNDARIES
COMPUTE TYPICAL COLLISIONS APPROPRIATE TO TIME Δtm
MOVE ALL MOLECULES THROUGH DISTANCES APPROPRIATE TO Δtm AND COMPUTE INTERACTIONS WITH BOUNDARIES AS REQUIRED
TIME INTERVAL > Δt1
NO
YES
FOR UNSTEADY FLOW REPEAT RUN UNTIL REQUIRED SAMPLE SIZE IS OBTAINED
SAMPLE CELL DENSITIES AND OTHER PROPERTIES AS REQUIRED
TIME > t1
NO
YES
AVERAGE RUNS FOR UNSTEADY FLOW OR AVERAGE SAMPLES TAKEN AFTER ESTABLISHMENT OF A STEADY FLOW
PRINT FINAL RESULTS
STOP

Table II

that the decoupling causes only small distortion of a typical molecular path. Since the change of the flow over the width of a cell is small, the molecules in a cell at any instant may be regarded as a sample of the molecules at the position of the cell. The relative positions of the molecules within the cell can then be disregarded and any randomly chosen pair within the cell is a possible collision pair. The actual collision probability of the pair depends on the molecular interaction model. For example in the case of ES in a gas obeying any inverse repulsive power law with a distance cut off, the collision probability is proportional to the relative velocity g between the random pair of molecules, and they are retained or rejected to this velocity. Random impact parameters are generated for the computation of each collision and the old velocities are replaced by the new ones. At each collision the time counter for the cell where it occured is advanced by

$$\Delta t = \frac{2}{N_c} \frac{1}{\sigma n g} \tag{111}$$

where N_c is the number of simulated molecules in a cell, σ the fixed collision cross section, n the number density and g the relative velocity.

Collisions are computed in the cell until the time counter has advanced through Δt_m. This procedure is repeated for every cell. After the completion of this collision procedure each molecule is moved through a distance equal to the product of its velocity and the time Δt_m and then the overall time is advanced by Δt_m. At time intervals $\Delta t_s \gg \Delta t_m$ the number density and other flow parameters may be sampled. For steady flows the establishment of stationary flow conditions with reduced statistical fluctuations is necessary before obtaining the sought HT quantities. Bird's [44] results of MC calculation indicating the scatter are shown on fig. 39 and the influence of the interaction potential on the density profile of the sw is shown on fig. 40.

The effect of high sw M number on density profiles for ES, MM and intermolecular repulsive forces proportional to the 8 power of the inverse distance calculated by the MC [44] method are shown on fig. 41. The strong influence of the softness of the molecule interaction force is very marked. The comparison of higher moments calculated by the MC method for a realistic intermolecular interaction with those obtained from the Ch.E. procedure at M = 8 given on fig. 42 shows the inadequacy of this last method at high M. The temperature profiles are similar to those obtained by Cheremisin. The heat flux vector q_x in units of upstream density times the most probable velocity shows the greatest percentage difference at the leading edge of the sw and a very great difference around $x/\lambda_1 = 2$ where the overall temperature gradient is very small but the gradients of the lateral and longitudinal temperatures are appreciable. Three components of the viscous stress tensor τ_{xn} as the mean of τ_{xy} and τ_{xz}, τ_{nn} as the mean of τ_{yy} and τ_{zz} and τ_{xx} are given. The NS value of $\tau_{xx} = 4/3\mu\, du/dx$ is also shown on the same figure. According to the NS description τ_{nn} is zero and the largest difference between the MC and NS results is in this stress component. The τ_{xn} MC results $\neq 0$ may be due to scatter.

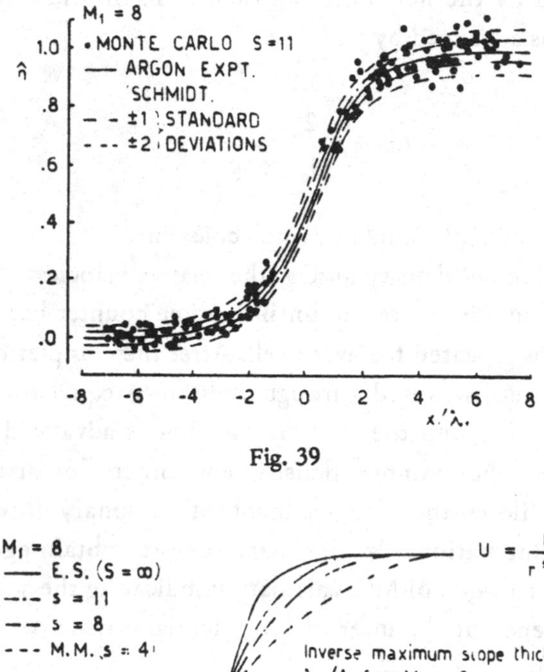

Fig. 39

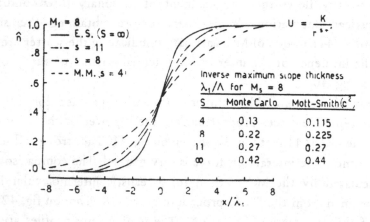

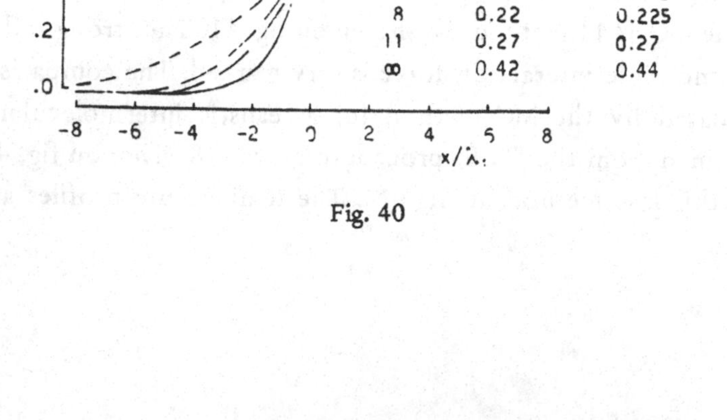

Fig. 40

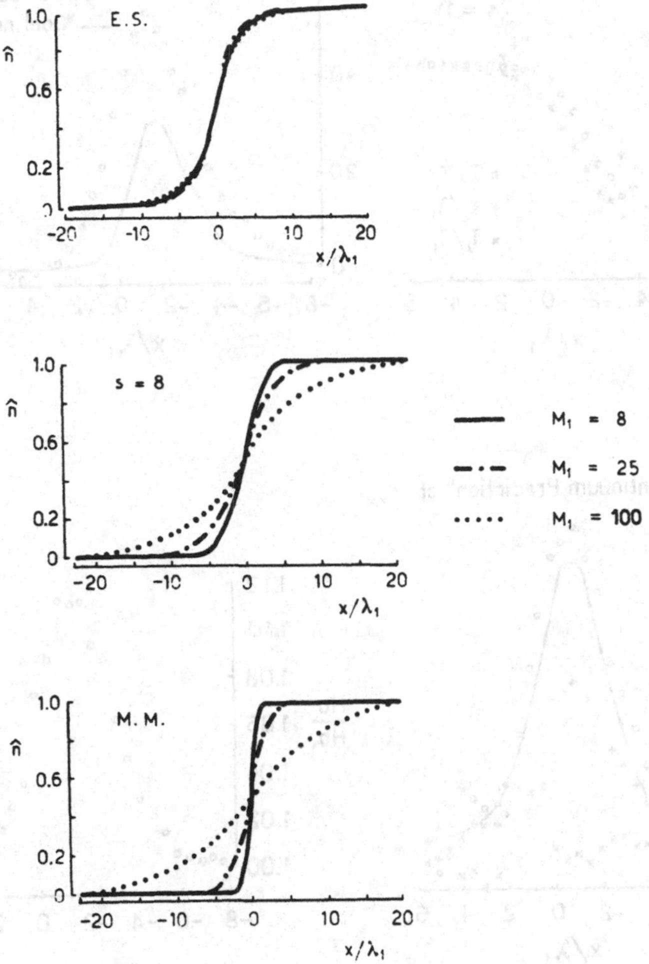

Fig. 41

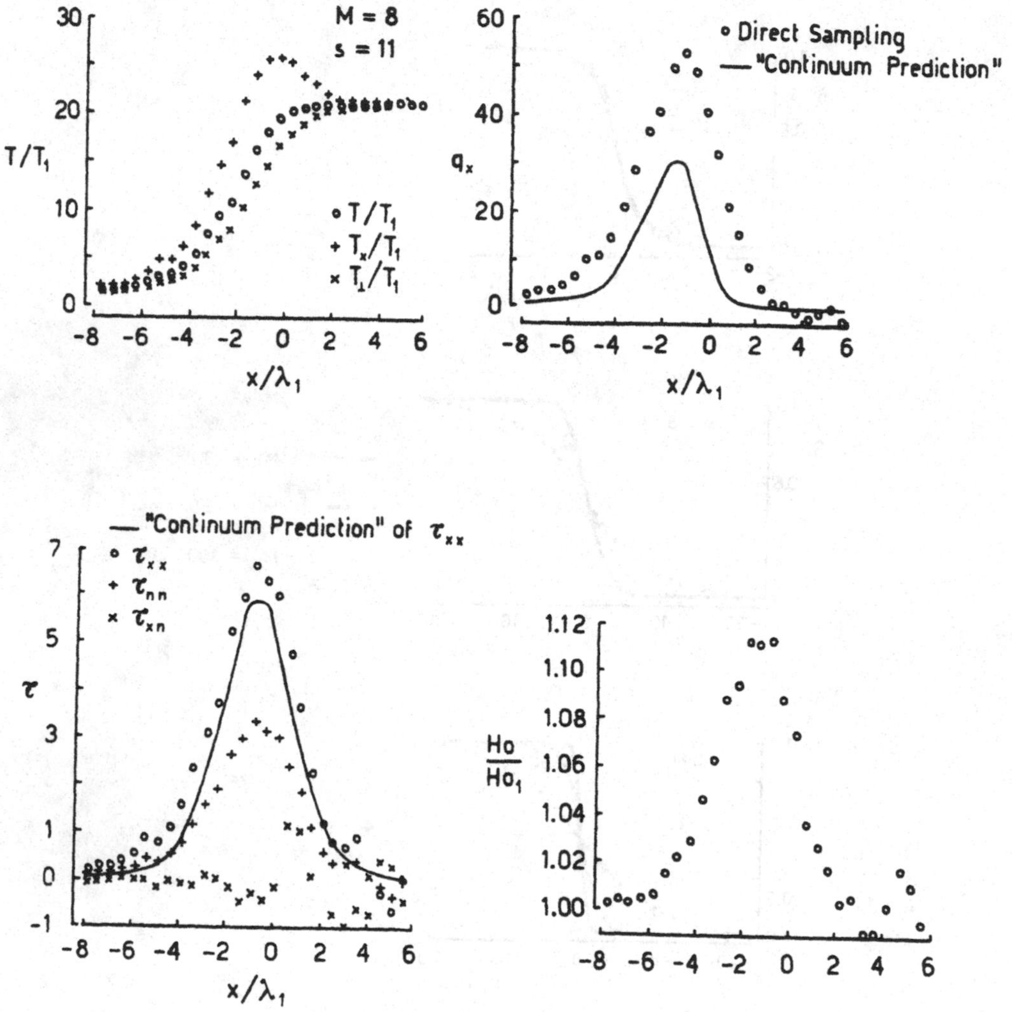

M = 8
s = 11
T/T1
x/λ1
T/T1
Tx/T1
T⊥/T1
Direct Sampling
"Continuum Prediction"
qx
"Continuum Prediction" of τxx
τxx
τnn
τxn
τ
Ho/Ho1

Fig. 42

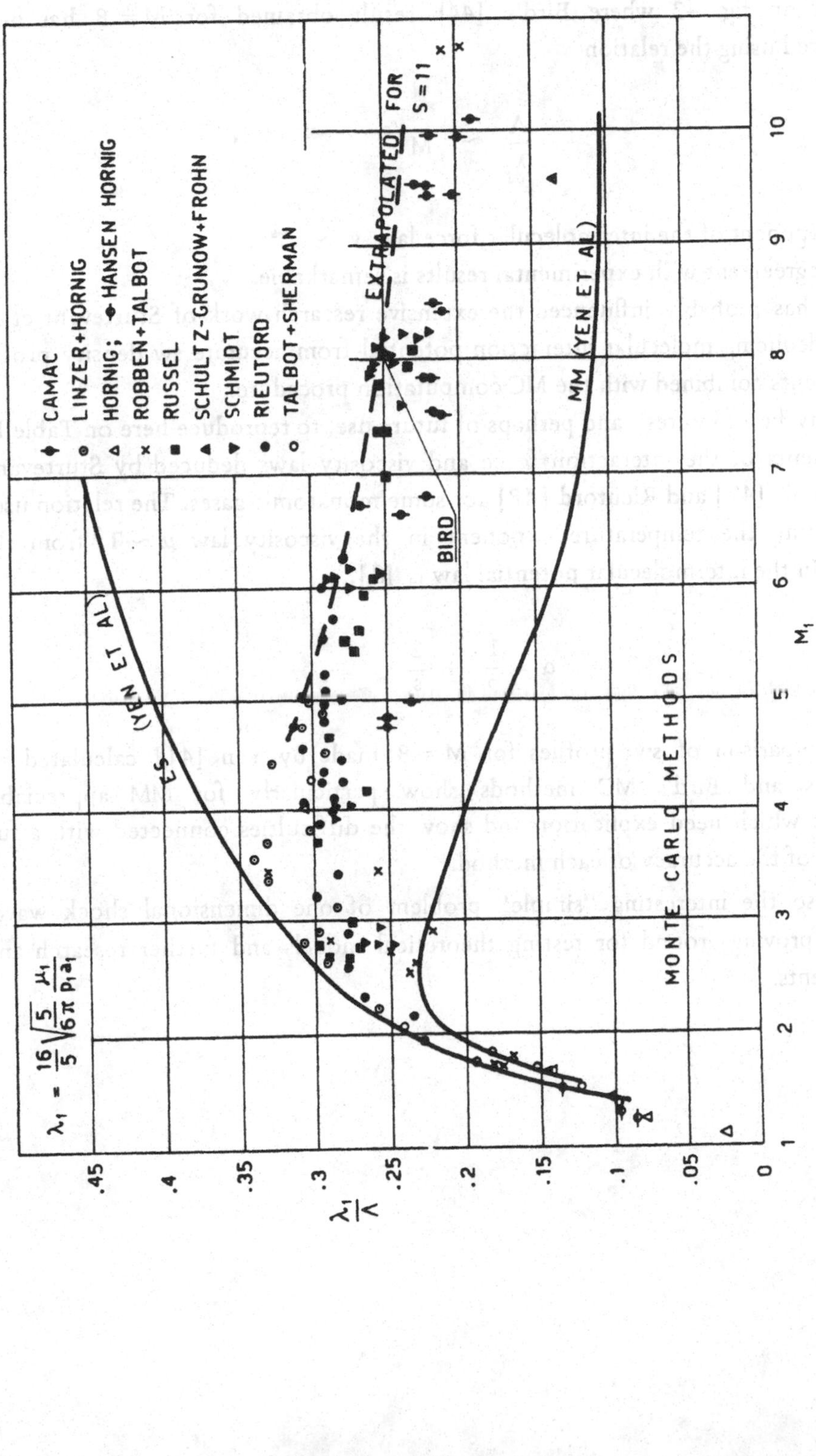

$\lambda_1 = \frac{16}{5}\sqrt{\frac{5}{6\pi}}\frac{\mu_1}{p_1 a_1}$
CAMAC
LINZER+HORNIG
HORNIG; HANSEN HORNIG
ROBBEN+TALBOT
RUSSEL
SCHULTZ-GRUNOW+FROHN
SCHMIDT
RIEUTORD
TALBOT+SHERMAN
ES (YEN ET AL)
MM (YEN ET AL)
BIRD
EXTRAPOLATED FOR S=11
MONTE CARLO METHODS
$\frac{\lambda_1}{\Lambda}$
0
.05
.1
.15
.2
.25
.3
.35
.4
.45
1
2
3
4
5
6
7
8
9
10
M_1

Fig. 43

Shock wave thicknesses calculated using the MC method and measured are compared on fig. 43 where Bird's [44] result obtained for M = 8 has been extrapolated using the relation

$$\frac{\Lambda}{\lambda_1} \sim M^{4/s} \tag{112}$$

s is the exponent of the intermolecular force law $v \sim r^{-s}$.

The agreement with experimental results is remarkable.

This has probably influenced the extensive research work of Sturtevant et al. [47] on deducing molecular interaction potential from accurate sw density profile measurements combined with the MC computation procedure.

It may be of interest, and perhaps of future use, to reproduce here on Table III the exponents of the interaction force and viscosity laws deduced by Sturtevant, Barcello et al. [47] and Rieutord [48] for some monatomic gases. The relation used in calculating the temperature exponent in the viscosity law $\mu \sim T^{\alpha}$ from the exponent in the intermolecular potential law is [8].

$$\alpha = \frac{1}{2} + \frac{2}{s} \tag{113}$$

A comparison of sw profiles for M = 8 made by Yen [41] calculated by Nordsieck's and Bird's MC methods show particularly for MM appreciable differences which need explication and show the difficulties connected with a full estimation of the accuracy of each method.

And so the interesting "simple" problem of one dimensional shock waves remains a proving ground for testing theoretical models and further research and improvements.

Gas \ Author	Mol. Weight	s, $\left(U \sim \frac{1}{r^s}\right)$ Barcelo	s, $\left(U \sim \frac{1}{r^s}\right)$ Rieutord	α, $(\mu \sim T^{\alpha})$ Barcelo	α, $(\mu \sim T^{\alpha})$ Rieutord
Helium	4.003		13.6		0.647
Neon	20.14	12	12	0.67	0.666
Argon	39.95	11	11 ÷ 9	0.68	0.682 ÷ 0.7
Krypton	83.80	10	(deduced from α)	0.70	
Xenon	131.30	9	(deduced from α)	0.72	

Table III

REFERENCES

[1] HAYES, W.D., Gasdynamic Discontinuities, Princeton 1960.

[2] KOGAN, M.N., Rarefied Gas Dynamics, Kinetic Theory. ed. Nauka 1967 (in Russian).

[3] CERCIGNANI, C., Mathematical Methods in Kinetic Theory, Plenum Press 1969.

[4] LIEPMANN, H.W., ROSHKO, A. Elements of Gasdynamics, J. Wiley 1957.

[5] THOMPSON, P.A., Compressible Fluid Dynamics, McGraw 1972.

[6] ZELDOVICH, Ya. B., RAIZER, Yu. P., Physics of Shock Waves, Nauka (in Russian) 1963, Academic Press 1967.

[7] GILBARG, D., PAOLUCCI, D., J. Rat. Mech. Anal. 2, 1953, (617-642).

[8] CHAPMAN, S., COWLING, T., The Mathematical Theory of Nonuniform Gases, Cambridge Univ. Press, 1939.

[9] GRAD, H., Comm. Pure Appl. Math., 2, 1949 (331-407), 5, 1952 (257-300).

[10] TRUESDELL, G., J.math.pures et appl. 30, 1951 (111-155).

[11] TAYLOR, G.I., Collected Papers, Cambridge University Press, 1963.

[12] MUNTZ, E.P., HARNETT, L.W., Phys. of Fluids, 12 (1969) 2027-2035.

[13] SMOLDEREN, J., Structure des chocs et théorie cinématique de gas, Choc et Ondes de Choc, ed. A.L. Jaumotte, Pub. Masson, Paris, 1971.

[14] LIEPMANN, H.W., NARASIMHA, R., CHAHINE, M.T., Phys. Fluids, 5, 1313-1324 (1962).

[15] FOCH, J.D., Acta Phys. Austr. X. Supp. (123-140) 1973.

[16] BHATNAGAR, P.L., GROSS, E.P., KROOK, M., Phys. Rev., 94, 511 (1964).

[17] HOLWAY, L.H., Phys. Fluids, 9, 1658-1673 (1966).

[18] KOGAN, M.N., Prikl. Mat. Mech. 22, 425 (1958).

[19] CHAHINE M.T., and NARASIMHA, R., IV RGD Transactions 140-159, Chahine III RGD Trans. 260-273.

[21] MUNTZ, E.P., HARNETT, L.W., Phys. FC, 12, 2027-2035 (1969).

[22] ABRAMOWITZ, M., J. Math. Phys., 32, 188 (1953).

[23] ANDERSON, D.G., MACOMBER, H.K., J. Mathem. Phys. (1962).

[24] SEGAL, B.M., FERZIGER, J.M., Phys. of Fl., 15, 1233-1247 (1972).

[25] SHAKHOV, E.M., Mech. Zhidk. Gaza 1969, 69-75; ZHUK, W.J., RIKOV, W.A., SHAKHOV, E.M., Mech. Zhidk., Gaza 1973, 135-141.

[26] SEGAL, B.M., Ph.D. Thesis, Stanford Univ. 1971.

[27] CHAHINE, M.T., NARASIMHA, R., J. Math. Phys., 43, 163 (1964).

[28] WANG-CHANG, C.S., UHLENBECK, G.E., Studies in Statistical Mechanics, Vol. 5, North Holland, 1970.

[29] GROSS, E.P., JACKSON, E.A., Phys. Fl., 2, 432 (1959).

[30] CERCIGNANI, C., TIRONI, G., RGD (1967) Proceedings, Acad. Press.

[31] MOTT-SMITH, H.M., Phys. Rev., 82, 885-892 (1951).

[32] TAMM, I.E., A.S. USSR, Trudy Inst. Leb. XXIX, 239 (1965).

[33] RODE, D.L., TANENBAUM, B.S., Phys. Fluids, 10, 1352-1353 (1967).

[34] MUCKENFUSS, Ch., Phys. Fluids, 5, 1325-1336 (1962).

[35] OBERAI, M.M., J. Mécanique, 6, 317-326 (1967).

[36] NARASIMHA, R., et al. VI RGD, Nar. Bull. de la Classe des Sciences, Ac.Roy. Belge 5e Serie T LVI 1970.

[37] SALVEN, H., GROSCH, Ch.E., ZIERING, S., Phys. Fluids, 7, 180-189 (1964).

[38] FISZDON, W., HERCZYNSKI, R., WALENTA, Z., IX RGD Symposium, DVLR-Press, W. Germany, 1974.

[39] NORDSIECK, A., HICKS, B.L., RGD (ed. C.L. Brundin) Acad. Press, 1967, vol. I, 695-710.

[40] HICKS, B.L., YEN, S.M., RGD (ed. I. Trilling and H.Y. Wachman), Acad. Press, 1969, vol. I, 313-317.

HICKS, B.L., SMITH, M.A., J. Comp. Phys. 3, 58-79 (1968).

HICKS, B.L., YEN. S.M., REILLY, B.J., J. Fluid, Mech. 53, 85-111 (1972).

[41] HICKS, B.L., YEN, S.M., Phys. Fluids, 10, 458-460 (1967).

YEN, S.M., Int.J.Heat Mass Transfer, 14, 1865-1869 (1971).

YEN, S.M., W. Ng. - to be published in J. Fluid Mech. (1974).

[42] MUNTZ, E.P., HARNETT, L.W., Phys. Fluids, 12, 2027-2035 (1969).

[43] CHEREMISIN, F.G., J. Comp. Math. Phys., 10, 654-665 (1970) in Russian.

[44] BIRD, G.A., RGD 1965, RGD 1969, May 1970.

[45] BIRD, G.A., Phys. Fluids, 13, 2676-2681 (1970 Nov.).

[46] SCHMIDT, B., J. Fluid Mech., 39, 361-373 (1969).

[47] STURTEVANT, B., STEINHILPER, E.A., Ca. Tech. to be published (1974).

BARCELO, B.T., Phys. Fluids, Cal. Inst. Techn. Pasadena (1970).

[48] RIEUTORD, E., Thése, Lyon (1970).

Printed in the United States
By Bookmasters